Study & Master

MATHEMATICS 7

NEW SYLLABUS

Karin Loots

in collaboration with
Linda Loots
Claris de Lange

ROEDURICO
trust

Somerset West

Roedurico Trust
P.O. Box 810 Somerset West 7129
Cover page: The graphic shop
Caricature created by Rufus Papenfus
Set in 10 pt Times
Litographed and bound by
National Book Printers, Drukkery Street
Goodwood, Western Cape

ISBN 0-947465-36-7

First edition
First impression 1996

PREFACE

This book was written to consolidate and expand the basic Mathematical knowledge as required by the syllabus, and to assist in the preparation of the pupil for tests and examinations. It provides a rich source of information for classwork, homework and revision.
The main aim of the book is to present the knowledge and concepts as prescribed by the Junior Secondary Syllabus, in such a clear and precise way that the pupil will find it easy to understand and assimilate. It also provides numerous questions which teachers may use for continuous evaluation, tests and examinations.

The topics are presented logically to assist both teachers and pupils:

- Explanation
- Examples
- Exercises
- Solutions

The complete solutions at the end of each chapter also serve as further examples.

We trust that both pupils and teachers will find this book useful and challenging.

The Authors

Index

BOLPY
Exercises
Examples
Solutions

1 *Real numbers*

The number system

N $= \{1; 2; 3; \ldots\}$ – Natural numbers

No $= \{0; 1; 2; 3; \ldots\}$ – Counting numbers

Z $= \{\ldots -3; -2; -1; 0; 1; 2; \ldots\}$ – Integers

Q = Rational numbers $= \dfrac{\text{Integer}}{\text{Integer}}$

R = Real numbers = Rational numbers + Irrational numbers (Q^1)

Natural numbers (N):

Subdivision:

1. Even numbers : 2; 4; 6; 8; . . . . .
2. Odd numbers : 1; 3; 5; 7; . . . . .
3. Prime numbers : 2; 3; 5; 7; 11; 13 . . . . .
4. Composite numbers : 4; 6; 8; 9; 10; 12; 14; 15; 16 . . . . .
5. Squares : 1; 4; 9; 16; 25; 36; 49; . . . . .
6. Cubes : 1; 8; 27; 64; 125; . . . . .

Rational numbers (Q):

All integers
All fractions
All finite decimal fractions
All recurring decimal fractions

Examples: Rational numbers

1. $3 = \dfrac{3}{1}$
2. $-8 = \dfrac{-8}{1}$
3. $-\dfrac{3}{4}$
4. $0{,}1 = \dfrac{1}{10}$
5. $0{,}03 = \dfrac{3}{100}$
6. $\sqrt{9} = 3 = \dfrac{3}{1}$
7. $0{,}58 = \dfrac{58}{100}$
8. $\sqrt[3]{8} = 2 = \dfrac{2}{1}$
9. $0{,}\dot{4}\dot{5} = \dfrac{5}{9}$
10. $0{,}\dot{3} = \dfrac{1}{3}$

N.B. All the examples above can be expressed as:

$\dfrac{\textbf{Integer}}{\textbf{Integer}}$ = **Rational number.**

Common fraction → Decimal fraction

Examples:

1. $\frac{41}{100} = 0,41$

2. $\frac{3}{5} = \frac{6}{10} = 0,6$

3. $\frac{3}{8} = \frac{3,000}{8}$

$= 0,375$

$= \frac{3,{}^{3}0{}^{6}0{}^{4}0}{8}$

$= 0,375$

4. $\frac{2}{9} = \frac{2,000}{9}$

$= 0,222 \; .....$

$= 0,\dot{2}$

(Recurring decimal fraction)

5. $\frac{2}{11} = \frac{2,000}{11}$

$= 0,181818 \; .....$

$= 0,\dot{1}\dot{8}$

6. $1\frac{1}{6} = \frac{7}{6}$

$= \frac{7,000}{6}$

$= 1,1666$

$= 1,1\dot{6}$

(the recurring pattern starts with the 6)

7. $\frac{3}{7} = \frac{3,000}{7}$

$= 0,\dot{4}\dot{2}\dot{8}\dot{5}\dot{7}\dot{1}$

Recurring decimal fraction → Common fraction

Examples:

1.

$0,\dot{6} = 0,666 \; .....$

Let $x = 0,666 \; ..... \; (1)$

$10x = 6,666 \; ..... \; (2)$

$(2)-(1) \quad 9x = 6$

$x = \frac{6}{9}$

$x = \frac{2}{3}$

$0,\dot{6} = \frac{2}{3}$

2.

$0,\dot{4}\dot{5} = 0,454545 \; .....$

Let $x = 0,454545 \; ..... \; (1)$

$\therefore \; 100x = 45,454545 \; ...... \; (2)$

$(2)-(1)\text{:} \; 99x = 45$

$x = \frac{45}{99}$

$= \frac{5}{9}$

$\therefore \; 0,0\dot{4}\dot{5} = \frac{5}{9}$

3. $2,\dot{1}\dot{8} = 2,181818 \; .....$

Let $x = 2,181818 \; ..... \; (1)$

$100x = 218,181818 \; ..... \; (2)$

$99x = 216$

$x = \frac{216}{99}$

$x = \frac{24}{11}$

$\therefore \; 2,\dot{1}\dot{8} = 2\frac{2}{11}$

Irrational numbers (Q^1):

All **infinite, non-recurring** decimal numbers are ***Irrational.***

$\sqrt{\text{Positive non–square}}$ and $\sqrt[3]{\text{non–cube}}$ ***are Irrational.***

Examples: Irrational numbers

1. 0,4381497 . . . . .

2. $\pi \approx \frac{22}{7}$ is only an approximate value, and is therefore also ***irrational.***

3. $\pi = 3,1415927$ . . . . .

4. $\sqrt{10} = 3,162277$ . . . . .

5. $\sqrt[3]{20} = 2,171441$ . . . . .

Note: **1.** All the deicmal numbers in the examples above are infinite; they are ***non*-recurring.**

2. $\sqrt{10} = 3,16$ has been **rounded off** to 2 decimals. Therefore $\sqrt{10}$: Irrational

N.B. **3.** $\sqrt{5} \Rightarrow$ Irrational number.

but $(\sqrt{5})^2 = 5 \Rightarrow$ Rational number.

Summary

1. Rational numbers: $\frac{3}{4}$; -2; $\sqrt{25}$; $0,\dot{3}$; $0,15$; $\sqrt{\frac{100}{49}}$;

$(\sqrt{10})^2$; $\sqrt[3]{8} + (\sqrt{2})^2$; . . . . .

2. Irrational numbers: $\sqrt{3}$; 213, 497 . . . . .; π; 5π $\sqrt[3]{9}$; . . . . .

3. Rational numbers + Irrational numbers = Real numbers

4. Division by 0: undefined

$\frac{1}{0}$ = *undefined*, but $\frac{0}{1} = 0$

5. $\sqrt{4}$; $\sqrt{9}$; $\sqrt{\frac{1}{25}}$; . . . . . Real, Rational

$\sqrt[3]{-8}$; $\sqrt[3]{1}$; $\sqrt[3]{-27}$ . . . . . Real, Rational

$\sqrt{3}$; $\sqrt{10}$; $\sqrt{50}$; . . . . . Real, Irrational

$\sqrt[3]{9}$; $\sqrt[3]{-11}$; $\sqrt[3]{30}$ . . . . . Real, Irrational

$\sqrt{-9}$; $\sqrt{-10}$; $\sqrt{-1}$ . . . . . Non-Real (Imaginary)

6. $(\sqrt{5})^2 \Rightarrow$ Rational

$2 + \sqrt{3} \Rightarrow$ Irrational

$\frac{\sqrt{6}}{2} \Rightarrow$ Irrational

$\pi + 3 \Rightarrow$ Irrational

Irrational numbers on the number line:

Irrational numbers like $\sqrt{2}$; $\sqrt{26}$; $\sqrt{10}$; etc. can be shown on the **number line** by using the **Theorem of Pythagoras** (Std 6).

Example:

Represent $\sqrt{26}$ on the number line.

STEP 1: Express the number 26 as the **sum** of 2 squares.
eg. $26 = 25 + 1$

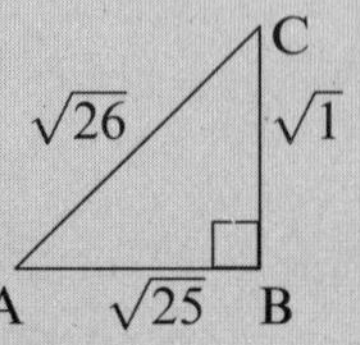

Now do the following geometry construction:

STEP 2: On the number line: mark off AB $= \sqrt{25} = 5$ (from 0)

STEP 3: Construct $\triangle$ ABC with $\hat{B} = 90°$, and BC $= \sqrt{1} = 1$

Note: same scale on AB and BC.

STEP 4: Join AC.

STEP 5: Using your pair of compasses, place the sharp point on A and the pencil on point C. Draw an arc to cut the number line in D. AD $= \sqrt{26}$; D represents $\sqrt{26}$ on the number line.

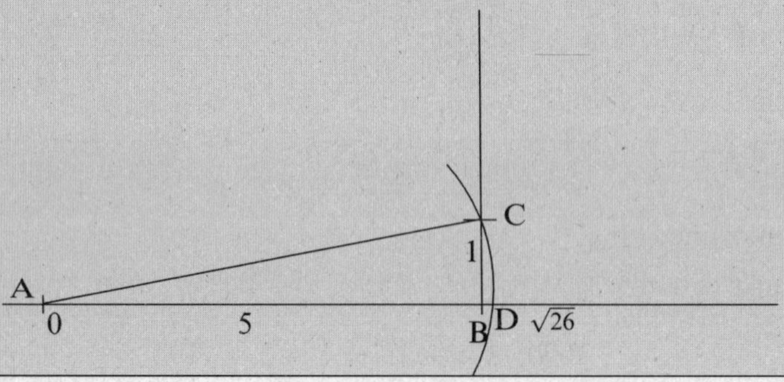

Inequalities on the Number line:

N; N_0; Z: represented by: 
R: represented by: 

Notation.

1. $\{1 \leq x < 6;\ x \in R\}$ / $[1; 6)$ } means: x is any real number between 1 and 6; 1 included; 6 excluded.

2. [ ]: Included / ( ): Excluded } applicable only to the **real numbers**.

3. •: **Included** / o: **Excluded** / ∞: **infinity** } on number line

Hint: Draw the number line and calibration in **pencil**, and then mark your interval in **pen**.

Examples on the number line:

Represent each of the following on the number line:

1. $-2 < x \leq 3;\quad x \in$ N:

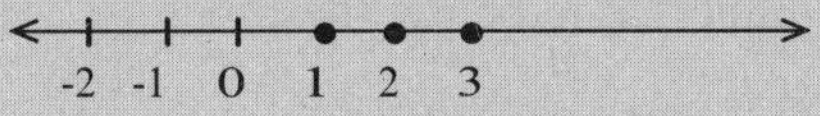

2. $-2 < x \leq 3;\quad x \in$ Z:

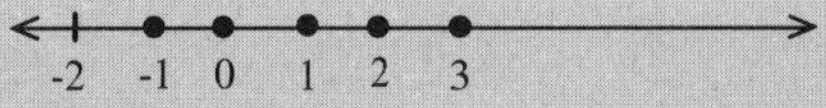

3. $-2 < x \leq 3;\quad x \in R$:

4. $-5 \leq x < 2;\quad x \in$ Z:

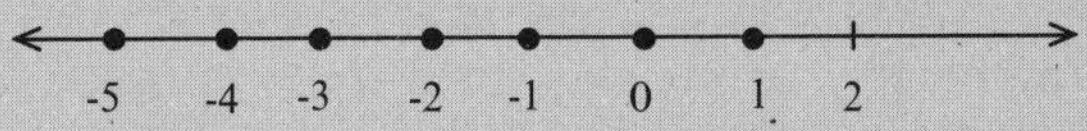

5. $x < -3$ *of* $x > 1;\quad x \in R$:

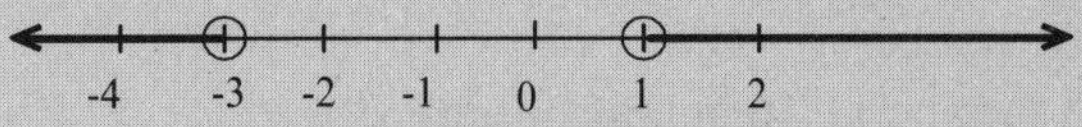

6. $0 < x \leq 3;\quad x \in R$:

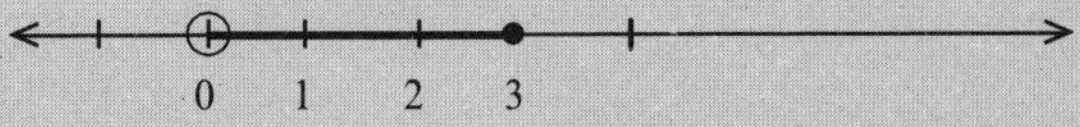

7. $4 > y \geq -2;\quad y \in$ Z:

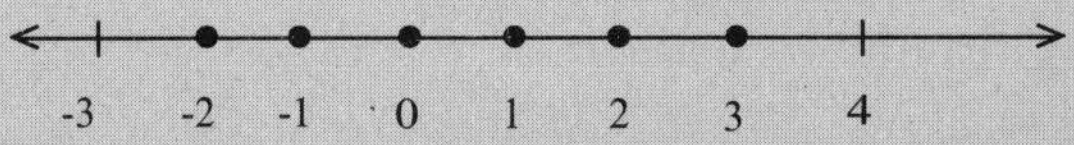

Examples:

Write down the interval shown on each of the number lines:

Solutions

1. 

$-3 < x \leq 2; \quad x \in R \quad or \quad (-3;\ 2];\ x \in R$

2. 

$x \leq 1$ *or* $x \geq 3;\ x \in Z$ *or* $x \neq 2;\ x \in Z$

3. 

$x \leq \sqrt{3}; \quad x \in R$

or $(-\infty;\ \sqrt{3}\,]$

4. 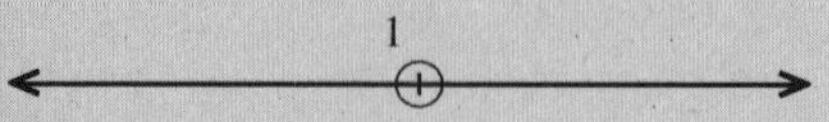

$x \neq 1; \quad x \in R$

Exercise:

1. List the following numbers one below the other:

$\frac{1}{7}; \quad -3; \quad \sqrt{\frac{9}{16}}; \quad -\sqrt[3]{8}; \quad 0,\dot{3}\dot{5}; \quad 2\pi; \quad \sqrt{-3};$

$\sqrt[3]{8}; \quad 1\frac{1}{6}; \quad 4; \quad \left(\sqrt{3}\right)^2; \quad \sqrt[3]{9}; \quad 4,5; \quad 0,\dot{1}2\dot{3}; \quad \sqrt{10}$

$2\sqrt{8}; \quad \sqrt{-2}; \quad 5,1897\ .\ .\ .\ .\ .$

Next to each number, write down the category to which it belongs. (N.B.: There could be more than one category eg. $-\sqrt[3]{8} = -2$: Real, Rational; Integer.)

The possible categories are:

Natural numbers (N)
Integers (Z)
Rational numbers (Q)
Irrational numbers (Q^1)
Non-real numbers. (Imaginary numbers)

2. Choose from the list: $-2\frac{1}{3}; \quad 9\pi; \quad \sqrt{20}; \quad 1,23\ .\ .\ .\ .\ .; \quad -3; \quad \sqrt{16}$

All the (a) Integers
(b) Rational numbers
(c) Irrational numbers

3. Construct each of the following irrational numbers on the number line.

3.1 $\sqrt{10}$ **3.2** $\sqrt{29}$ **3.3** $\sqrt{40}$

3.4 $\sqrt{136}$ **3.5** $\sqrt{45}$ **3.6** $\sqrt{13}$

4. Express each of the following common fractions as a decimal fraction:

4.1 $\frac{3}{4}$ **4.2** $\frac{5}{8}$ **4.3** $\frac{4}{3}$ **4.4** $-\frac{4}{9}$

4.5 $-\frac{1}{9}$ **4.6** $2\frac{5}{6}$ **4.7** $\frac{3}{11}$ **4.8** $\frac{2}{7}$

4.9 $1\frac{2}{3}$ **4.10** $-4\frac{7}{11}$

5. Express each of the following recurring decimal fractions as common fractions (show all calculations).

5.1 $0{,}\dot{6}$ **5.2** $0{,}\dot{3}\dot{5}$ **5.3** $0{,}\dot{1}\dot{5}$

5.4 $0{,}\dot{4}$ **5.5** $1{,}\dot{1}2\dot{3}$ **5.6** $2{,}\dot{3}\dot{6}$

6. Represent each of the following inequalities graphically (on the number line):

6.1 $-2 < x \leq 5$; $x \in N$

6.2 $-5 \leq x < 1$; $x \in Z$

6.3 $x \leq 4$; $x \in N$

6.4 $0 < x \leq 4$; $x \in R$

6.5 $x \leq -1$ *or* $x > 3$; $x \in R$

6.6 $x < 2$ *or* $x \geq 4$; $x \in R$

6.7 $-5 < x < -1$; $x \in R$

6.8 $-2\frac{1}{4} \leq x < 3\frac{1}{2}$; $x \in R$

6.9 $x \leq -3$ *or* $x > 2$; $x \in Z$

6.10 (3: 5]

6.11 $(-5; \infty)$

6.12 $(-\infty; 2]$

6.13 $-3 \leq x < 0$ *or* $x > 3$; $x \in R$

6.14 $-2\frac{1}{2} \leq x < 1$ *or* $x \geq 4\frac{1}{2}$; $x \in Z$

7. Write down an inequality (*in x*) for each of the following graphic representations:

7.1

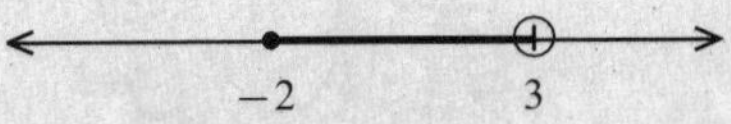

7.2

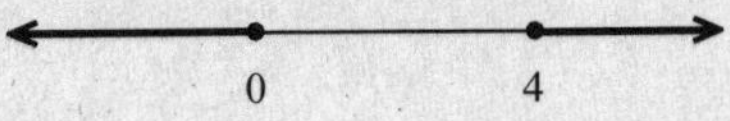

7.3

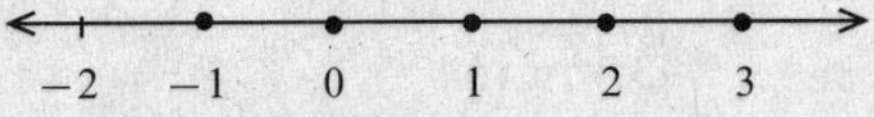

7.4

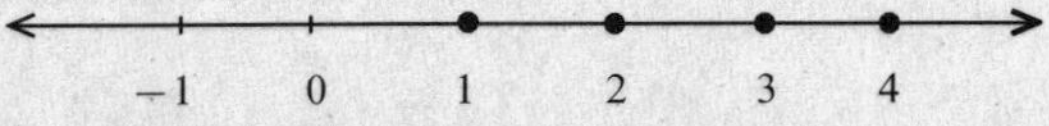

7.5

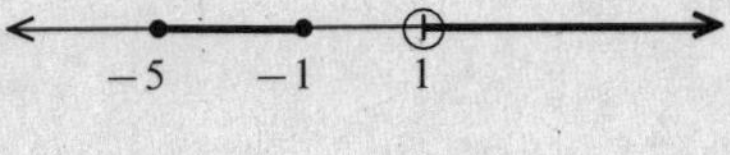

7.6

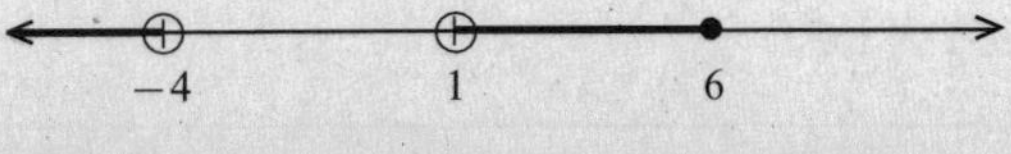

SOLUTIONS

R: Real

Q: Rational

Z: Integer

N: Natural number

Q^1: Irrational

1. $\frac{1}{7}$: Real; Rational

-3 : Real, Rational; Integer

$\sqrt{\frac{9}{16}} = \frac{3}{4}$: Real, Rational

$-\sqrt[3]{8} = -2$: Real, Rational; Integer

$0,\dot{3}\dot{5}$: Real, Rational

2π : Real, Q^1

$\sqrt{-3}$: Non-Real

$\sqrt[3]{8} = 2$: *R; Q; Z; N*

$1\frac{1}{6} = \frac{7}{6}$: *R; Q*

4 : *R; Q; Z; N*

$\left(\sqrt{3}^2\right) = 3$: *R; Q; Z; N*

$\sqrt[3]{9}$: *R*; Q^1

$4,5$: *R; Q*

$0,\dot{1}\dot{2}\dot{3}$: *R; Q*

$\sqrt{10}$: *R*; Q^1

$2\sqrt{8}$: *R*; Q^1

$\sqrt{-2}$: Non-Real (Imaginary)

$5,1897\ldots\ldots$: *R*; Q^1

2. (a) Integers : -3; $\sqrt{16} = 4$

(b) Rational numbers : $-2\frac{1}{3}$; -3; $\sqrt{16} = 4$

(c) Irrational numbers : 9π; $\sqrt{20}$; $1,23\ \ldots\ldots$

3.1 $\sqrt{10}$

$10 = 9 + 1$

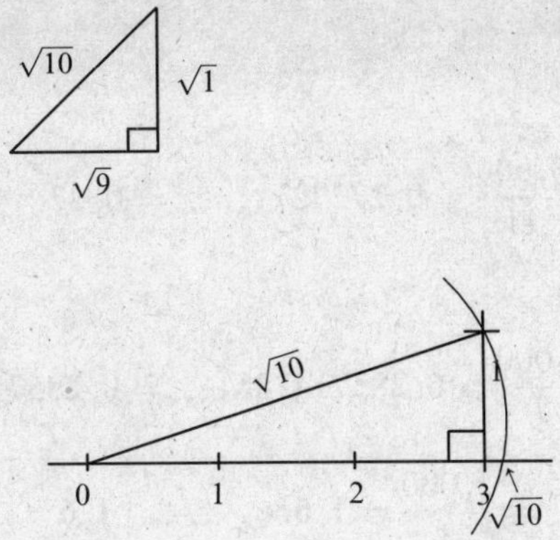

3.4 $\sqrt{136}$

$136 = 100 + 36$

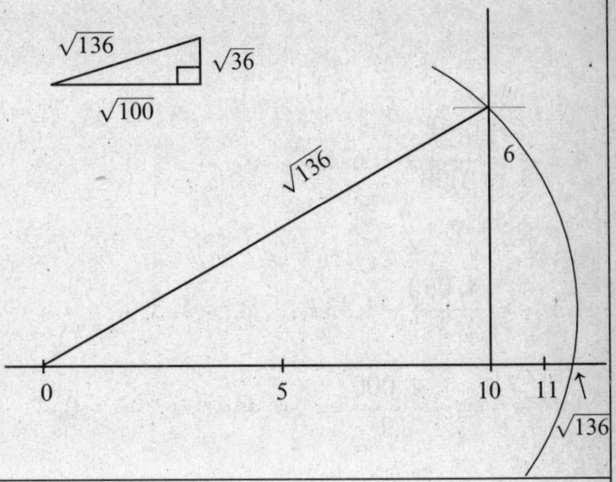

3.2 $\sqrt{29}$

$29 = 25 + 4$

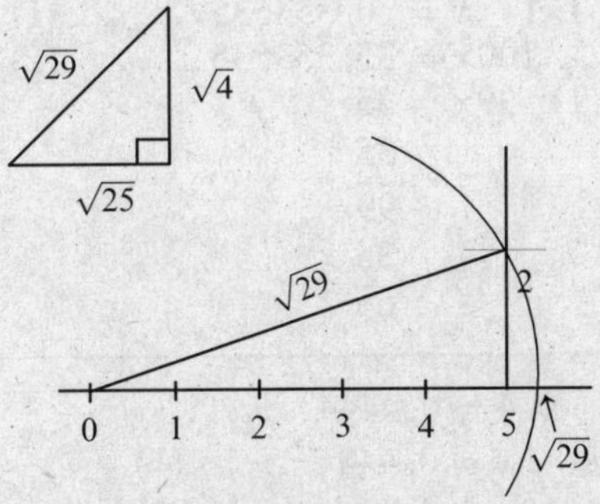

3.5 $\sqrt{45}$

$45 = 36 + 9$

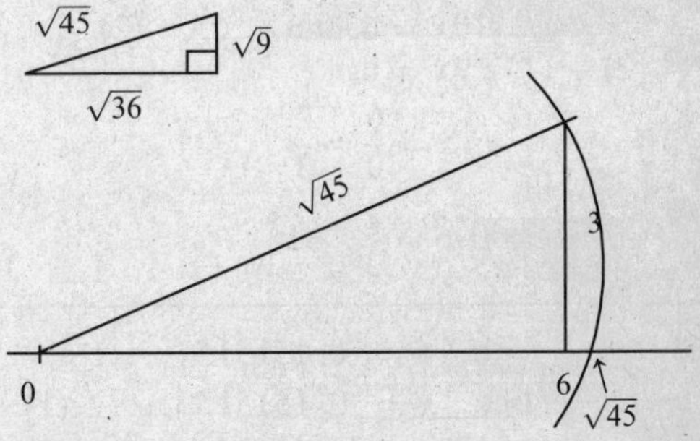

3.3 $\sqrt{40}$

$40 = 36 + 4$

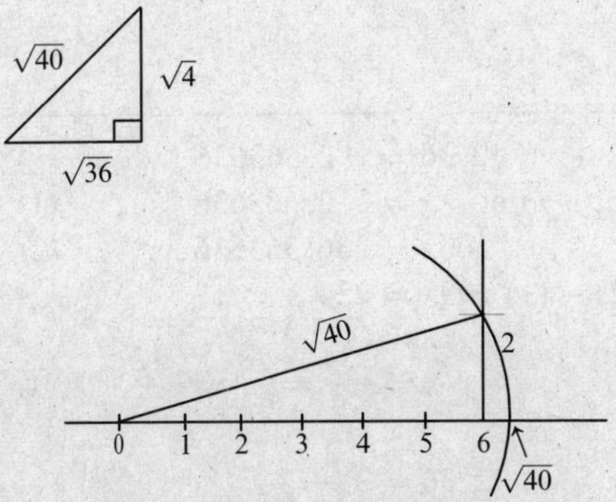

3.6 $\sqrt{13}$

$13 = 9 + 4$

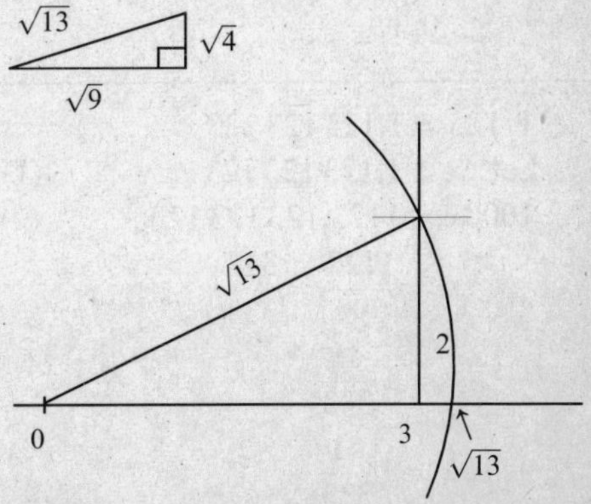

4.1 $\frac{3}{4}=\frac{75}{100}=0,75$ (numerator $\times 25$, denominator $\times 25$)

4.2 $\frac{5}{8}=\frac{625}{1000}=0{,}625$ (numerator $\times 125$, denominator $\times 125$)

4.3 $\frac{4}{3}=\frac{4,000}{3}=1{,}333\ldots\ldots=1,\dot{3}$

4.4 $-\frac{4}{9}=-\frac{4,000}{9}=-0,444\ldots\ldots=-0,\dot{4}$

4.5 $-\frac{1}{9}=-\frac{1,000}{9}=-0,111\ldots\ldots\ -0,\dot{1}$

4.6 $2\frac{5}{6}=\frac{17}{6}=\frac{17,000}{6}=2,8333\ldots\ldots=2,8\dot{3}$

4.7 $\frac{3}{11}=\frac{3,000}{11}=0,272727\ldots\ldots=0,\dot{2}\dot{7}$

4.8 $\frac{2}{7}=\frac{2,000}{7}=0,285714285\ldots.=0,\dot{2}\dot{8}\dot{5}\dot{7}\dot{1}\dot{4}$

4.9 $1\frac{2}{3}=\frac{5}{3}=\frac{5,000}{3}=1,666\ldots\ldots=1,\dot{6}$

4.10 $-4\frac{7}{11}=\frac{-51}{11}=\frac{-51,00}{11}=-4{,}6363=-4,\dot{6}\dot{3}$

5.1

$0,\dot{6}=0,666\ldots\ldots$

Let $x=0,666\ldots\ldots$ (1)

$\therefore\ 10x=6,666\ldots\ldots$ (2)

$(2)-(1):9x=6$

$x=\frac{6}{9}=\frac{2}{3}$

$\therefore\ 0,\dot{6}=\frac{2}{3}$

5.2

$0,\dot{3}\dot{5}=0,353535\ldots\ldots$

Let $x=0,353535\ldots\ldots$ (1)

$\therefore\ 100x=35,353535\ldots\ldots$ (2)

$(2)-(1):\ 99x=35$

$x=\frac{35}{99}$

$0,\dot{3}\dot{5}=\frac{35}{99}$

5.3

$0,\dot{1}\dot{5}=0,151515\ldots\ldots$

Let $x=0,151515\ldots\ldots$ (1)

$\therefore\ 100x=15,151515\ldots\ldots$ (2)

$(2)-(1):99x=15$

$x=\frac{15}{99}$

$=\frac{5}{33}$

$\therefore\ 0,\dot{1}\dot{5}=\frac{5}{33}$

5.4

$0,\dot{4}=0,4444\ldots\ldots$

Let $x=0,444\ldots\ldots$ (1)

$\therefore\ 10x=4,444\ldots\ldots$ (2)

$(2)-(1):9x=4$

$x=\frac{4}{9}$

$\therefore\ 0,\dot{4}=\frac{4}{9}$

5.5

$1,\dot{1}2\dot{3}=1{,}123\,123\,123\ldots\ldots$

Let $x=1{,}123\,123\,123\ldots\ldots$ (1)

$\therefore\ 1000x=1123{,}123\,123\,123\ldots\ldots$ (2)

$\therefore\ x=\frac{1122}{999}=\frac{374}{333}$

$=1\frac{41}{333}$

$\therefore\ 1,\dot{1}2\dot{3}=1\frac{41}{333}$

5.6

$2,\dot{3}\dot{6}=2,363636\ldots\ldots$

Let $x=2,363636\ldots\ldots$ (1)

$\therefore\ 100x=236,363636\ldots\ldots.$ (2)

$(2)-(1):99x=234$

$x=\frac{234}{99}=\frac{78}{33}=2\frac{12}{33}$

$\therefore\ 2,\dot{3}\dot{6}=2\frac{4}{11}$

6.1 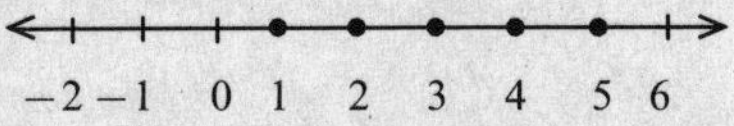

$-2 < x \leq 5; \quad x \in N$

6.2 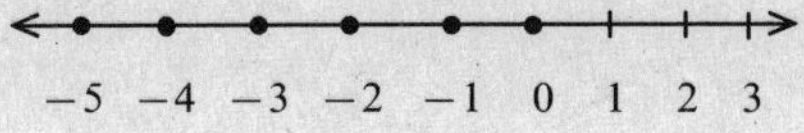

$-5 \leq x < 1; \quad x \in Z$

6.3 

$x \leq 4; \quad x \in N;$

6.4 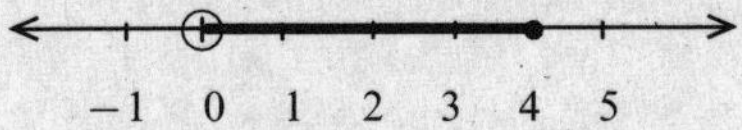

$0 < x \leq 4; \quad x \in R$

6.5 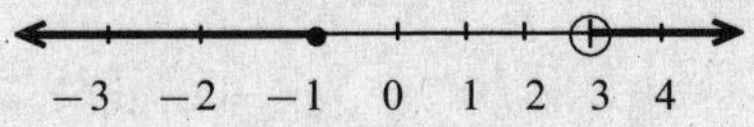

$x \leq -1$ *or* $x > 3; \quad x \in R.$

6.6 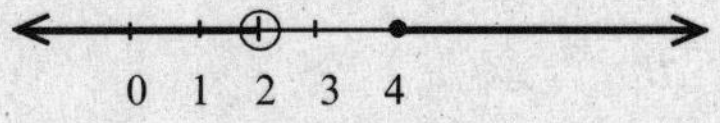

$x < 2$ *or* $x \geq 4; \quad x \in R$

6.7 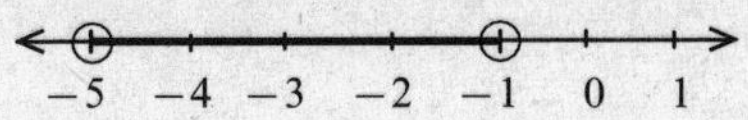

$-5 < x < -1; \quad x \in R$

6.8 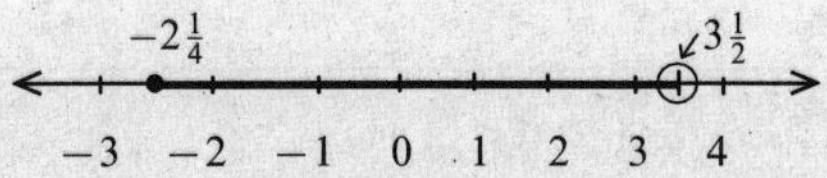

$-2\frac{1}{4} \leq x < 3\frac{1}{2}; \quad x \in R$

6.9 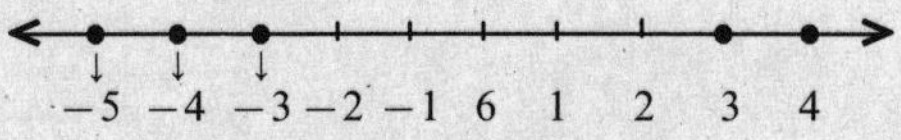

$x \leq -3$ *or* $x > 2; \quad x \in Z$

6.10 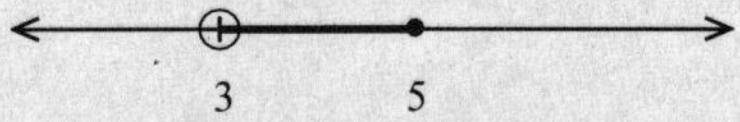

(3; 5]

6.11 

$(-5; \infty)$

6.12 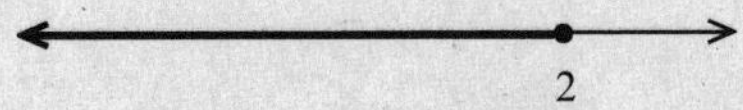

$(-\infty; 2]$

6.13 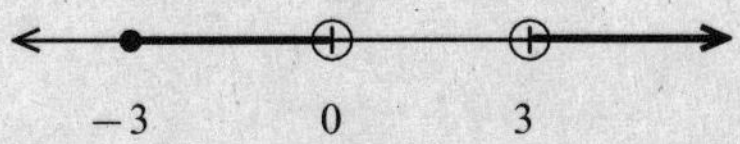

$-3 \leq x < 0$ *or* $x > 3; \quad x \in R$

6.14 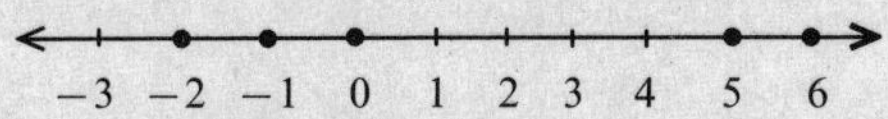

$-2\frac{1}{2} \leq x < 1$ *or* $x \geq 4\frac{1}{2}; \quad x \in Z$

7.1 $-2 \leq x < 3; \quad x \in R$

7.2 $x \leq 0$ *or* $x \geq 4; \quad x \in R$

7.3 $-1 \leq x \leq 3; \quad x \in Z$

7.4 $x \geq 1; \quad x \in Z$ [alternatively $x \in N$]

7.5 $-5 \leq x \leq -1$ *or* $x > 1; \quad x \in R$

7.6 $x < -4$ *or* $0 < x \leq 6; \quad x \in R$

2 *Indices*

Meaning $3^5 = 3 \times 3 \times 3 \times 3 \times 3 = 243$

$2x^3$ (circled; arrow labelled "power" pointing to it)

2 : Coefficient
x : Base
3 : Index

N.B.: The index 3 applies only to x.

Eg. $2(3)^3 = 2 \times 3 \times 3 \times 3 = 54$

$(2x)^3 = 2x \times 2x \times 2x = 8x^3$

Something to learn

$2^2 = 4$
$2^3 = 8$
$2^4 = 16$
$2^5 = 32$
$2^6 = 64$
$2^7 = 128$
$2^8 = 256$

$3^2 = 9$
$3^3 = 27$
$3^4 = 81$

$4^2 = 16$
$4^3 = 64$

$5^2 = 25$
$5^3 = 125$
$5^4 = 625$

$6^2 = 36$
$6^3 = 216$

$7^2 = 49$
$7^3 = 343$

$8^2 = 64$

$9^2 = 81$

$10^2 = 100$
$10^3 = 1000$

Remember:

$1^3 = 1$
$1^{10} = 1$
$1^{50} = 1$

$(-1)^3 = -1$
$(-1)^{10} = 1$
$(-1)^{51} = -1$

Laws of Indices and Deductions	Applications
A. Multiplication $a^m \times a^n = a^{m+n}$	1. $x^2x^3 = x^5$ 2. $2^2 \times 2^4 = 2^6 = 64$ 3. $-4^3 \times -4^4 = 4^7$ $(- \times - = +)$ 4. $(-2)^3 \times (-2)^2 = (-2)^5 = -32$
B. Division $\frac{a^m}{a^n} = a^{m-n}$	1. $\frac{x^7}{x^5} = x^2$ 2. $\frac{a^3}{a} = a^2$ 3. $\frac{b^7}{b^9} = b^{7-9} = b^{-2} = \frac{1}{b^2}$ (see D)
C. Power of O $a^o = 1$	1. $(3x)^o = 1$ 2. $y^o = 1$ 3. $4b^o = 4.1 = 4$ 4. $-5a^ob^2 = -5b^2$
D. Negative index $a^{-n} = \frac{1}{a^n}$	1. $x^{-1} = \frac{1}{x}$ 2. $\frac{1}{y^{-2}} = y^2$ 3. $\frac{x^{-1}}{y^{-1}} = \frac{y}{x}$ 4. $\frac{4a^{-2}}{b} = \frac{4}{a^2b}$ 5. $\left(\frac{2}{3}\right)^{-1} = \frac{3}{2}$ 6. $\left(\frac{3}{4}\right)^{-2} = \left(\frac{4}{3}\right)^2 = \frac{16}{9}$
E. Raise to a power $(a^m)^n = a^{m \times n}$ $= a^{mn}$	1. $(x^2)^3 = x^{2\times 3} = x^6$ 2. $\frac{(a^5)^3}{(a^2)^4} = \frac{a^{15}}{a^8} = a^7$
F. The Power of a Product $(ab)^m = a^mb^m$	1. $(2x^3)^2 = 2^2x^6 = 4x^6$ 2. $(a^2b)^4 = a^8b^4$

Exercise 1: Direct applications of laws of indices

1. Laws: $a^m a^n = a^{m+n}$ *and* $\dfrac{a^m}{a^n} = a^{m-n}$

Simplify:

1.1 $x^8 . x^3$

1.2 $x^4 . x^5 . x^6$

1.3 $\dfrac{a^3}{a}$

1.4 $\dfrac{a^5 . b^6 c^2}{ab^4}$

1.5 $2 . y^3 . y^5 . y^2 . x^3$

1.6 $2^3 . 2^4$

1.7 $x^3 . x^5 . x^2 \div x^4$

1.8 $2x^3 y^5 z \times - 5x^5 y^6$

1.9 $\dfrac{3x^{17}}{24x^{13}}$

1.10 $a^{2n} . y . a^{5n} . y^3$

1.11 $2^b . 2^a . 2$

2. Laws: $(a^m)^n = a^{mn}$ *and* $(ab)^m = a^m b^m$

2.1 $(xy)^3$

2.2 $(a^3)^2 . (a^2)^4$

2.3 $(3x^2)^2$

2.4 $\left(\dfrac{2}{x^3}\right)^2$

2.5 $(4x^5 y^3)^2$

2.6 $(2a^3 b)^3$

2.7 $(-3a^4 b^2 c)^3$

2.8 $\left(\dfrac{-2x}{y^3}\right)^3$

2.9 $(a^2)^5 \times (a^5)^2$

2.10 $\left(\dfrac{1}{2} a^2 b^3\right)^2$

3. Deduction: $a^0 = 1$

3.1 6^0

3.2 $2x^0$

3.3 $-3x^2 y^0$

3.4 $-a^0$

3.5 $(-a)^0$

3.6 $4a^0 b^2 c^0$

3.7 $\dfrac{x^0}{3}$

3.8 $\left(\dfrac{2}{3}\right)^0$

4. Deduction: $a^{-n} = \dfrac{1}{a^n}$

4.1 $\dfrac{a^{-2}}{b^{-3}}$

4.2 $\dfrac{mn}{x^{-1}}$

4.3 $(x^{-3})^2$

4.4 $2^{-2} . 3^2$

4.5 $(2a)^{-2}$

4.6 $5a^{-2} b^3$

4.7 $\left(\dfrac{2}{3}\right)^{-1}$

4.8 $\dfrac{x^{-2} y^0}{2^{-3}}$

Sometimes no unknown (variable) occurs in the expression. However, without using a calculator, the index laws may be used to simplify the expression.

Examples:

1. $2^{-5} = \frac{1}{2^5} = \frac{1}{32}$

2. $\frac{10^{-1}}{2.3^{-1}} = \frac{3}{2.10} = \frac{3}{20}$

3. $\frac{3^5.2^2}{2^5.3^2} = \frac{3^3}{2^3} = \frac{27}{8}$

4. $5^0.3^{-2} = \frac{1}{3^2} = \frac{1}{9}$

Mixed Examples:

1. $4x^3y^2 \times 3x^4y^2 = 12x^7y^4$

2. $\frac{5a^0b^5}{b} = 5b^4$

3. $\frac{4a^0+1}{(4a)^0+1} = \frac{4+1}{1+1} = \frac{5}{2}$

4. $(3a^3)^2 - a^{-4} = 9a^6 \times a^{-4} = 9a^2$

5. $2x^{-2} \times 3x = 6x^{-1} = \frac{6}{x}$

6. $\frac{5a^0b^5}{b^{-2}} = 5.b^5b^2 = 5b^7$

7. $3(x^4)^2 + 4x^8y^0 + \frac{9(x^4)^3}{18x^4}$

$= 3x^8 + 4x^8 + \frac{9x12}{18x^4}$

$= 3x^8 + 4x^8 + \frac{1}{2}x^8$

$= 7\frac{1}{2}x^8$

8. $\frac{(2a^2b^3)^4 \times (8ab^{-2})^2}{4ab^5 \times 12a^9b^7}$

$= \frac{\overset{4}{\cancel{16}}a^8b^{12}}{\underset{1}{\cancel{4}}ab^5} \times \frac{\overset{16}{\cancel{64}}a^2b^{-4}}{\underset{3}{\cancel{12}}a^9b^7}$

$= \frac{64a^{10}b^8}{3a^{10}b^{12}} = \frac{64}{3b^4}$

Exercise 2: Mixed

Simplify:

1. $ab^2 \times a^3b^4$

2. $(3x^{-1})^3$

3. $2x^3 . 3x^4$

4. $5^0 + 5^2 + 5^{-1}$

5. $2x^3(2x^3)^3$

6. $(a^2)^5 . a^2 . a^5$

7. $2x^0 + (2x)^0$

8. $\dfrac{x^5y^8}{xy^{10}}$

9. $2x^2 . 2x . (2x)^2$

10. 2^{-3}

11. $2^{-2}x^3$

12. $(2^3x^4)^2$

13. $-3(-2a^4)^3$

14. $(2x^2y)^3 \times \dfrac{5x^0y}{y^5}$

15. $-(-3x)^2$

16. $\left(\dfrac{3}{x^{-2}}\right)^2$

17. $(5x)^0 + 2^{-1}$

18. $\dfrac{(a^2b^3c)^2 . (2ac^2)^4}{2a^4 \times (b^2c)^3}$

19. $2xy^2 \div (4x^2y^3)^2 \times 8x^4y^5$ (be careful!)

20. $\dfrac{16a^5b}{12a^3b^6}$

21. $2p^0 + (3x)^0 + p$

22. $(x^2y^3)^4 \div x^2y^4$

23. $\left(\dfrac{3a^3b^4}{4c^5}\right)^2$

24. $\dfrac{(4a^2b^3)^3}{c^4a^4b}$

25. $\dfrac{15p^{14}r^5}{25p^9r^6}$

26. $(m^6n^4)^5 \div (m^3n^2)^2$

27. $2(2a^2b^3)^2(-2a^2b)$

28. $-2(4a)^2(-a)^2$

29. $\dfrac{ab^3 \times a^4b^2}{(ab^3)^5}$

30. $\dfrac{(x^2y)^3 \times (x^4y^2)^5}{y^7 \times x^6y^6}$

31. $\dfrac{(a^2b^3)^4 \times (8ab^{-2})^2}{4ab^5 \times 12a^9b^7}$

32. $\dfrac{3^{2n-1}}{3^{n+5}}$

33. $\dfrac{7^0x^3y^{-2}}{x^{-2}}$

34. $\dfrac{-15m^9n^2}{5m^3n^6}$

35. $(2x^3)^2 + (3x^2)^3$

36. $-2x^2 + 2(-x)^2$

37. $\dfrac{(x^6)^2}{-x^{12}}$

38. $(\frac{1}{2}x^3)^2$

39. $\dfrac{x^{6-p}}{x^{10-2p}}$

40. $(a^{-2})(-3a^0)$

41. $(4x^{-2}y^0z)^3$

42. $a^{x-1} \times a^x \times a^{1-x}$

43. $(3x^{n-1})^3$

44. $\dfrac{x^{pq}}{x^p}$

45. $\dfrac{-x^0}{(-2x)^0}$

46. $(4x^{n-1})^2$

47. $(4^{n-1})^2$

48. $-2(-2a^2b)^4$

49. $4a^2 - 3a \times 4b^2 - \sqrt{36a^2b^4} + \dfrac{8a^3}{4a}$

50. $\dfrac{x^{-4}y^3}{2x^{-1}y}$

Substitution:

This means that a specific value may be assigned to each unknown namely a, b, x, y . . . etc. The expression will now have a numerical value.

Method:

Substitute numbers for the variables, (preferably in ( )).

Examples:

1. If $x = 2$, find: $(-2x)^4$

Solution: $(-2x)^4 = (-2(2))^4$

$= (-4)^4$

$= 256$

2. If $a = -2$, find: $-2a^3 - (-2a)^3$

Solution: $-2a^3 - (-2a)^3 = -2(-2)^3 - (-2(-2))^3$

$= -2(-8) - ({}^{+}4)^3$

$= 16 - 64$

$= -48$

3. If $x = -1$ and $y = +2$, find: $x^2 - (2xy)^3$

Solution: $x^2 - (2xy)^3 = (-1)^2 - (2(-1)(+2))^3$

$= 1 - (-4)^3$

$= 1 - (-64)$

$= 1 + 64$

$= 65$

Exercise 3: Substitution

1. Use $x = -1$; $y = -2$ and $z = 3$ to find the value of:

1.1 $4 - x^2 - z^2$

1.2 $4x^2y$

1.3 $-2 + (xy)^2$

1.4 $-(2x)^3$

1.5 $(\frac{1}{2}x)^3$

1.6 $3(y^2)^3$

1.7 $-(-y)^3$

1.8 $xy^0 - z^2$

1.9 $2x^3 - 3yz$

2. If $a = -2$; $b = 5$ and $c = 0$, calculate the value of:

2.1 $-a^7 - 2ab$

2.2 $(a + b)^c + (a + b)(a - b)$

3. If $a = -3$; $c = 1$; $d = -2$, calculate the value of:

3.1 $\sqrt{a^2 + 2d - c}$

3.2 $-3a^2 + d^2$

3.3 $\sqrt{a + d^2}$

Scientific Notation:

Scientific notation is a convenient way (notation) to express very large or very small numbers.

Examples:

Scientific notation requires 1 significant (not zero) digit in front of the comma.

1. $0,003\,18 = 3,18 \times 10^{-3}$

2. $1\,340\,000\,000 = 1,34 \times 10^{9}$

Comma is "moved" from: $\uparrow_1$ to $\uparrow_2$, that is 9 places.

3. $1,25 \times 10^{4} = 12\,500$

4. $5,068 \times 10^{-6} = 0,000\,005\,068$

Calculations using very large or very small numbers can be simplified without a calculator, by using **Scientific Notation and the Laws of Indices**.

Examples:

1. $2 \times 10^{4} \times 9,1 \times 10^{-8}$
$= 2 \times 9,1 \times 10^{4-8}$
$= 18,2 \times 10^{-4}$
$= 1,82 \times 10^{-3}$

2. $1,05 \times 10^{-2} + 3,8 \times 10^{-3}$
$= 1,05 \times 10^{-2} + \mathbf{0,38 \times 10^{-2}}$
$= (1,05 + 0,38) \times 10^{-2}$
$= 1,43 \times 10^{-2}$

Exercise 4: Scientific Notation

1. Write in scientific notation:

1.1 3 728 935
1.2 173,529
1.3 0,000 060 5
1.4 0,007 832
1.5 87 500 000
1.6 0,006
1.7 345 000 000
1.8 158,002 3

2. Write these as ordinary numbers:

2.1 $8,515 \times 10^{4}$
2.2 $6,12 \times 10^{-5}$
2.3 $3,206 \times 10^{6}$
2.4 $7,632 \times 10^{-2}$
2.5 $8,43 \times 10^{-1}$

3. Simplify the following without using a calcolator:

3.1 $3,6 \times 10^{3} \times 2 \times 10^{4}$
3.2 $2,34 \times 10^{-12} \times 5 \times 10^{7}$
3.3 $1,6 \times 10^{8} \times 3 \times 10^{7}$
3.4 $4,2 \times 10^{-4} + 5,1 \times 10^{-4}$
3.5 $7,23 \times 10^{-2} + 6,9 \times 10^{-2}$

SOLUTIONS

Exercise 1: Solution. (Direct application of the laws of indices)

1.1 $x^8x^3 = x^{11}$

1.2 $x^4x^5x^6 = x^{15}$

1.3 $\dfrac{a^3}{a} = a^2$

1.4 $\dfrac{a^5b^6c^2}{ab^4} = a^4b^2c^2$

1.5 $2y^3y^5y^2x^3 = 2y^{10}x^3$

1.6 $2^3 2^4 = 2^7$

1.7 $x^3x^5x^2 \div x^4 = \dfrac{x^{10}}{x^4} = x^6$

1.8 $2x^3y^5z \times -5x^5y^6 = -10x^8y^{11}z$

1.9 $\dfrac{3x^{17}}{24x^{13}} = \dfrac{x^4}{8}$

1.10 $a^{2n}.\,y\,.\,a^{5n}.\,y^3 = a^{7n}y^4$

1.11 $2^b.\,2^a.\,2 = 2^{b+a+1}$

2.1 $(xy)^3 = x^3y^3$

2.2 $(a^3)^2.\,(a^2)^4 = a^6.\,a^8 = a^{14}$

2.3 $(3x^2)^2 = 9x^4$

2.4 $\left(\dfrac{2}{x^3}\right)^2 = \dfrac{4}{x^6}$

2.5 $(4x^5y^3)^2 = 16x^{10}y^6$

2.6 $(2a^3b)^3 = 8a^9b^3$

2.7 $(-3a^4b^2c)^3 = -27a^{12}b^6c^3$

2.8 $\left(\dfrac{-2x}{y^3}\right)^3 = \dfrac{-8x^3}{y^9}$

2.9 $(a^2)^5 \times (a^5)^2 = a^{10} \times a^{10}$
$= a^{20}$

2.10 $\left(\dfrac{1}{2}a^2b^3\right)^2 = \dfrac{1}{4}a^4b^6$

3.1 $6^0 = 1$

3.2 $2x^0 = 2$

3.3 $-3x^2y^0 = -3x^2$

3.4 $-a^0 = -1$

3.5 $(-a)^0 = 1$

3.6 $4a^0b^2c^0 = 4b^2$

3.7 $\dfrac{x^0}{3} = \dfrac{1}{3}$

3.8 $\left(\dfrac{2}{3}\right)^0 = 1$

4.1 $\dfrac{a^{-2}}{b^{-3}} = \dfrac{b^3}{a^2}$

4.2 $\dfrac{mn}{x^{-1}} = mnx$

4.3 $(x^{-3})^2 = x^{-6} = \dfrac{1}{x^6}$

4.4 $2^{-2}.\,3^2 = \dfrac{9}{4}$

4.5 $(2a)^{-2} = \dfrac{1}{(2a)^2} = \dfrac{1}{4a^2}$

4.6 $5a^{-2}b^3 = \dfrac{5b^3}{a^2}$

4.7 $\left(\dfrac{2}{3}\right)^{-1} = \dfrac{3}{2}$

4.8 $\dfrac{x^{-2}y^0}{2^{-3}} = \dfrac{8}{x^2}$

Exercise 2: Solutions (Mixed)

1. $ab^2 3b^4 = a^4b^6$

2. $(3x^{-1})^3 = 27x^{-3} = \dfrac{27}{x^3}$

3. $2x^3 . 3x^4 = 6x^7$

4. $5^0 + 5^2 + 5^{-1} = 1 + 25 + \dfrac{1}{5} = 26\dfrac{1}{5}$

5. $2x^3(2x^3)^3 = 2x^3(8x^9) = 16x^{12}$

6. $(a^2)^5 . a^2 . a^5 = a^{10} . a^2 . a^5 = a^{17}$

7. $2x^0 + (2x)^0 = 2 + 1 = 3$

8. $\dfrac{x^5y^8}{xy^{10}} = \dfrac{x^4}{y^2}$

9. $2x^2 . 2x(2x)^2 = 2x^2 . 2x . 4x^2$

   $= 16x^5$

10. $2^{-3} = \dfrac{1}{2^3} = \dfrac{1}{8}$

11. $2^{-2}x^3 = \dfrac{x^3}{2^2} = \dfrac{x^3}{4}$

12 $(2^3x^4)^2 = 2^6x^8 = 64x^8$

13. $-3(-2a^4)^3 = -3(-8a^{12}) = 24a^{12}$

14. $(2x^2y)^3 \times \dfrac{5x^0y}{y^5} = 8x^6y^3 \times \dfrac{5}{y^4} = \dfrac{40x^6}{y}$

15. $-(-3x)^2 = -(9x^2) = -9x^2$

16. $\left(\dfrac{3}{x^{-2}}\right)^2 = \dfrac{9}{x^{-4}} = 9x^4$

17. $(5x)^0 + 2^{-1} = 1 + \dfrac{1}{2} = 1\dfrac{1}{2}$

18. $\dfrac{(a^2b^3c)^2(2ac^2)^4}{2a^4 \times (b^2c)^3} = \dfrac{a^4b^6c^2 \times 16a^4c^8}{2a^4 \times b^6c^3}$

    $= \dfrac{16a^8b^6c^{10}}{2a^4b^6c^3}$

    $= 8a^4c^7$

19. $2xy^2 \div (4x^2y^3)^2 \times 8x^4y^5$

    $= \dfrac{2xy^2 \times 8x^4y^5}{16x^4y^6} = \dfrac{16x^5y^7}{16x^4y^6} = xy$

20. $\dfrac{16a^5b}{12a^3b^6} = \dfrac{4a^2}{3b^5}$

21. $2p^0 + (3x)^0 + p = 2 + 1 + p = 3 + p$

22. $\dfrac{(x^2y^3)^4}{x^2y^4} = \dfrac{x^8y^{12}}{x^2y^4} = x^6y^8$

23. $\left(\dfrac{3a^3b^4}{4c^5}\right)^2 = \dfrac{9a^6b^8}{16c^{10}}$

24. $\dfrac{(4a^2b^3)^3}{c^4a^4b} = \dfrac{64a^6b^9}{c^4a^4b} = \dfrac{64a^2b^8}{c^4}$

25. $\dfrac{15p^{14}r^5}{25p^9r^6} = \dfrac{3p^5}{5r}$

26. $\dfrac{(m^6n^4)^5}{(m^3n^2)^2} = \dfrac{m^{30}n^{20}}{m^6n^4} = m^{24}n^{16}$

27. $2(2a^2b^3)^2(-2a^2b)$

    $= 2(4a^4b^6)(-2a^2b)$

    $= -16a^6b^7$

28. $-2(4a)^2(-a)^2 = -2(16a^2)(a^2)$

    $= -32a^4$

29. $\dfrac{ab^3 \times a^4b^2}{(ab^3)^5} = \dfrac{a^5b^5}{a^5b^{15}} = \dfrac{1}{b^{10}}$

30. $\dfrac{(x^2y)^3 \times (x^4y^2)^5}{y^7 \times x^6y^6} = \dfrac{x^6y^3 \times x^{20}y^{10}}{y^{13}x^6}$

    $= \dfrac{x^{26}y^{13}}{x^6y^{13}} = x^{20}$

31. $\dfrac{(a^2b^3)^4 \times (8ab^{-2})^2}{4ab^5 \times 12a^9b^7}$

    $= \dfrac{a^8b^{12} \times 64a^2b^{-4}}{4ab^5 \times 12a^9b^7}$

    $= \dfrac{64a^{10}b^8}{48a^{10}b^{12}}$

    $= \dfrac{4}{3b^4}$

32. $\dfrac{3^{2n-1}}{3^{n+5}} = 3^{2n-1-(n+5)}$

    $= 3^{2n-1-n-5}$

    $= 3^{n-6}$

33. $\dfrac{7^0x^3y^{-2}}{x^{-2}} = x^{3-(-2)}y^{-2}$

$= \dfrac{x^5}{y^2}$

34. $\dfrac{-15m^9n^2}{5m^3n^6} = \dfrac{-3m^6}{n^4}$

35. $(2x^3)^2 + (3x^2)^3$

$= 4x^6 + 27x^6$

$= 31x^6$

36. $-2x^2 + 2(-x)^2$

$= -2x^2 + 2x^2 = 0$

37. $\dfrac{(x^6)^2}{-x^{12}} = \dfrac{x^{12}}{-x^{12}} = -1$

38. $\left(\dfrac{1}{2}x^3\right)^2 = \dfrac{1}{4}x^6$

39. $\dfrac{x^{6-p}}{x^{10-2p}} = x^{6-p-(10-2p)}$

$= x^{6-p-10+2p}$

$= x^{p-4}$

40. $(a^{-2})(-3a^0) = \left(\dfrac{1}{a^2}\right)(-3)$

$= \dfrac{-3}{a^2}$

41. $(4x^{-2}y^0z)^3 = 64x^{-6}z^3 = \dfrac{64z^3}{x^6}$

42. $a^{x-1} \times a^x \times a^{1-x} = a^{x-1+x+1-x}$

$= a^x$

43. $(3x^{n-1})^3 = 27x^{3n-3}$

44. $\dfrac{x^{pq}}{x^p} = x^{pq-p}$

45. $\dfrac{-x^0}{(-2x)^0} = \dfrac{-1}{1} = -1$

46. $(4x^{n-1})^2 = 16x^{2n-2}$

47. $(4^{n-1})^2 = 4^{2n-2}$

48. $-2(-2a^2b)^4 = -2(16a^8b^4)$

$= -32a^8b^4$

49. $4a^2 - 3a \times 4b^2 - \sqrt{36a^2b^4} + \dfrac{8a^3}{4a}$

$= 4a^2 - 12ab^2 - 6ab^2 + 2a^2$

$= 6a^2 - 18ab^2$

50. $\dfrac{x^{-4}y^3}{2x^{-1}y} = \dfrac{xy^3}{2x^4y} = \dfrac{y^2}{2x^3}$

Exercise 3: Solutions (Substitution)

1.1 $4 - x^2 - z^2 = 4 - (-1)^2 - (3)^2$

$= 4 - 1 - 9 = -6$

1.2 $4x^2y = 4(-1)^2(-2)$

$= 4(1)(-2) = -8$

1.3 $-2 + (xy)^2 = -2 + ((-1)(-2))^2$

$= -2 + (2)^2$

$= -2 + 4 = 2$

1.4 $-(2x)^3 = -(2(-1))^3 = -(-2)^3$

$= -(-8) = +8$

1.5 $\left(\dfrac{1}{2}x\right)^3 = \left(\dfrac{1}{2}(-1)\right)^3 = \left(-\dfrac{1}{2}\right)^3 = -\dfrac{1}{8}$

1.6 $3(y^2)^3 = 3y^6 = 3(-2)^6$

$= 3 \times 64 = 192$

1.7 $-y^3 = -(-2)^3$

$= -(-8) = 8$

1.8 $xy^0 - z^2 = x - z^2$

$= (-1) - (3)^2 = -1 - 9$

$= -10$

1.9 $2x^3 - 3yz$

$= 2(-1)^3 - 3(-2)(3)$

$= 2(-1) + 18$

$= -2 + 18 = 16$

2.1 $-a^7 - 2ab = -(-2)^7 - 2(-2)(5)$
$= -(-128) + 20 = 148$

2.2 $(a+b)^c + (a+b)(a-b)$
$= (-2+5)^0 + (-2+5)(-2-5)$
$= 1 + (3)(-7) = 1 - 21$
$= -20$

3.1 $\sqrt{a^2 + 2d - c} = \sqrt{(-3)^2 + 2(-2) - 1}$
$= \sqrt{9 - 4 - 1}$
$= \sqrt{4} = 2$

3.2 $-3a^2 + d^2 = -3(-3)^2 + (-2)^2$
$= -3(9) + 4$
$= -27 + 4 = -23$

3.3 $\sqrt{a + d^2} = \sqrt{-3 + (-2)^2}$
$= \sqrt{-3 + 4}$

Exercise 4: Solutions (Scientific notation)

1.1 $3\ 728\ 935 = 3,728\ 935 \times 10^6$

1.2 $173,529 = 1,735\ 29 \times 10^2$

1.3 $0,000\ 060\ 5 = 6,05 \times 10^{-5}$

1.4 $0,007\ 832 = 7,832 \times 10^{-3}$

1.5 $87\ 500\ 000 = 8,75 \times 10^7$

1.6 $0,006 = 6 \times 10^{-3}$

1.7 $345\ 000\ 000 = 3,45 \times 10^8$

1.8 $158,002\ 3 = 1,580\ 023 \times 10^2$

2.1 $8,515 \times 10^4 = 85\ 150$

2.2 $6,12 \times 10^{-5} = 0,000\ 061\ 2$

2.3 $3,206 \times 10^6 = 3\ 206\ 000$

2.4 $7,632 \times 10^{-2} = 0,076\ 32$

2.5 $8,43 \times 10^{-1} = 0,843$

3.1 $3,6 \times 10^3 \times 2 \times 10^4$

$= 7,2 \times 10^7$

3.2 $2,34 \times 10^{-12} \times 5 \times 10^7$

$= (2,34 \times 5) \times 10^{-12+7}$

$= 11,70 \times 10^{-5}$

$= 1,17 \times 10^{-4}$

3.3 $1,6 \times 10^8 \times 3 \times 10^7$

$= 4,8 \times 10^{15}$

3.4 $4,2 \times 10^{-4} + 5,1 \times 10^{-4}$

$= (4,2 + 1,5) \times 10^{-4}$

$= 5,7 \times 10^{-4}$

3.5 $7,23 \times 10^{-2} + 6,9 \times 10^{-2}$

$= (7,23 + 6,9) \times 10^{-2}$

$= 14,13 \times 10^{-2}$

$= 1,413 \times 10^{-1}$

3 *Polynomials*

Definition:

A polynomial is an algebraic expression where the indices of the variables ($x; y$, etc.) are positive integers or O, and the variables only appear in the numerator of the terms.

Examples:

1. $3x^6 - 5x^4 + 2x^3 - \frac{1}{2}x^2 - 5x + 8$ is a polynomial in descending powers of x.

- The **degree** of the expression is: 6; (i.e. the highest power of x).
- The **coefficient** of the 2nd term is: -5
- The **coefficient** of the 4th term is: $-\frac{1}{2}$
- The **index** of the 1st term is: 6
- The **index** of the 5th term is: 1
- 8 is the **constant term**.
- If $x = -1$, the value of the expression is:

$$3(-1)^6 - 5(-1)^4 + 2(-1)^3 - \tfrac{1}{2}(-1)^2 - 5(-1) + 8$$
$$= 3(1) - 5(1) + 2(-1) - \tfrac{1}{2}(1) + 5 + 8$$
$$= 3 - 5 - 2 - \tfrac{1}{2} + 5 + 8$$
$$= 8\tfrac{1}{2}$$

2. $4b^3 - 2ab^2 + \frac{1}{3}a^2b - 5a^3$ is a polynomial in **descending powers of *b* and in ascending powers of *a*.**

Exercise 1:

1. In the expression: $-3x^4 + 5x^3 - 2x^2 - 8x + 2$,

(*a*) the coefficient of the 1st term is: ____________________
(*b*) the index of the 4th term is: ____________________
(*c*) the constant term is: ____________________
(*d*) the degree of the polynomial is: ____________________
(*e*) the number of terms is: ____________________
(*f*) the terms are arranged in ____________________ powers of x.
(*g*) if $x = -2$, the value of the polynomial is: ____________________

2. Given: $5x + 6x^3 + 2x^2 + 3 - 5x^3 - 2x^2 + 6x$.

(*a*) Simplify and write the expression in **ascending powers of *x***
Now find:
(*b*) the degree of the polynomial.
(*c*) the value of the polynomial if: $x = -1$.

3. Given: $27ab - 16a^4b^2 + 8(ab)^3 - 6 + 3a^2$.

(*a*) Simplify and write the polynomical in **decending powers of *a***.
Now find:
(*b*) the constant term.
(*c*) the numerical coefficient of the first term.
(*d*) the index of b in the second term.
(*e*) the value of the expression if: $a = -1$ and $b = \frac{1}{2}$

4. Determine the value of:

4.1 $2x + 5$ if $x = 3$

4.2 $-x + 3$ if $x = -5$

4.3 $-2x + 5$ if $x = \frac{1}{2}$

4.4 $x^2 + 4x$ if $x = -2$

4.5 $a^2 - 2ab + 5b^2$ if $a = -3$ and $b = 2$

4.6 $-3x^2 - x$ if $x = -\frac{1}{2}$

4.7 $-2x^2 - 4x + 8$ if $x = -3$

4.8 $6a^2 + 2a - 3$ if $a = \frac{1}{3}$

In the remaining part of the chapter we shall be adding, subtracting, multiplying and dividing polynomials.

1. Addition:

Example:

Add the following polynomials:
$-2a^2 - 3a + 5$; $-5a + 2$; $3 - a + 5a^2$

Solution:

Horizontal method: $-2a^2 - 3a + 5 - 5a + 2 + 3 - a + 5a^2 = 3a^2 - 9a + 10$
or
Vertical method:

$$\begin{array}{r} -2a^2 - 3a + 5 \\ -5a + 2 \\ \underline{5a^2 - \quad a + 3} \\ \underline{3a^2 - 9a + 10} \end{array}$$

Exercise 2:

Add:/Determine the sum of:

1. $-a^3 + 4a^2 + 5a + 2$; $-2a^3 - 3a^2 + 4a - 2$
2. $x^2 - 3x + 2$; $-2x^2 - 5x$; $3x^2 + 8x - 6$
3. $-2a + 5$; $-a^2 + 3a - 1$; $5a^2 - 2a + 5$
4. $-4a^2 - 2$; $-3a + 10$; $3 - 5a^2$
5. $-2x^2y + 5xy$; $2xy - 3x^2y$; $4xy^2 - 5xy - 2x^2y$

2. Subtraction:

Example:

Subtract $(4a^2 - 2ab + 8b^2)$ from $(-a^2 - 3ab + 5b^2)$

Solution:

Note which polynomial is written down first.

Horizontal:

$$\begin{aligned} & -a^2 - 3ab + 5b^2 - (4a^2 - 2ab + 8b^2) \\ &= -a^2 - 3ab + 5b^2 - 4a^2 + 2ab - 8b^2 \\ &= -5a^2 - ab - 3b^2 \end{aligned}$$

Vertical:

$$\begin{array}{rcrcr} -a^2 & - & 3ab & + & 5b^2 \\ {}^{(-)}4a^2 & \overset{(+)}{-} & 2ab & \overset{(-)}{+} & 8b^2 \\ \hline -5a^2 & - & ab & - & 3b^2 \end{array}$$

Exercise 3:

1. Subtract $4x^2 - 2x + 3$ from $-5x + 8 + 6x^2$
2. Subtract $5a^2b - 2ab^2 + 4$ from $3 - 6a^2b - 7ab^2$
3. From $8p^2 - 3p - 10$ subtract $4 - 2p + 6p^2$
4. From $3x$ subtract $5x^2 - 2x$
5. What must be added to $x^2y - 2xy$ to give $4 - 3x^2y + xy$?

3. Multiplication:

Type 1: $a(b + c) = ab + ac$

Type 2: $(a + b)(c + d) = ac + ad + bc + bd$

Type 3: $(a + b)(c + d + e) = ac + ad + ae + bc + bd + be$

Type 4: (Method 1): $(a + b)^2 = (a + b)(a + b) = a^2 + ab + ab + b^2$ (as in type 2)
$= a^2 + 2ab + b^2$

(Method 2): $(a + b)^2 = a^2 + 2ab + b^2$ (square a → a^2; square b → b^2; $2ab$ = $a \times b \times 2$)

Examples: Type 1:

1. $5xy(3x - 2y^3) = 15x^2y - 10xy^4$
2. $x(x + 1) - (2x + 3) = x^2 + x - 2x - 3 = x^2 - x - 3$

Exercise 4:

Simplify:

1. $2a(3a - b - 4)$
2. $3x^2y(-2xy + 4y^2)$
3. $4x(x + 2y) + 5(x^2 - 2xy) - (3x^2 - 4xy)$
4. $4x - 3x(2x - \frac{1}{3})$
5. $\frac{2}{3}x(9x - 6y + 4)$

Example: Type 2:

1. $(2x + 3y)(4x + y) = 8x^2 + 2xy + 12xy + 3y^2$
$\quad = 8x^2 + 14xy + 3y^2$

2. $(4x^2 - y)(2x^2 + \frac{1}{2}y) = 8x^4 + 2x^2y - 2x^2y - \frac{1}{2}y^2$
$\quad = 8x^4 - \frac{1}{2}y^2$

Exercise 5:

Simplify:

1. $(x + 3)(x + 5)$
2. $(2x + 3)(4x - 5)$
3. $(4x - 3y)(x + y)$
4. $(5x - 2y)(5x + 2y)$
5. $(3x^2 - 2y)(8x^2 + y)$
6. $(x + \frac{1}{2})(x - \frac{1}{2})$
7. $(x^2 + 1)(x^2 - 1)$
8. $-2(3x + y)(5x - 4y)$

Example: Type 3:

$(2x + y)(x^2 + 3xy - 5y^2)$

$= 2x^3 + 6x^2y - 10xy^2 + x^2y + 3xy^2 - 5y^3$
$= 2x^3 + 7x^2y - 7xy^2 - 5y^3$

Exercise 6:

Simplify:

1. $(3x - 4)(2x^2 - 3x + 2)$
2. $(4a - b)(a^2 + 4ab - 3b^2)$
3. $(4x - 5y)(2x^2 - 10xy + 7y^2)$
4. $(a^2 - 3ax + 5x^2)(2a - 3x)$
5. $(4x - \frac{1}{2}y)(x^2 - 2xy + 6y^2)$

Example: Type 4:

Choose either method 1, which is exactly the same as type 2, or method 2, which is a short cut. To master method 2 requires practise.

Method 1: 1. $(2x - 5y)^2 = (2x - 5y)(2x - 5y)$
$\quad = 4x^2 - 10xy - 10xy + 25y^2$
$\quad = 4x^2 - 20xy + 25y^2$

2. $2(4x - 3y)^2 = 2(4x - 3y)(4x - 3y)$
$\quad = 2(16x^2 - 12xy - 12xy + 9y^2)$
$\quad = 32x^2 - 24xy - 24xy + 18y^2$
$\quad = 32x^2 - 48xy + 18y^2$

Method 2: 1. $(2x - 5y)^2 = 4x^2 + 2(2x)(-5y) + 25y^2$
$\quad = 4x^2 - 20xy + 25y^2$

2. $2(4x - 3y)^2 = 2(16x^2 + 2(4x)(-3y) + 9y^2)$
$\quad = 32x^2 - 48xy + 18y^2$

To find the middle term multiply the product of the two terms by 2.

Exercise 7:

Simplify:

1. $(4x - 3y)^2$
2. $(2a + 5b)^2$
3. $(c + d)^2$
4. $(5x - 6y)^2$
5. $(4a - 5b)^2$
6. $(2a^2 - 3b)^2$
7. $(4x^2 - y^3)^2$
8. $-2(3x - 5y)^2$
9. $[2a(4 - 3a)]^2$

Mixed example:

$(3x - 2y)^2 - 4(x + 2y)(8x - y) - (x - y)^2$

$= (9x^2 - 12xy + 4y^2) - 4(8x^2 - xy + 16xy - 2y^2) - (x^2 - 2xy + y^2)$

$= 9x^2 - 12xy + 4y^2 - 32x^2 + 4xy - 64xy + 8y^2 - x^2 + 2xy - y^2$

$= -24x^2 - 70xy + 11y^2$

Exercise 8: Mixed

Simplify:

1. $(2a + 5b)(2a - 5b)$
2. $(5x - 3y)(3x + 2y)$
3. $(3s - 4t)(s^2 - 2)$
4. $-7a(-a + 1) - 3a(5 - a)$
5. $(2x^2 - 5y)(3x^2 + 4y)$
6. $3x - 3x(6x^2 - 7x + 1)$
7. $2a(4a - 5b)(3a + 8b)$
8. $\left(\frac{2}{x} + 1\right)\left(\frac{3}{x} + 3\right)$
9. $\frac{a^2}{b^3}\left(b^2 - \frac{b^3}{a^2}\right)$
10. $(x - 2)(x + 3)(2x + 1)$
11. $x(x + 1) - (x + 1)(x + 3)$
12. $\left(a + \frac{2}{a}\right)^2$
13. $3(2x - 5y)^2$
14. $(3x + y)(3x - y) - (2x - y)^2$
15. $[2(x - y)]^2$
16. $(3x - 2y)^2 - 3x(5x - y)$
17. $(3a - 2b)^2 - 3(4a^2 + 2) + (a + b)(3a - b)$
18. $(x + 2y)(x^2 + 3xy - 5y^2)$
19. $(x + 5)^2 - (x - 5)^2$
20. $2(x - 3)^2 - 4(2x - 3) - (x + 4)(2x - 1)$

4. Division:

(The syllabus only requires division by a monomial, i.e. one term)

Examples:

1. $\dfrac{16a^4 - 32a^3 + 8a^2}{-8a^2} = -2a^2 + 4a - 1$

2. $\dfrac{24a^3b - 12a^2b^2 + 3a^2}{-3a^2b} = -8a + 4b - \dfrac{1}{b}$

3. $\dfrac{4x^3 - 8x + 1}{-4x^2} = -x + \dfrac{2}{x} - \dfrac{1}{4x^2}$

Exercise 9:

Simplify:

1. $\dfrac{9a^3b^5 + 24a^2b^8 - 3ab^3}{3ab^3}$

2. $\dfrac{14x^5 - 20x^3y^2 + 8x^6y^3}{-2x^2}$

3. $\dfrac{-15xy^3 - 5x^2y^5 + 30x^8y^2}{-5xy^3}$

4. $\dfrac{20xy^2 + xy - 2x^2y^5}{4x^3y^2}$

5. $\dfrac{-9x^3y^2 + 6xy^5 - 8x^5}{6x^3y^3}$

Division by two terms (optional)

Example:

$(6x^3 + 11x^2 - 19x + 6) \div (3x - 2)$

Solution:

Step 2 Step 3

$6x^3 + 11x^2 - 19x + 6 = (3x - 2)(2x^2 + 5x - 3)$ ($-4x^2$)

Step 4

Step 1: Write down a 2nd bracket with $(\ldots x^2 \ldots x \ldots)$

Step 2: Determine the coefficient of x^2, i.e. $3x \times ?x^2 = 6x^3$; which is $\mathbf{2x^2}$

Step 3: Determine the constant term in the 2nd bracket, i.e. $-2 \times ? = +6$; which is $\mathbf{-3}$

Step 4: Now try to find: $11x^2$

You already have: $-2 \times 2x^2 = -4x^2$

What do you need to have a total of $11x^2$? $+\ 15x^2$

And $3x \times .?. = 15x^2$, therefore $\mathbf{5x}$

Exercise 10:

Simplify:

1. $(8x^3 - 18x^2 + 5x + 6) \div (2x - 3)$

2. $(6x^3 + 19x^2 + 6x - 10) \div (2x + 5)$

3. $(10x^3 - 21x^2 + 5x + 6) \div (2x - 3)$

4. $(12k^3 - 4k^2 - 17k + 10) \div (6k - 5)$

5. $(30 - 3p^2 + p^3 - 23p) \div (p - 6)$
(First rearrange in the correct order)

6. $(x^3 + 2x^2y - 17xy^2 + 6y^3) \div (x - 3y)$

(In this case the second bracket will be: $(...x^2 ...xy ...y^2)$)

7. $(2a^3 - 7a^2b + 7ab^2 - 2b^3) \div (a - 2b)$

8. $(8x^3 - 2x^2y + 3xy^2 + y^3) \div (4x + y)$

Expressions with multiple brackets:

Simplify the expression by always removing the innermost brackets.

Example:

$3\{2 - 4[x - 2(-3x + 1)] - 5x\}$

$= 3\{2 - 4[x + 6x - 2 - 5x]\}$

$= 3\{2 - 4x - 24x + 8 + 20x\}$

$= 6 - 12x - 72x + 24 + 60x$

$= -18x + 24$

Exercise 11:

Simplify:

1. $4x^2 + 2[4x - 2x\{3 - (x - 2)\} + 3]$

2. $5x - [5y - (2x - 2y) - 5x]$

3. $4a - 3[a + 2(-b + a) - (2a - 3b)]$

4. $2a - 3[4a - \{2(a + 1) - 7\}]$

5. $[-3a - 4(2 + a)][-(-2a)^2 - 6(a - 1)]$

Squares and square roots:

Examples:

1. $\sqrt{a^2} = a$ N.B. Divide the indices of a, b, ... by 2

2. $\sqrt{64a^2y^6} = 8ay^3$

3. $\sqrt{25(x-y)^2} = 5(x-y)$

4. $\sqrt{5^2 4^2} = 5 \times 4 = 20$

5. $\sqrt{5^2 - 4^2} = \sqrt{25 - 16} = \sqrt{9} = 3$

6. $\sqrt{10 \times 15 \times 6} = \sqrt{\underbrace{2 \times 5}_{10} \times \underbrace{5 \times 3}_{15} \times \underbrace{3 \times 2}_{6}}$
   $= \sqrt{2^2.5^2.3^2} = 2 \times 5 \times 3 = 30$

7. $\sqrt{\dfrac{18a^{16}b^7}{50b^9}} = \sqrt{\dfrac{9a^{16}}{25b^2}} = \dfrac{3a^8}{5b}$

8. $(\sqrt{3})^2 = 3$

9. $(2\sqrt{3})^2 + (3\sqrt{5})^2 = 4 \times 3 + 9 \times 5$
   $= 12 + 45$
   $= 57$

10. $2\sqrt{7} \times 3\sqrt{7} = 6 \times 7 = 42$

11. $2\sqrt{7} + 3\sqrt{7} = 5\sqrt{7}$

Exercise 12:

Simplify:

1. $\sqrt{25a^4b^2}$

2. $\sqrt{(a+b)^2}$

3. $\sqrt{16x^{16}}$

4. $\sqrt{9x^4 + 16x^4}$

5. $\sqrt{\dfrac{m^2(a+b)^4}{121}}$

6. $\sqrt{\dfrac{32a^3b^4}{8a}}$

7. $\sqrt{\dfrac{20a^5}{5a}}\sqrt{\dfrac{9a^3b^2}{a^5}}$

8. $\sqrt{\dfrac{50a^6b}{98b^9}}$

9. $3\sqrt{3}\ \sqrt{3}$

10. $-3\sqrt{12}\ \sqrt{3}$

11. $a\sqrt{a^4}\ a^2$

12. $\sqrt{x^2}\ \sqrt{x^6}$

13. $(2\sqrt{3})^2 + (4\sqrt{2})^2$

14. $3(\sqrt{12})^2 + (\sqrt{5})^2$

15. $\sqrt{2\frac{1}{4}a^6}$

SOLUTIONS

Exercise 1: Solutions

1. (*a*) coefficient of 1st term: -3
(*b*) index of 4th term: 1
(*c*) constant term: 2
(*d*) degree: 4
(*e*) number of terms: 5
(*f*) descending
(*g*) $-3x^4 + 5x^3 - 2x^2 - 8x + 2$
$= -3(-2)^4 + 5(-2)^3 - 2(-2)^2 - 8(-2) + 2$
$= -3(16) + 5(-8) - 2(4) + 16 + 2$
$= -48 - 40 - 8 + 16 + 2 = -78$

2. (*a*) $3 + 11x + x^3$
(*b*) degree: 3
(*c*) $x + 11x + 3$
$= (-1)^3 + 11(-1) + 3$
$= -1 - 11 + 3 = -9$

3. (*a*) $-16a^4b^2 + 8a^3b^3 + 3a^2 + 27ab - 6$
(*b*) constant term: -6
(*c*) coefficient of 1st term: -16
(*d*) index of b in 2nd term: 3
(*e*) $-16a^4b^2 + 8a^3b^3 + 3a^2 + 27ab - 6$
$= -16(-1)^4\left(\frac{1}{2}\right)^2 + 8(-1)^3\left(\frac{1}{2}\right)^3$
$+3(-1)^2 + 27(-1)\left(\frac{1}{2}\right) - 6$
$= -16(1)\left(\frac{1}{4}\right) + 8(-1)\left(\frac{1}{8}\right)$
$+3(1) + 27(-1)\left(\frac{1}{2}\right) - 6$
$= -4 - 1 + 3 - 13\frac{1}{2} - 6$
$= -21\frac{1}{2}$

4.

4.1 $2x + 5 = 2(3) + 5 = 6 + 5 = 11$

4.2 $-x + 3 = -(-5) + 3 = 5 + 3 = 8$

4.3 $-2x + 5 = -2\left(\frac{1}{2}\right) + 5 = -1 + 5 = 4$

4.4 $x^2 + 4x = (-2)^2 + 4(-2) = 4 - 8 = -4$

4.5 $a^2 - 2ab + 5b^2 = (-3)^2 - 2(-3)(2) + 5(2)^2$
$= 9 + 12 + 20$
$= 41$

4.6 $-3x^2 - x = -3\left(-\frac{1}{2}\right)^2 - \left(-\frac{1}{2}\right)$
$= -3\left(\frac{1}{4}\right) + \frac{1}{2}$
$= -\frac{3}{4} + \frac{1}{2}$
$= -\frac{1}{4}$

4.7 $-2x^2 - 4x + 8 = -2(-3)^2 - 4(-3) + 8$
$= -2(9) + 12 + 8$
$= -18 + 12 + 8$
$= 2$

4.8 $6a^2 + 2a - 3 = 6\left(\frac{1}{3}\right)^2 + 2\left(\frac{1}{3}\right) - 3$
$= 6\left(\frac{1}{9}\right) + \frac{2}{3} - 3$
$= \frac{6}{9} + \frac{2}{3} - 3$
$= \frac{2}{3} + \frac{2}{3} - 3$
$= \frac{4}{3} - 3$
$= -1\frac{2}{3}$

Exercise 2: Solutions

1. $-a^3 + 4a^2 + 5a + 2 - 2a^3 - 3a^2 + 4a - 2 = -3a^3 + a^2 + 9a$

2. $x^2 - 3x + 2 - 2x^2 - 5x + 3x^2 + 8x - 6 = 2x^2 - 4$

3. $-2a + 5 - a^2 + 3a - 1 + 5a^2 - 2a + 5 = 4a^2 - a + 9$

4.
$$\begin{array}{rrr} -4a^2 & & -2 \\ & -3a & +10 \\ -5a^2 & & +3 \\ \hline -9a^2 & -3a & +11 \end{array}$$

5.
$$\begin{array}{rrr} -2x^2y & +5xy & \\ -3x^2y & +2xy & \\ -2x^2y & -5xy & +4xy^2 \\ \hline -7x^2y & +2xy & +4xy^2 \end{array}$$

Exercise 3: Solutions

1. $(-5x + 8 + 6x^2) - (4x^2 - 2x + 3)$
$= -5x + 8 + 6x^2 - 4x^2 + 2x - 3$
$= 2x^2 - 3x + 5$

2. $(3 - 6a^2b - 7ab^2) - (5a^2b - 2ab^2 + 4)$
$= 3 - 6a^2b - 7ab^2 - 5a^2b + 2ab^2 - 4$
$= -11a^2b - 5ab^2 - 1$

3. $(8p^2 - 3p - 10) - (4 - 2p + 6p^2)$
$= 8p^2 - 3p - 10 - 4 + 2p - 6p^2$
$= 2p^2 - p - 14$

4. $3x - (5x^2 - 2x)$
$= 3x - 5x^2 + 2x$
$= -5x^2 + 5x$

5. $(4 - 3x^2y + xy) - (x^2y - 2xy)$
$= 4 - 3x^2y + xy - x^2y + 2xy$
$= 4 + 3xy - 4x^2y$

Exercise 4: Solutions

1. $2a(3a - b - 4) = 6a^2 - 2ab - 8a$

2. $3x^2y(-2xy + 4y^2) = -6x^3y^2 + 12x^2y^3$

3. $4x(x + 2y) + 5(x^2 - 2xy) - (3x^2 - 4xy)$
$= 4x^2 + 8xy + 5x^2 - 10xy - 3x^2 + 4xy$
$= 6x^2 + 2xy$

4. $4x - 3x(2x - \frac{1}{3})$
$= 4x - 6x^2 + x$
$= 5x - 6x^2$

5. $\frac{2}{3}x(9x - 6y + 4)$
$= 6x^2 - 4xy + \frac{8}{3}x$

Exercise 5: Solutions

1. $(x + 3)(x + 5)$
$= x^2 + 5x + 3x + 15$
$= x^2 + 8x + 15$

2. $(2x + 3)(4x - 5)$
$= 8x^2 - 10x + 12x - 15$
$= 8x^2 + 2x - 15$

3. $(4x - 3y)(x + y)$
$= 4x^2 + 4xy - 3xy - 3y^2$
$= 4x^2 + xy - 3y^2$

4. $(5x - 2y)(5x + 2y)$
$= 25x^2 + 10xy - 10xy - 4y^2$
$= 25x^2 - 4y^2$

5. $(3x^2 - 2y)(8x^2 + y)$
$= 24x^4 + 3x^2y - 16x^2y - 2y^2$
$= 24x^4 - 13x^2y - 2y^2$

6. $(x + \frac{1}{2})(x - \frac{1}{2})$
$= x^2 - \frac{1}{2}x + \frac{1}{2}x - \frac{1}{4}$
$= x^2 - \frac{1}{4}$

7. $(x^2 + 1)(x^2 - 1) = x^4 - x^2 + x^2 - 1$
$= x^4 - 1$

8. $-2(3x + y)(5x - 4y)$
$= -2(15x^2 - 12xy + 5xy - 4y^2)$
$= -30x^2 + 24xy - 10xy + 8y^2$
$= -30x^2 + 14xy + 8y^2$

Exercise 6: Solutions

1. $(3x - 4)(2x^2 - 3x + 2) = 6x^3 - 9x^2 + 6x - 8x^2 + 12x - 8$
$= 6x^3 - 17x^2 + 18x - 8$

2. $(4a - b)(a^2 + 4ab - 3b^2) = 4a^3 + 16a^2b - 12ab^2 - a^2b - 4ab^2 + 3b^3$
$= 4a^3 + 15a^2b - 16ab^2 + 3b^3$

3. $(4x - 5y)(2x^2 - 10xy + 7y^2) = 8x^3 - 40x^2y + 28xy^2 - 10x^2y + 50xy^2 - 35y^3$
$= 8x^3 - 50x^2y + 78xy^2 - 35y^3$

4. $(a^2 - 3ax + 5x^2)(2a - 3x) = 2a^3 - 3a^2x - 6a^2x + 9ax^2 + 10ax^2 - 15x^3$
$= 2a^3 - 9a^2x + 19ax^2 - 15x^3$

5. $(4x - \frac{1}{2}y)(x^2 - 2xy + 6y^2) = 4x^3 - 8x^2y + 24xy^2 - \frac{1}{2}x^2y + xy^2 - 3y^3$
$= 4x^3 - 8\frac{1}{2}x^2y + 25xy^2 - 3y^3$

Exercise 7: Solutions

Method 1:

1. $(4x - 3y)^2 = (4x - 3y)(4x - 3y)$
$= 16x^2 - 12xy - 12xy + 9y^2$
$= 16x^2 - 24xy + 9y^2$

2. $(2a + 5b)^2 = (2a + 5b)(2a + 5b)$
$= 4a^2 + 10ab + 10ab + 25b^2$
$= 4a^2 + 20ab + 25b^2$

3. $(c + d)^2 = (c + d)(c + d)$
$= c^2 + cd + cd + d^2$
$= c^2 + 2cd + d^2$

4. $(5x - 6y)^2 = (5x - 6y)(5x - 6y)$
$= 25x^2 - 30xy - 30xy + 36y^2$
$= 25x^2 - 60xy + 36y^2$

5. $(4a - 5b)^2 = (4a - 5b)(4a - 5b)$
$= 16a^2 - 20ab - 20ab + 25b^2$
$= 16a^2 - 40ab + 25b^2$

Method 2:

6. $(2a^2 - 3b)^2 = 4a^4 - 12a^2b + 9b^2$
(middle term: $2(2a^2)(-3b)$)

7. $(4x^2 - y^3)^2 = 16x^4 - 8x^2y^3 + y^6$
(middle term: $2(4x^2)(-y^3)$)

8. $-2(3x - 5y)^2 = -2(9x^2 - 30xy + 25y^2)$
$= -18x^2 + 60xy - 50y^2$

9. $[2a(4 - 3a)]^2 = 4a^2(16 - 24a + 9a^2)$
$= 64a^2 - 96a^3 + 36a^4$

Exercise 8: Solutions: Mixed

1. $(2a + 5b)(2a - 5b)$
$= 4a^2 - 10ab + 10ab - 25b^2$
$= 4a^2 - 25b^2$

2. $(5x - 3y)(3x + 2y)$
$= 15x^2 + 10xy - 9xy - 6y^2$
$= 15x^2 + xy - 6y^2$

3. $(3s - 4t)(s^2 - 2)$
$= 3s^3 - 6s - 4ts^2 + 8t$

4. $-7a(-a + 1) - 3a(5 - a)$
$= 7a^2 - 7a - 15a + 3a^2$
$= 10a^2 - 22a$

5. $(2x^2 - 5y)(3x^2 + 4y)$
$= 6x^4 + 8x^2y - 15x^2y - 20y^2$
$= 6x^4 - 7x^2y - 20y^2$

6. $3x - 3x(6x^2 - 7x + 1)$
$= 3x - 18x^3 + 21x^2 - 3x$
$= -18x^3 + 21x^2$

7. $2a(4a - 5b)(3a + 8b)$
$= 2a(12a^2 + 32ab - 15ab - 40b^2)$
$= 24a^3 + 34a^2b - 80b^2a$

8. $\left(\frac{2}{x} + 1\right)\left(\frac{3}{x} + 3\right)$
$= \frac{6}{x^2} + \frac{6}{x} + \frac{3}{x} + 3$
$= \frac{6}{x^2} + \frac{9}{x} + 3$

9. $\frac{a^2}{b^3}\left(b^2 - \frac{b^3}{a^2}\right) = \frac{a^2}{b} - 1$

10. $(x-2)(x+3)(2x+1)$
$= (x-2)(2x^2+x+6x+3)$
$= (x-2)(2x^2+7x+3)$
$= 2x^3+7x^2+3x-4x^2-14x-6$
$= 2x^3+3x^2-11x-6$

11. $x(x+1)-(x+1)(x+3)$
$= x^2+x-(x^2+3x+x+3)$
$= x^2+x-x^2-3x-x-3$
$= -3x-3$

12. $\left(a+\frac{2}{a}\right)^2 = a^2+4+\frac{4}{a^2}$

(middle term: $2(a)\left(\frac{2}{a}\right) = 4$)

13. $3(2x-5y)^2$
$= 3(4x^2-20xy+25y^2)$
$= 12x^2-60xy+75y^2$

14. $(3x+y)(3x-y)-(2x-y)^2$
$= 9x^2-3xy+3xy-y^2-(4x^2-4xy+y^2)$
$= 9x^2-y^2-4x^2+4xy-y^2$
$= 5x^2+4xy-2y^2$

15. $[2(x-y)]^2$
$= 4(x^2-2xy+y^2)$
$= 4x^2-8xy+4y^2$

16. $(3x-2y)^2-3x(5x-y)$
$= 9x^2-12xy+4y^2-15x^2+3xy$
$= -6x^2-9xy+4y^2$

17. $(3x-2b)^2-3(4a^2+2)+(a+b)(3a-b)$
$= 9a^2-12ab+4b^2-12a^2-6$
$+3a^2-ab+3ab-b^2$
$= -10ab+3b^2$

18. $(x+2y)(x^2+3xy-5y^2)$
$= x^3+3x^2y-5xy^2+2x^2y+6xy^2-10y^3$
$= x^3+5x^2y+xy^2-10y^3$

19. $(x+5)^2-(x-5)^2$
$= x^2+10x+25-(x^2-10x+25)$
$= x^2+10x+25-x^2+10x-25$
$= 20x$

20. $2(x-3)^2-4(2x-3)-(x+4)(2x-1)$
$= 2(x^2-6x+9)-8x+12$
$-(2x^2-x+8x-4)$
$= 2x^2-12x+18-8x+12$
$-2x^2+x-8x+4$
$= -27x+34$

Exercise 9: Solutions

1. $\frac{9a^3b^5+24a^2b^8-3ab^3}{3ab^3}$

$= 3a^2b^2+8ab^5-1$

2. $\frac{14x^5-20x^3y^2+8x^6y^3}{-2x^2}$

$= 7x^3+10xy^2-4x^4y^3$

3. $\frac{-15xy^3-5x^2y^5+30x^8y^2}{-5xy^3}$

$= 3+xy^2-\frac{6x^7}{y}$

4. $\frac{20xy^2+xy-2x^2y^5}{4x^3y^2}$

$= \frac{5}{x^2}+\frac{1}{4x^2y}-\frac{y^3}{2x}$

5. $\frac{-9x^3y^2+6xy^5-8x^5}{6x^3y^3}$

$= \frac{-3}{2y}+\frac{y^2}{x^2}-\frac{4x^2}{3y^3}$

Exercise 10: *Solutions*

1. $8x^3 - 18x^2 + 5x + 6$
$= (2x - 3)(4x^2 - 3x - 2)$

2. $6x^3 + 19x^2 + 6x - 10$
$= (2x + 5)(3x^2 + 2x - 2)$

3. $10x^3 - 21x^2 + 5x + 6$
$= (2x - 3)(5x^2 - 3x - 2)$

4. $12k^3 - 4k^2 - 17x + 10$
$= (6k - 5)(2k^2 + k - 2)$

5. $30 - 3p^2 + p^3 - 23p$
$= p^3 - 3p^2 - 23p + 30$
$= (p - 6)(p^2 + 3p - 5)$

6. $x^3 + 2x^2y - 17xy^2 + 6y^3$
$= (x - 3y)(x^2 + 5xy - 2y^2)$

7. $2a^3 - 7a^2b + 7ab^2 - 2b^3$
$= (a - 2b)(2a^2 - 3ab + b^2)$

8. $8x^3 - 2x^2y + 3xy^2 + y^3$
$= (4x + y)(2x^2 - xy + y^2)$

Exercise 11: *Solutions*

1. $4x^2 + 2\Big[4x - 2x\{3 - (x - 2)\} + 3\Big]$

$= 4x^2 + 2\Big[4x - 2x\{3 - x + 2\} + 3\Big]$ remove ( )

$= 4x^2 + 2\Big[4x - 6x + 2x^2 - 4x + 3\Big]$ remove { }

$= 4x^2 + 8x - 12x + 4x^2 - 8x + 6$ remove [ ]

$= 8x^2 - 12x + 6$

2. $5x - [5y - (2x - 2y) - 5x]$
$= 5x - [5y - 2x + 2y - 5x]$ remove ( )
$= 5x - 5y + 2x - 2y + 5x$ remove [ ]
$= 12x - 7y$

3. $4a - 3[a + 2(-b + a) - (2a - 3b)]$
$= 4a - 3[a + 2b + 2a - 2a + 3b]$
$= 4a - 3a + 6b - 6a + 6a - 9b$
$= a - 3b$

4. $2a - 3\Big[4a - \{2(a + 1) - 7\}\Big]$

$= 2a - 3\Big[4a - \{2a + 2 - 7\}\Big]$ remove ( )

$= 2a - 3\Big[4a - 2a - 2 + 7\Big]$ remove { }

$= 2a - 12a + 6a + 6 - 21$ remove [ ]

$= -4a - 15$

5. $[-3a - 4(2 + a)][-(-2a)^2 - 6(a - 1)]$

$= [-3a - 8 - 4a][-4a^2 - 6a + 6]$

$= [-7a - 8][-4a^2 - 6a + 6]$

$= 28a^3 + 42a^2 - 42a + 32a^2 + 48a - 48$

$= 28a^3 + 74a^2 + 6a - 48$

Exercise 12: Solutions

1. $\sqrt{25a^4b^2} = 5a^2b$

2. $\sqrt{(a+b)^2} = a+b$

3. $\sqrt{16x^{16}} = 4x^4$

4. $\sqrt{9x^4 + 16x^4} = \sqrt{25x^4} = 5x^2$

5. $\sqrt{\dfrac{m^2(a+b)^4}{121}} = \dfrac{m(a+b)^2}{11}$

6. $\sqrt{\dfrac{32a^3b^4}{8a}} = \sqrt{4a^2b^4} = 2ab^2$

7. $\sqrt{\dfrac{20a^5}{5a}} . \sqrt{\dfrac{9a^3b^2}{a^5}}$

$= \sqrt{4a^4} . \sqrt{\dfrac{9b^2}{a^2}}$

$= 2a^2 . \dfrac{3b}{a} = 6ab$

8. $\sqrt{\dfrac{50a^6b}{98b^9}} = \sqrt{\dfrac{25a^6}{49b^8}} = \dfrac{5a^3}{7b^4}$

9. $3\sqrt{3}.\sqrt{3} = 3 \times 3 = 9$

10. $-3\sqrt{12}\sqrt{3} = -3\sqrt{36} = -3 \times 6 = -18$

11. $a\sqrt{a^4}. a^2 = a. a^2. a^2 = a^5$

12. $\sqrt{x^2}.\sqrt{x^6} = x. x^3 = x^4$

13. $(2\sqrt{3})^2 + (4\sqrt{2})^2 = 4 \times 3 \times 16 \times 2$
$= 12 + 32 = 44$

14. $3(\sqrt{12})^2 + (\sqrt{5})^2 = 3 \times 12 + 5$
$= 36 + 5 = 41$

15. $\sqrt{2\frac{1}{4}a^6} = \sqrt{\frac{9}{4}a^6} = \dfrac{3a^3}{2}$

4 *Factorisation*

Find the product : $3 \times 2 = 6$
Faktorize : $6 = 3 \times 2$

So: Faktorization is the reverse process of determining the product. (Chapter 2)
Thus: Faktorization of a polynomial means breaking it up into 2 or more factors, namely (. . .)(. . . .)(. . .)
In Std 7 we do **3 types of factorisation**.

Recognizing the different types is very important.

Type 1: Common Factor

Take out **the greatest common factor** occuring in all the different terms of the expression.

Examples:

1. $2ab + 3ac + 4ad + 5ae = a(2b + 3c + 4d + 5e)$

2. $3a + 3b - 3 = 3(a + b - \mathbf{1})$

3. $7x - 14x^2 + 21x^3 = 7x(1 - 2x + 3x^2)$

4. $12a^2b - 16a^3b^2 + 20a^4b^3 = 4a^2b(3 - 4ab + 5a^2b^2)$

5. $(a + b)x - (a + b)y = (a + b)(x - y)$
N.B. (1) The **whole bracket** is the common factor.
(2) **Keep the bracket** if it is "taken out" as a common factor.

6. $2x(a + b)^2 + y(a + b) = (a + b)(2x(a + b) + y)$

7. $q(a - c) + m(c - a) = q(a - c) - m(a - c)$
$= (a - c)(q - m)$

Exercise 1:

Faktorise:

1. $2x^2 - 4$
2. $18m^2 - 24m^3$
3. $4ab - 8ac + 12ad$
4. $3(k + 1) - k(k + 1)$
5. $(m - 2) + p(m - 2)$
6. $2x(a + 2b) - (a + 2b)$
7. $6x^3(a + b) + x(b + a)$
8. $(x - 3)^2 + (x - 3)$
9. $3x(x - 3) + 2(3 - x)$
10. $2x(a - 2b) - y(2b - a)$

Type 2: Difference of two squares

Identification: 1. **Only 2 terms**
2. **There must be a – (minus) between the 2 terms**
3. **Both terms must be squares**

Examples of squares: 1; 4; 9; $\frac{16}{25}$; $\frac{1}{4}$; ......
$4a^2$; $9x^2$; $\frac{16a^4}{25b^6}$; $4(a-b)^2$ ......

Thus: the indices must be **EVEN** numbers.

Examples:

1. $x^2 - y^2 = (x+y)(x-y)$

2. $4a^2 - 9b^2 = (2a-3b)(2a+3b)$

3. $9x^2 - \frac{1}{4} = (3x - \frac{1}{2})(3x + \frac{1}{2})$

4. $k^4 - 1 = (k^2-1)(k^2+1)$
$= (k+1)(k-1)(k^2+1)$

5. $x^2 - (m+n)^2$
$= [x + (m+n)][x - (m+n)]$

Exercise 2:

Faktorise:

1. $a^2 - b^2$

2. $25a^2 - 16b^2$

3. $9a^2 - 4b^2$

4. $9x^4 - 1$

5. $4x^6 - \frac{1}{9}$

6. $1 - 16x^4$

7. $2\frac{1}{4}x^2 - 9y^4$

8. $(a+b)^2 - 1$

9. $(x+y)^2 - 4x^2$

10. $\frac{9x^2}{25y^2} - 4a^4$

Type 3: Trinomial

Identification: **Always consists of 3 terms in a specific order.** (Sometimes it is necessary to rearrange the terms in the correct order).

$x^2 \ldots x \ldots\ldots$ i.e. $(x^2 - 3x - 10)$

or: $x^2 \ldots xy \ldots\ldots y^2$ i.e. $(x^2 - 7xy - 10y^2)$

Examples:

1. $x^2 + 7x + 12$

Step 1. Write down all the **factors** of 12, $\left\{\begin{array}{c} 12 \times 1 \\ 6 \times 2 \\ 4 \times 3 \end{array}\right.$

Step 2. The (+) in front of 12, means that the two factors of 12, when added, must give 7 (middle term's coefficient); therefore 3×4 is the correct combination.

Step 3. If the 3rd term, namely 12, is (+), the signs in the brackets must be **the same**; and has to be the sign of the middle term, i.e. + and +.
thus $\mathbf{x^2 + 7x + 12 = (x + 3)(x + 4)}$

2. $x^2 - 5x + 6$
$= (x - 3)(x - 2)$

- (+6 →) 1. Factors of 6 must be (+) to give 5, i.e. 3 and 2.
  2. Signs in the brackets must be **the same**.
- (−5x →) indicates that the signs in the brackets are −

3. $x^2 - 5x - 6$
$= (x - 6)(x + 1)$

- (−6 →) indicates 1. Find factors of 6 which, when subtracted (−) will give five, i.e., 6 and 1.
  2. Signs in the brackets **differ**.
- (−5x →) 3. The **larger** number of 6 and 1 must be (−).

4. $x^2 + 5x - 6 = (x + 6)(x - 1)$

5. $y^2 - 8y - 20 = (y - 10)(y + 2)$

6. $b^2 + b - 12 = (b + 4)(b - 3)$

7. $a^2 - 13a + 30 = (a - 10)(a - 3)$

8. $x^4 + 8x^2 + 15 = (x^2 + 3)(x^2 + 5)$

9. $x^2 + 7 - 8x = x^2 - 8x + 7 = (x - 7)(x - 1)$

Exercise 3:

Factorise:

1. $x^2 + 5x + 6$
2. $x^2 + 8x + 12$
3. $x^2 + 7x + 10$
4. $x^2 - 8x + 15$
5. $y^2 - 5y + 6$
6. $a^2 + 5a - 14$
7. $p^2 + 3p - 10$
8. $x^2 + x - 12$
9. $x^2 - 2xy - 15y^2$
10. $a^2 - 3a - 18$

Mixed Factorisation:

It is often necessary to use a combination of the different types of factorisation in order to factorise an expression completely.

Hint: 1. **Always try to find a common factor first.**
2. **Check whether the 2nd bracket needs further factorisation.**

Examples:

1. $2k^2 - 32 = 2(k^2 - 16)$
$= 2(k-4)(k+4)$

2. $2x^2 - 12x + 18 = 2(x^2 - 6x + 9)$
$= 2(x-3)(x-3)$

3. $(x+y)^2 - 25 = (x+y-5)(x+y+5)$

4. $(x^2 - 1) + (x+1) = (x+1)(x-1) + (x+1)$
$= (x+1)(x-1+1)$
$= (x+1)(x)$

5. $32a^3 - 50ab^6 = 2a(16a^2 - 25b^6)$
$= 2a(4a - 5b^3)(4a + 5b^3)$

6. $4a^2(a-2) - 36(a-2) = 4(a-2)(a^2 - 9)$
$= 4(a-2)(a-3)(a+3)$

7. $(x-2)^2 - (x-2)(x-6) - 2x(x-2) = (x-2)[(x-2) - (x-6) - 2x]$
$= (x-2)[x - 2 - x + 6 - 2x]$
$= (x-2)(4-2x)$
$= (x-2)2(2-x)$
$= -2\,(x-2)^2$ [optional]

Exercise 4: Mixed

Faktorise completely:

1. $2x^3 - 18x$
2. $2x^3 + 2x^2 - 12x$
3. $x(a-c) + y(c-a)$
4. $3x^2 - 21x + 36$
5. $12x^3 - 24x^2 + 12x$
6. $x^8 - 16$
7. $9a^3 - 3a$
8. $8a^4b - 18a^2b^3$
9. $x^2 - 12x + 27$
10. $2b^2(a-c) + 18(c-a)$
11. $x^2(p-q) + (q-p)$
12. $9x^2 - 1$
13. $a^2x^2 - 2ax - 15$
14. $25a^2 - \dfrac{9x^4}{16y^6}$
15. $(2x-3y)(2c+d) - 2x + 3y$
16. $3x^3 - 6x^2 + 3x$
17. $x^2 - 5xy - 24y^2$
18. $2q(m-n) - 3(n-m)$
19. $m(m-1) - 12$
20. $(2a-b)(x-y) - cx + cy$
21. $9x^2 - \dfrac{1}{4}$
22. $2x(a-b) - 3(b-a)$
23. $9(a+b) - x^2(a+b)$
24. $9(x+y)^2 - 4$
25. $2b^2(x-y) - 2x + 2y$
26. $x^4 - 13x^2 + 36$

SOLUTIONS

Exercise 1: Solutions

1. $2x^2 - 4 = 2(x^2 - 2)$

2. $18m^2 - 24m^3 = 6m^2(3 - 4m)$

3. $4ab - 8ac + 12ad$
$= 4a(b - 2c + 3d)$

4. $3(k+1) - k(k+1)$
$= (k+1)(3-k)$

5. $(m-2) + p(m-2)$
$= (m-2)(1+p)$

6. $2x(a+2b) - (a+2b)$
$= (a+2b)(2x-1)$

7. $6x^3(a+b) + x(b+a)$
$= \mathbf{x}(a+b)(6x^2+1)$

8. $(x-3)^2 + (x-3)$
$= (x-3)(x-3+1)$
$= (x-3)(x-2)$

9. $3x(x-3) + 2(3-x)$
$= 3x(x-3) - \mathbf{2(x-3)}$
$= (x-3)(3x-2)$

10. $2x(a-2b) - y(2b-a)$
$2x(a-2b) + \mathbf{y(a-2b)}$
$(a-2b)(2x+y)$

Exercise 2 Solutions

1. $a^2 - b^2 = (a+b)(a-b)$

2. $25a^2 - 16b^2 = (5a+4b)(5a-4b)$

3. $9a^2 - 4b^2 = (3a+2b)(3a-2b)$

4. $9x^4 - 1 = (3x^2-1)(3x^2+1)$

5. $4x^6 - \frac{1}{9} = \left(2x^3 - \frac{1}{3}\right)\left(2x^3 + \frac{1}{3}\right)$

6. $1 - 16x^4$
$= (1-4x^2)(1+4x^2)$
$= (1-2x)(1+2x)(\mathbf{1+4x^2})$

7. $2\frac{1}{4}x^2 - 9y^4$
$= \frac{9}{4}x^2 - 9y^4$
$= \left(\frac{3}{2}x - 3y^2\right)\left(\frac{3}{2}x + 3y^2\right)$

8. $(a+b)^2 - 1 = (a+b+1)(a+b-1)$

9. $(x+y)^2 - 4x^2$
$= (x+y-2x)(x+y+2x)$
$= (y-x)(3x+y)$

10. $\frac{9x^2}{25y^2} - 4a^4$
$= \left(\frac{3x}{5y} - 2a^2\right)\left(\frac{3x}{5y} + 2a^2\right)$

Exercise 3: Solutions

1. $x^2 + 5x + 6 = (x+3)(x+2)$

2. $x^2 + 8x + 12 = (x+6)(x+2)$

3. $x^2 + 7x + 10 = (x+5)(x+2)$

4. $x^2 - 8x + 15 = (x-5)(x-3)$

5. $y^2 - 5y + 6 = (y-3)(y-2)$

6. $a^2 + 5a - 14 = (a+7)(a-2)$

7. $p^2 + 3p - 10 = (p+5)(p-2)$

8. $x^2 + x - 12 = (x+4)(x-3)$

9. $x^2 - 2xy - 15y^2 = (x-5y)(x+3y)$

10. $a^2 - 3a - 18 = (a-6)(a+3)$

Exercise 4: Mixed. Solutions

1. $2x^3 - 18x$
$= 2x(x^2 - 9)$
$= 2x(x + 3)(x - 3)$

2. $2x^3 + 2x^2 - 12x$
$= 2x(x^2 + x - 6)$
$= 2x(x + 3)(x - 2)$

3. $x(a - c) + y(c - a)$
$= x(a - c) - y(a - c)$
$= (a - c)(x - y)$

4. $3x^2 - 21x + 36$
$= 3(x^2 - 7x + 12)$
$= 3(x - 4)(x - 3)$

5. $12x^3 - 24x^2 + 12x$
$= 12x(x^2 - 2x + 1)$
$= 12x(x - 1)(x - 1)$

6. $x^8 - 16$
$= (x^4 - 4)(x^4 + 4)$
$= (x^2 - 2)(x^2 + 2)(x^4 + 4)$

7. $9a^3 - 3a = 3a(3a^2 - 1)$

8. $8a^4b - 18a^2b^3$
$= 2a^2b(4a^2 - 9b^2)$
$= 2a^2b(2a - 3b)(2a + 3b)$

9. $x^2 - 12x + 27 = (x - 9)(x - 3)$

10. $2b^2(a - c) + 18(c - a)$
$= 2b^2(a - c) - 18(a - c)$
$= 2(a - c)(b^2 - 9)$
$= 2(a - c)(b - 3)(b + 3)$

11. $x^2(p - q) + (q - p)$
$= x^2(p - q) - (p - q)$
$= (p - q)(x^2 - 1)$
$= (p - q)(x + 1)(x - 1)$

12. $9x^2 - 1 = (3x + 1)(3x - 1)$

13. $a^2x^2 - 2ax - 15 = (ax - 5)(ax + 3)$

14. $25a^2 - \frac{9x^4}{16y^6}$
$= \left(5a - \frac{3x^2}{4y^3}\right)\left(5a + \frac{3x^2}{4y^3}\right)$

15. $(2x - 3y)(2c + d) - 2x + 3y$
$= (2x - 3y)(2c + d) - (2x - 3y)$
$= (2x - 3y)(2c + d - 1)$

16. $3x^3 - 6x^2 + 3x = 3x(x^2 - 2x + 1)$
$= 3x(x - 1)(x - 1)$

17. $x^2 - 5xy - 24y^2 = (x - 8y)(x + 3y)$

18. $2q(m - n) - 3(n - m)$
$= 2q(m - n) + 3(m - n)$
$= (m - n)(2q + 3)$

19. $m(m - 1) - 12$
$= m^2 - m - 12$
$= (m - 4)(m + 3)$

20. $(2a - b)(x - y) - cx + cy$
$= (2a - b)(x - y) - c(x - y)$
$= (x - y)(2a - b - c)$

21. $9x^2 - \frac{1}{4}$
$= \left(3x - \frac{1}{2}\right)\left(3x + \frac{1}{2}\right)$

22. $2x(a - b) - 3(b - a)$
$= 2x(a - b) + 3(a - b)$
$= (a - b)(2x + 3)$

23. $9(a + b) - x^2(a + b)$
$= (a + b)(9 - x^2)$
$= (a + b)(3 - x)(3 + x)$

24. $9(x + y)^2 - 4$
$= [3(x + y) - 2][3(x + y) + 2]$

25. $2b^2(x - y) - 2x + 2y$
$= 2b^2(x - y) - 2(x - y)$
$= 2(x - y)(b^2 - 1)$
$= 2(x - y)(b + 1)(b - 1)$

26. $x^4 - 13x^2 + 36$
$= (x^2 - 4)(x^2 - 9)$
$= (x - 2)(x + 2)(x - 3)(x + 3)$

5 *Algebraic fractions*

Simplification, multiplication and division:

Factorise all **denominators** and **numerators** and "cancel" identical **factors**.

eg. $\dfrac{y^2 - 3y}{y^2 - 9} = \dfrac{y\cancel{(y-3)}}{\cancel{(y-3)}(y+3)} = \dfrac{y}{y+3}$

and **not** $\dfrac{\cancel{y^2} - 3y}{\cancel{y^2} - 9}$ **(take care:** don't do this!)

When **dividing** by a fracton: multiply by the reciprocal (inverse).
Work carefully through the following examples.

Examples:

1. $\dfrac{x^2 - 9}{x^2 - 3x - 18} = \dfrac{(x-3)\cancel{(x+3)}}{(x-6)\cancel{(x+3)}} = \dfrac{x-3}{x-6}$

2. $\dfrac{a^2 - b^2}{2a^2 + 2ab} \div \dfrac{a^2 - 2ab + b^2}{2a^2 - ba}$

$= \dfrac{a^2 - b^2}{2a^2 + 2ab} \times \dfrac{2a^2 - 6a}{a^2 - 2ab + b^2}$

$= \dfrac{\cancel{(a-b)}\cancel{(a+b)}}{\cancel{2a}\cancel{(a+b)}} \times \dfrac{\cancel{2a}(a-3)}{\cancel{(a-b)}(a-b)}$

$= \dfrac{a-3}{a-b}$

3. $\dfrac{(a-x)^2}{a^2 - x^2} = \dfrac{\cancel{(a-x)}(a-x)}{\cancel{(a-x)}(a+x)} = \dfrac{a-x}{a+x}$

4. $\dfrac{x^2 - xy}{y^2 + xy} \div \dfrac{y^2 - xy}{x^2 + xy} \times \dfrac{2x^2 - 2xy}{2x^2}$

$= \dfrac{x^2 - xy}{y^2 + xy} \times \dfrac{x^2 + xy}{y^2 - xy} \times \dfrac{2x(x-y)}{2x^2}$

$= \dfrac{x(x-y)}{\cancel{y(y+x)}} \times \dfrac{\cancel{x(x+y)}}{y(y-x)} \times \dfrac{2x(x-y)}{2x^2}$

$= \dfrac{2x^3(x-y)(x-y)}{2x^2y^2(y-x)}$

$= -\dfrac{\cancel{2}x^3\cancel{(x-y)}(x-y)}{\cancel{2}x^2y^2\cancel{(x-y)}}$

$= -\dfrac{x(x-y)}{y^2}$

N.B.
1. **(y − x) = −(x − y)**
2. **(2a − 3b) = −(3b − 2a)**
3. **(−a − b) = − (a + b)**
4. **a + b = b + a**

Addition and subtraction of fractions:

Determine the L.C.M. of the **denominators**.

Examples:

1. $\frac{x+2}{3}-\frac{x-4}{2}$ LCM of 3 and 2 is 6

$=\frac{2(x+2)-3(x-4)}{6}$

$=\frac{2x+4-3x+12}{6}$

$=\frac{-x+16}{6}$

2. $\frac{a-2}{4}-\frac{a}{3}-\frac{a-4}{12}$

$=\frac{3(a-2)-4a-(a-4)}{12}$ **NB: Remember the brackets**

$=\frac{3a-6-4a-a+4}{12}$

$=\frac{-2a-2}{12}=\frac{-2(a+1)}{12}=\frac{-(a+1)}{6}$

3. $\frac{2x-3}{3x}-\frac{4x+1}{2x}-\frac{1}{6}$

$=\frac{2(2x-3)-3(4x+1)-x}{6x}$

$=\frac{4x-6-12x-3-x}{6x}$

$=\frac{-9x-9}{6x}=\frac{9(-x-1)}{6x}=\frac{3(-x-1)}{2x}$

4. $\frac{a-1}{3a^2}+\frac{3}{4a}-\frac{4+a}{6a^2}$

$=\frac{4(a-1)+(3a)3-2(4+a)}{12a^2}$

$=\frac{4a-4+9a-8-2a}{12a^2}$

$=\frac{11a-12}{12a^2}$

Exercise:

Simplify:

1. $\dfrac{21x^7}{7x^5}$

2. $\dfrac{3x^2-6x}{3x}$

3. $\dfrac{x^2-9}{x^2-3x-18}$

4. $\dfrac{(a-x)^2}{a-x}$

5. $\dfrac{(a-b)^2}{3a^2-3b^2}$

6. $\dfrac{x^2-1}{(x-1)^2}$

7. $\dfrac{6x+6y}{3x+3y}$

8. $\dfrac{a^2-9}{a^2-a-6} \times \dfrac{a^2+2a}{a^2}$

9. $\dfrac{3m-3p}{a^2+3a+2} \div \dfrac{m-p}{a+2}$

10. $\dfrac{x^2}{x^2-4} \div \dfrac{3x-x^2}{x^2-5x+6}$

11. $\dfrac{4x^2-9}{6xa-9a} \times \dfrac{ax+a}{2x-3}$

12. $\dfrac{2x^2-8x}{x^2-6x+8} \times \dfrac{x^2-4}{x^2+2x}$

13. $\dfrac{x^4-81x^2}{2x} \div \dfrac{x^3-9x}{6}$

14. $\dfrac{2x-1}{3x} - \dfrac{x-1}{2x} - \dfrac{1}{2}$

15. $\dfrac{3}{2x} - \dfrac{1}{x} + \dfrac{3}{x^2}$

16. $\dfrac{m-4}{2} + \dfrac{2}{3}(m-1) - \dfrac{m-1}{4}$

17. $\dfrac{3x+y}{2} - \dfrac{2x-y}{3} + 2x$

18. $\dfrac{2x-1}{x} + \dfrac{x+1}{2x} - \dfrac{3x^2-x+1}{x^2}$

19. $\dfrac{1}{a^2bc} + \dfrac{2a}{b^2c} + \dfrac{3}{ac}$

20. $\dfrac{2}{3p^2q} - \dfrac{3}{8pq^2} + \dfrac{5}{12pq}$

General guidelines for fractions:

1. $a \div b \times c = \dfrac{a}{1} \times \dfrac{1}{b} \times \dfrac{c}{1} = \dfrac{ac}{b}$

2. $a \div (b \times c) = a \div (bc) = \dfrac{a}{1} \times \dfrac{1}{bc} = \dfrac{a}{bc}$

3. $a \div b \div c = \dfrac{a}{1} \times \dfrac{1}{b} \times \dfrac{1}{c} = \dfrac{a}{bc}$

4. $a \div (b \div c) = a \div \left(\dfrac{b}{1} \times \dfrac{1}{c}\right) = a \div \dfrac{b}{c} = a \times \dfrac{c}{b} = \dfrac{ac}{b}$

Examples:

1. $\dfrac{4}{\frac{1}{2}} = 4 \times \dfrac{2}{1} = 8$

2. $\dfrac{\frac{3}{4}}{5} = \dfrac{3}{4} \div \dfrac{5}{1} = \dfrac{3}{4} \times \dfrac{1}{5} = \dfrac{3}{20}$

3. $\dfrac{2x^2y}{3ax} \div \dfrac{y^2}{9xa} = \dfrac{2x^2y}{3ax} \times \dfrac{9xa}{y2} = \dfrac{18x^3ay}{3ay^2x} = \dfrac{6x^2}{y}$

SOLUTIONS

1. $\dfrac{21x^7}{7x^5} = 3x^2$

2. $\dfrac{3x^2 - 6x}{3x} = \dfrac{\cancel{3x}(x-2)}{\cancel{3x}} = x - 2$

3. $\dfrac{x^2 - 9}{x^2 - 3x - 18} = \dfrac{\cancel{(x+3)}(x-3)}{(x-6)\cancel{(x+3)}} = \dfrac{x-3}{x-6}$

4. $\dfrac{(a-x)^2}{a-x} = a - x$

5. $\dfrac{(a-b)^2}{3a^2 - 3b^2} = \dfrac{(a-b)^2}{3(a^2 - b^2)}$

$= \dfrac{\cancel{(a-b)}(a-b)}{3\cancel{(a-b)}(a+b)}$

$= \dfrac{a-b}{3(a+b)}$

6. $\dfrac{x^2 - 1}{(x-1)^2} = \dfrac{(x+1)\cancel{(x-1)}}{(x-1)\cancel{(x-1)}} = \dfrac{x+1}{x-1}$

7. $\dfrac{6x + 6y}{3x + 3y} = \dfrac{6\cancel{(x+y)}}{3\cancel{(x+y)}} = 2$

8. $\dfrac{a^2 - 9}{a^2 - a - 6} \times \dfrac{a^2 + 2a}{a^2}$

$= \dfrac{\cancel{(a-3)}(a+3)}{\cancel{(a-3)}\cancel{(a+2)}} \times \dfrac{a\cancel{(a+2)}}{a^2} = \dfrac{a+3}{a}$

9. $\dfrac{3m - 3p}{a^2 + 3a + 2} \div \dfrac{m-p}{a+2}$

$= \dfrac{3\cancel{(m-p)}}{\cancel{(a+2)}(a+1)} \times \dfrac{\cancel{(a+2)}}{\cancel{m-p}} = \dfrac{3}{a+1}$

10. $\dfrac{x^2}{x^2 - 4} \div \dfrac{3x - x^2}{x^2 - 5x + 6}$

$= \dfrac{x^2}{(x-2)(x+2)} \times \dfrac{x^2 - 5x + 6}{3x - x^2}$

$= \dfrac{x^2}{\cancel{(x-2)}(x+2)} \times \dfrac{(x-3)\cancel{(x-2)}}{x(3-x)}$

$= -\dfrac{x\cancel{(x-3)}}{(x+2)\cancel{(x-3)}} = -\dfrac{x}{x+2}$

11. $\dfrac{4x^2 - 9}{6xa - 9a} \times \dfrac{ax + a}{x^2 - 2x - 3}$

$= \dfrac{\cancel{(2x-3)}(2x+3)}{3\cancel{a}\cancel{(2x-3)}} \times \dfrac{\cancel{a}\cancel{(x+1)}}{(x-3)\cancel{(x+1)}}$

$= \dfrac{2x+3}{3(x-3)}$

12. $\dfrac{2x^2 - 8x}{x^2 - 6x + 8} \times \dfrac{x^2 - 4}{x^2 + 2x}$

$= \dfrac{2\cancel{x}\cancel{(x-4)}}{\cancel{(x-4)}\cancel{(x-2)}} \times \dfrac{\cancel{(x-2)}\cancel{(x+2)}}{\cancel{x}\cancel{(x+2)}} = 2$

13. $\dfrac{x^1 - 81x^2}{2x} \div \dfrac{x^3 - 9x}{6}$

$= \dfrac{\cancel{x^2}\cancel{(x^2-9)}}{\cancel{2x}} \times \dfrac{\cancel{6}^{3}}{\cancel{x}\cancel{(x^2-9)}} = 3$

14. $\dfrac{2x-1}{3x} - \dfrac{x-1}{2x} - \dfrac{1}{2}$

$= \dfrac{2(2x-1) - 3(x-1) - 3x}{6x}$

$= \dfrac{4x - 2 - 3x + 3 - 3x}{6x} = \dfrac{-2x + 1}{6x}$

15. $\frac{3}{2x}-\frac{1}{x}+\frac{3}{x^2}$

$=\frac{3x-2x+6}{2x^2}=\frac{x+6}{2x^2}$

16. $\frac{m-4}{2}+\frac{2}{3}(m-1)-\frac{m-1}{4}$

$=\frac{6(m-4)+8(m-1)-3(m-1)}{12}$

$=\frac{6m-24+8m-8-3m+3}{12}$

$=\frac{11m-29}{12}$

17. $\frac{3x+y}{2}-\frac{2x-y}{3}+2x$

$=\frac{3(3x+y)-2(2x-y)+12x}{6}$

$=\frac{9x+3y-4x+2y+12x}{6}=\frac{17x+5y}{6}$

18. $\frac{2x-1}{x}+\frac{x+1}{2x}-\frac{3x^2-x+1}{x^2}$

$=\frac{2x(2x-1)+x(x+1)-2(3x^2-x+1)}{2x^2}$

$=\frac{4x^2-2x^2+x-6x^2+2x-2}{2x^2}$

$=\frac{-x^2+x-2}{2x^2}$

19. $\frac{1}{a^2bc}+\frac{2a}{b^2c}+\frac{3}{ac}$

$=\frac{b+2a^3+3ab^2}{a^2b^2c}$

20. $\frac{2}{3p^2q}-\frac{3}{8pq^2}+\frac{5}{12pq}$

$=\frac{8q-9p+10pq}{24p^2q^2}$

6 *Equations*

To solve a linear equation in x *e.g.* $x + 8 = 10$, you need to find the value of x (only one), namely $x = 2$, so that the LHS : $x + 8$ = the RHS : 10.
We say $x = 2$ **satisfies** the equation.
In any equation the left hand side (LHS) must **always be equal** to the right hand side (RHS).

On **both sides** of the =		If $x = y$
you may: add the same number	$\Rightarrow$	$x + 6 = y + 6$
subtract the same number	$\Rightarrow$	$x - 3 = y - 3$
multiply by the same number	$\Rightarrow$	$6x = 6y$
divide by the same number.	$\Rightarrow$	$\dfrac{x}{-4} = \dfrac{y}{-4}$

We are now going to solve equations, which have only one variable (unknown) eg. x; a; . . . *etc.*

Example:

Solve for x:

	Goal	**Method**
1. $\mathbf{-2x + 4 = 16}$		
$\mathbf{-2x + 4 - 4 = 16 - 4}$	Only x-terms on LHS	Add (-4) to both sides.
$\therefore\ \mathbf{-2x = 12}$		
$\dfrac{\mathbf{-2x}}{\mathbf{-2}} = \dfrac{\mathbf{12}}{\mathbf{-2}}$	Isolate x	Divide both sides by (-2)
$\mathbf{x = -6}$		

2. $6x - 12 = 2x + 4$

$6x - 12 - 2x = 2x + 4 - 2x$ (All x-terms to LHS)

$4x - 12 = 4$

$4x - 12 + 12 = 4 + 12$ (All constant terms to RHS)

$4x = 16$

$\dfrac{4x}{4} = \dfrac{16}{4}$ (Isolate x)

$x = 4$

After much practice, you discover the following:

3. $6x + 16 = 8 - 12x$

$$6x + 12x = 8 - 16$$
$$18x = -8$$
$$x = \frac{-8}{18}$$
$$x = -\frac{4}{9}$$

4. $8 - 2x - 5 + 6x = 3x - 4x + 10$

$$\therefore\ -2x + 6x - 3x + 4x = 10 - 8 + 5$$
$$5x = 7$$
$$x = \frac{7}{5}$$

Exercise 1:

Solve for x:

1. $2x = 6$
2. $3x = 12$
3. $4x = 7$
4. $5x = 3$
5. $-x = 12$
6. $-4x = -16$
7. $3x + 5 = 14$
8. $7 + 2x = 9$
9. $3x + 8 = 17$
10. $4x - 5 = 2$
11. $3x - 7 = 5x - 13$
12. $5x - 14 = 2x + 1$
13. $10x - 9 = 6x - 7$
14. $3x - 27 = 8x - 22$
15. $5x + 8 = 8x + 14$

Equations with brackets:

Remove the brackets

Examples:

1. $3(x - 4) = x - 6$
$$3x - 12 = x - 6$$
$$3x - x = -6 + 12$$
$$2x = 6$$
$$x = 3$$

2. $2(x - 3) = 3(2x - 7)$
$$2x - 6 = 6x - 21$$
$$2x - 6x = -21 + 6$$
$$-4x = -15$$
$$x = \frac{-15}{-4}$$
$$\therefore x = \frac{15}{4} = 3\frac{3}{4}$$

3. $4(2x + 3) - (3x + 8) = 5 - 3(4x - 2)$
$$\therefore 8x + 12 - 3x - 8 = 5 - 12x + 6$$
$$8x - 3x + 12x = 5 + 6 - 12 + 8$$
$$17x = +7$$
$$x = \frac{7}{17}$$

Exercise 2:

Solve for x:

1. $5(2x - 3) = 3(2x - 9)$
2. $2(x - 5) = 4(2x + 3)$
3. $3(3x + 4) = 2(2x - 4)$
4. $2(4x - 5) - (3x + 6) = -2(x + 3)$
5. $-4(3x - 1) = 3(5x + 2) - (8x - 3)$
6. $5(x + 3) = 2(10x - 1) - 4(2x + 6)$
7. $2(3x + 1) = 7 - 4x$
8. $2(2x - 1) - 3x = 4 - x$
9. $2(x + 3) + x + 7 = 2(2x + 9) + 5(4 - x) + 3$
10. $8(x - 1) + 2(3x - 2) = -7(x + 2) + 3(x - 6) - 2x$

Equations with fractions:

Eliminate fractions by multiplying both sides by the LCM of the denominators.
NB. We multiply both sides: **every term** by the **same number**, namely the LCM.

Example:

1. $\frac{a}{7} + 2 = 4$

(LCM: 7)

$$\therefore a + 14 = 28$$
$$a = 28 - 14$$
$$a = 14$$

2. $\frac{3x}{8} = \frac{1}{4}$

(LCM: 8)

$$3x = \tfrac{1}{4} \times 8$$
$$\therefore 3x = 2$$
$$x = \tfrac{2}{3}$$

3. $\dfrac{x-6}{2} = -1$

(LCM: 2)

$\therefore x - 6 = -2$

$x = -2 + 6$

$x = 4$

4. $\dfrac{2x-1}{3} - \dfrac{x+1}{4} = 3$

(LCM: 12) **(Remember!)**

$4(2x-1) - 3(x+1) = 36$

$8x - 4 - 3x - 3 = 36$

$8x - 3x = 36 + 4 + 3$

$5x = 43$

$x = \dfrac{43}{5}$

5. $\dfrac{x}{2} - \dfrac{x-2}{6} = \dfrac{5}{3} + \dfrac{4-x}{2} + 2$

(LCM: 6)

$\therefore 3x - (x-2) = 10 + 3(4-x) + (12)$

$\therefore 3x - x + 2 = 10 + 12 - 3x + 12$

$3x - x + 3x = 10 + 12 + 12 - 2$

$5x = 32$

$x = \dfrac{32}{5}$

$\dfrac{1}{2} + \dfrac{1}{3x} = \dfrac{2-x}{6x}$

(LCM: $6x$)

$\therefore 3x + 2 = 2 - x$

$3x + x = 2 - 2$

$4x = 0$

$x = 0$ but $x \neq 0$

thus the equation has **no solution**

{ You may not divide by 0.

7. $\dfrac{3}{a} - 2 = \dfrac{1}{2} - \dfrac{9}{2a}$

(LCM: 2a)

$\therefore 6 - 4a = a - 9$

$\therefore -4a - a = -9 - 6$

$\therefore -5a = -15$

$a = \dfrac{-15}{-5}$

$= 3$

6. If x appears in the DENOMINATOR of a fraction remember that this denominator may not be 0! Check your answers and reject any solution which makes the denominator 0.

Exercise 3:

Solve the following equations:

1. $\dfrac{a}{2} + \dfrac{a}{3} = 1$
2. $\dfrac{2p+3}{4} - 2 = -\dfrac{3p-1}{8}$
3. $\dfrac{2x-3}{2} - \dfrac{3x+1}{4} = 1$
4. $\dfrac{3a+1}{5} - \dfrac{2a-1}{10} = 2$
5. $\dfrac{2k}{5} = \dfrac{3k-1}{2}$
6. $\dfrac{x}{7} - 2 = x + \dfrac{4-3x}{7}$
7. $\dfrac{2(x-1)}{3} + 3 = \dfrac{3}{4} + \dfrac{3(2x+3)}{2}$
8. $\dfrac{x}{5} + \dfrac{x-9}{3} = \dfrac{-x}{6} + x - 1 - \dfrac{x}{2}$
9. $\dfrac{3}{8}(a-2) = \dfrac{5}{12}(a+3) - \dfrac{2}{3}$
10. $\dfrac{5x+2}{3} - \dfrac{x}{5} = x - \dfrac{3x-5}{15}$

Summary

To solve a linear equation:

1. Eliminate fractions by multiplying every term by the LCM of the denominators.
2. Remove all brackets.
3. Move all x-terms to LHS and all the constant terms to the RHS.
4. Divide by the coefficient of x.

Exercise 4: Mixed

Solve the following equations:

1. $\frac{x}{2} = \frac{11}{8}$

2. $\frac{2x}{3} + 1 = 5$

3. $6x - 4 + x - 9x = 4x - 5$

4. $2(x + 2) - (x - 3) = 5$

5. $7 - x = \frac{2x - 7}{5}$

6. $x - \frac{x - 1}{2} = 3$

7. $\frac{2x + 1}{2} - 3 = \frac{x - 1}{4}$

8. $\frac{2x + 1}{5} = \frac{x + 1}{2} + 1$

9. $\frac{3(2x - 1)}{4} = 1 + \frac{2(x + 2)}{3}$

10. $\frac{a + 1}{2} - 4 = \frac{a - 1}{4}$

11. $\frac{a}{3} - \frac{a + 1}{2} = 2$

12. $\frac{3p - 2}{4} + \frac{4p - 1}{5} = \frac{p + 1}{20}$

13. $\frac{a + 3}{4} - \frac{a + 2}{8} = \frac{a}{2} - 1$

14. $\frac{3(y + 6)}{4} - \frac{y}{3} = 2 - \frac{2y - 3}{6}$

15. $\frac{1}{2}(2x + 4) + 2\frac{1}{4} = \frac{3}{4}(3x - 5)$

16. $\frac{2}{a} + \frac{3}{2a} = 7$

17. $\frac{5}{4} + \frac{2}{3a} = 4 - \frac{a - 3}{12a}$

18. $5 + \frac{8}{x} = \frac{2(x + 4)}{x}$

19. $\frac{x - 1}{3x} - \frac{1}{9} = \frac{-3 + x}{9x}$

20. $\frac{4}{3x} + \frac{1}{5} = 1\frac{1}{3} + \frac{1}{5x}$

Identities:

If **any real value** of x satisfies a linear equation in x, then such an equation is called **an identity**.

Example:

1. $2x + 3 = 2x + 3$...........(1)
try to solve it and you find:
$2x - 2x = 3 - 3$
$0 = 0$
$\therefore x \in R$

you could already see in (1) that **any value** of x would satisfy the equation.

2. $2(x + 2) + x - 9 = 3(x - 1) - 2$
$\therefore 2x + 4 + x - 9 = 3x - 3 - 2$
$\therefore \underbrace{3x - 5 = 3x - 5}_{\text{Identity}}$
$\therefore x \in R$

Solution of problems:

The whole object of the solution of equations is to be able to solve practical problems. The so-called "story sum" is reduced to an equation which can then be solved. This section will cover easy examples. Work carefully through each one.

Guidelines:

1. **Read! Read! Read!** the problem.
2. **Decide what x will be**, eg. the number; the width; the boy's age; the distance; etc. ...... Then begin with:
   "Let the ...... be x" ←NB
   Further guidelines for this step:
   If there are 2 unknowns, let the **smaller one** be x.
   Or **check** what the required quantity is (i.e. what is **the actual question**: Find the distance ......) and let that be x
3. **If 2 items are involved, express the one in terms of the other using x**
   Eg. son's age is x ⇒ father's age is $3x$
   width is x ⇒ length is $x + 10$
   one number is x ⇒ the other number is $50 - x$ or $\frac{20}{x}$; etc.
4. **Now set up an equation.**
5. **Solve the equation.**
6. **Again read the actual question.** Often the x that you have chosen and solved, does not represent the answer to the question.
7. **Ensure that you give a logical answer.**
   impossible if: son: -3 years old (negative)
   breadth: -5 cm (negative)

Remember:
7 times a number is $7x$
7 more than a number is $x + 7$

Example:

1. 3 less than seven times a number, is equal to 11. Find the number.

Solution: **Let the number be x**

$$\therefore 7x - 3 = 11$$
$$7x = 14$$
$$x = 2$$

∴ the number is 2.

2. The length of a rectangle is 4 cm more than the width. If the perimeter is 32 cm, find the length and the width.

Solution: Read (Step 1)

Let the width be x (Step 2)

$\therefore$ **the length is $x + 4$** (Step 3)

2 lengths + 2 widths = perimeter

$$\therefore\ x+4+x+4+x+x = 32$$
$$\therefore\ 4x+8 = 32$$
$$4x = 32-8$$
$$4x = 24$$
$$x = 6$$

$\therefore$ Width is 6 cm

and the length is 10 cm

3. A grandfather is 50 years older than his grandson. Five years ago he was 6 times as old as his grandson. How old is the grandfather now?

Solution: Let the boy be **now**: x (choose the smaller x)

the grandfather is **now**: $x + 50$

5 years ago: boy: $x - 5$

grandfather: $x + 50 - 5$

$$\therefore\ 6(x-5) = x+50-5$$
$$6x-30 = x+50-5$$
$$6x-x = 50-5+30$$
$$5x = 75$$
$$x = 15 \text{ years old}$$

$\therefore$ Grandfather is **now**: $x + 50$: 65 years old.

4. A batsman scored 72 runs off fours, sixes and singles. He hit an equal amount of fours and sixes, and twice as many singles as fours. How many of each did he hit?

Solution: Say he hit: x fours $\Rightarrow 4x$ runs.

x sixes $\Rightarrow 6x$ runs.

and $2x$ singles $\Rightarrow 2x$ runs.

$$\therefore\ 4x+6x+2x = 72$$
$$12x = 72$$
$$x = 6$$

$\therefore$ 6 fours; 6 sixes and 12 singles

5. Zukile walks at 4 km/h and runs at 6 km/h. If he runs all the way, he reaches the station 15 minutes earlier. Find the distance to the station.

Solution:	Walk	Run
Distance (D)	x	x
Speed (S)	4 km/h	6 km/h
Time (T)	$\frac{x}{4}$ hour	$\frac{x}{6}$ hour

Tips:

1. Fill in what is given, namely the speed.
2. Check what is asked, namely distance. Distance has to be x.
3. Complete table by using your formula: Time $= \frac{\text{Distance}}{\text{Speed}}$
4. Now set up an equation, using the remaining information.

$\therefore$ Let the distance to the station be x km.

Now the equation:

$$\therefore \frac{x}{4} = \frac{x}{6} + \frac{15}{60} \text{ (15 min earlier } = \frac{15}{60} \text{ hour.)}$$

$$\times 60: \quad 15x = 10x + 15$$

$$15x - 10x = 15$$

$$5x = 15$$

$$x = 3$$

$\therefore$ distance to station is 3 km.

Remember: *Distance = Speed × Time*

$$Time = \frac{Distance}{Speed}$$

$$Speed = \frac{Distance}{Time}$$

Exercise 5:

1. If you multiply a certain number by 5, and then subtract 3, the answer is 17. Find the number.
2. Find 2 numbers which have a sum of 60, and a difference of 36.
3. A number increased by three times the number, is 28. Find the number.
4. The length of a rectangle is 10 cm longer than its breadth. The perimeter is 48 cm. Determine the area of the rectangle.
5. The denominator of a fractionis is 1 more than the numerator. If 4 is added to the numerator and 6 to the denominator, the new fraction will be equal to the original fraction. Determine the original fraction.
6. The perimeter of a rectangle is 30 cm. If the breadth is doubled (length unchanged) it becomes a square. Determine the dimensions of the rectangle.

7. A salesman knows he has 36 tables in his store-room, some of which have 3 legs and other 4 legs. He discovers by counting that there are 124 legs. How many tables with 4 legs does he have?

8. The joint ages of A and B is 82 years. In 6 years time A will be twice as old as B was 4 years ago. Determine their ages now.

9. The ages of the 3 children in our family total 32 years. I am twice as old as my sister, and my brother is 2 years older than I. How old am I?

10. A tank is $\frac{2}{3}$ full of water. If 35 liters are tapped off, the tank is $\frac{1}{6}$ full. Determine the volume of water in a full tank.

11. If one side of a square is extended by 8 m and the adjacent side is reduced by 6 m, a rectangle is obtained which has a perimeter of 100 m. Determine the area of the original square.

12. There were 400 coins in the container after a street collection. If there were only 5c- and 10c-coins, with a total value of R27,50, how many 5c-coins and how many 10c-coins were there?

13. A man may fire 30 shots at a target. Each time he hits the target, he gets 25c, but for every miss, he has to pay in 15c. If he makes a profit of R3,50, how many times did he hit the target?

14. In a test, consisting of multiple choice questions, one mark is awarded for each correct answer and two marks are deducted for every incorrect answer. If a pupil answers 30 questions and obtains 21 marks, how many correct answers were there?

15. I bought 25 CD's. Some were R50, others R80 each. Altogether I paid R1 430. How many of each did I buy?

16. A CD costs R3 more than last year. I have to pay R496 for 12 CD's at the current price, after being given a discount of R80. What did a CD cost last year?

17. A bottle contains 120 ml medicine. A second bottle has twice the amount. Determine how much of the first bottle's contents must be added to the second bottle, so that the second bottle will have 3 times as much as the first.

18. A man has to complete a journey of 520 km in 6 hours. Having traveled for 4 hours, he realises that will have to increase his speed by 20 km/h in order to complete the journey on time. What was his original speed?

19. A cyclist rides against the wind for 36 km at a speed of 12 km/h. On his return journey with the wind behind him, his speed is 18 km/h. **What was his average speed**?

$$\text{Remember: average speed} = \frac{\text{Total Distance}}{\text{Total Time}}$$

20. 120 people attended a concert. Entrance fees were R12 per adult and R8 per child. If the concert raised R1 280, how many adults and how many children attended the concert?

SOLUTIONS

Exercise 1: Solutions

1. $2x = 6$
$\therefore x = 3$

2. $3x = 12$
$x = 4$

3. $4x = 7$
$x = \frac{7}{4}$

4. $5x = 3$
$x = \frac{3}{5}$

5. $-x = 12$
$x = -12$

6. $-4x = -16$
$x = \frac{-16}{-4} = 4$

7. $3x + 5 = 14$
$3x = 14 - 5$
$3x = 9$
$x = 3$

8. $7 + 2x = 9$
$2x = 9 - 7$
$2x = 2$
$x = 1$

9. $3x + 8 = 17$
$3x = 17 - 8$
$3x = 9$
$x = 3$

10. $4x - 5 = 2$
$4x = 2 + 5$
$4x = 7$
$x = \frac{7}{4}$

11. $3x - 7 = 5x - 13$
$3x - 5x = -13 + 7$
$-2x = -6$
$x = \frac{-6}{-2} = 3$

12. $5x - 14 = 2x + 1$
$5x - 2x = 14 + 1$
$3x = 15$
$x = 5$

13. $10x - 9 = 6x - 7$
$10x - 6x = -7 + 9$
$4x = 2$
$x = \frac{2}{4} = \frac{1}{2}$

14. $3x - 27 = 8x - 22$
$3x - 8x = -22 + 27$
$-5x = 5$
$x = -1$

15. $5x + 8 = 8x + 14$
$5x - 8x = 14 - 8$
$-3x = 6$
$x = \frac{6}{-3}$
$= -2$

Exercise 2: Solutions

1. $5(2x-3)=3(2x-9)$
$10x-15=6x-27$
$10x-6x=-27+15$
$4x=-12$
$x=-3$

2. $2(x-5)=4(2x+3)$
$2x-10=8x+12$
$2x-8x=12+10$
$-6x=22$
$x=\frac{22}{-6}=\frac{11}{3}$

3. $3(3x+4)=2(2x-4)$
$9x+12=4x-8$
$9x-4x=-8-12$
$5x=-20$
$x=\frac{-20}{5}=-4$

4. $2(4x-5)-(3x+6)=-2(x+3)$
$8x-10-3x-6=-2x-6$
$8x-3x+2x=-6+10+6$
$7x=10$
$x=\frac{10}{7}$

5. $-4(3x-1)=3(5x+2)-(8x-3)$
$-12x+4=15x+6-8x+3$
$-12x-15x+8x=6+3-4$
$-19x=5$
$x=\frac{5}{-19}$

6. $5(x+3)=2(10x-1)-4(2x+6)$
$5x+15=20x-2-8x-24$
$5x-20x+8x=-2-24-15$
$-7x=-41$
$x=\frac{-41}{-7}=\frac{41}{7}$

7. $2(3x+1)=7-4x$
$6x+2=7-4x$
$6x+4x=7-2$
$10x=5$
$x=\frac{1}{2}$

8. $2(2x-1)-3x=4-x$
$4x-2-3x=4-x$
$4x-3x+x=+2+4$
$2x=6$
$x=3$

9. $2(x+3)+x+7=2(2x+9)+5(4-x)+3$
$2x+6+x+7=4x+18+20-5x+3$
$2x+x-4x+5x=18+20+3-6-7$
$4x=28$
$x=7$

10. $8(x-1)+2(3x-2)=-7(x+2)+3(x-6)-2x$
$8x-8+6x-4=-7x-14+3x-18-2x$
$8x+6x+7x-3x+2x=-14-18+8+4$
$20x=-20$
$x=-1$

Exercise 3: Solutions

1. $\frac{a}{2}+\frac{a}{3}=1$
×6: $3a+2a=6$
$5a=6$
$a=\frac{6}{5}$

2. $\frac{2p+3}{4}-2=-\frac{3p-1}{8}$
×8: $2(2p+3)-16=-(3p-1)$
$4p+6-16=-3p+1$
$4p+3p=1-6+16$
$7p=11$
$p=\frac{11}{7}$

3. $\frac{2x-3}{2}-\frac{3x+1}{4}=1$
×4: $2(2x-3)-(3x+1)=4$
$4x-6-3x-1=4$
$4x-3x=4+6+1$
$x=11$

4. $$\frac{3a+1}{5}-\frac{2a-1}{10}=2$$

$\times$ **10:** $2(3a+1)-(2a-1)=20$

$6a+2-2a+1=20$

$6a-2a=20-2-1$

$4a=17$

$a=\frac{17}{4}$

5. $$\frac{2k}{5}=\frac{3k-1}{2}$$

$\times$ **10:** $4k=5(3k-1)$

$4k=15k-5$

$4k-15k=-5$

$-11k=-5$

$k=\frac{-5}{11}=\frac{5}{11}$

6. $$\frac{x}{7}-2=x+\frac{4-3x}{7}$$

$\times$ **7:** $x-14=7x+4-3x$

$x-7x+3x=4+14$

$-3x=18$

$x=-6$

7. $$\frac{2(x-1)}{3}+3=\frac{3}{4}+\frac{3(2x+3)}{2}$$

$\times$ **12:** $8(x-1)+36=9+18(2x+3)$

$8x-8+36=9+36x+54$

$8x-36x=54+9-36+8$

$-28x=35$

$x=\frac{35}{-28}=-\frac{5}{4}$

8. $$\frac{x}{5}+\frac{x-9}{3}=\frac{-x}{6}+x-1-\frac{x}{2}$$

$\times$ **30:** $6x+10(x-9)=-5x+30x-30-15x$

$6x+10x-90=-5x+30x-30-15x$

$6x+10x+5x-30x+15x=+90-30$

$6x=60$

$x=10$

9. $$\frac{3}{8}(a-2)=\frac{5}{12}(a+3)-\frac{2}{3}$$

$\times$ **24:** $9(a-2)=10(a+3)-16$

$9a-18=10a+30-16$

$9a-10a=30-16+18$

$-a=32$

$a=-32$

10. $$\frac{5x+2}{3}-\frac{x}{5}=x-\frac{3x-5}{15}$$

$\times$ **15:** $5(5x+2)-3x=15x-(3x-5)$

$25x+10-3x=15x-3x+5$

$25x-3x-15x+3x=5-10$

$10x=-5$

$x=-\frac{1}{2}$

Exercise 4: Mixed Solutions

1. $$\frac{x}{2}=\frac{11}{8}$$

$\times$ **8:** $4x=11$

$x=\frac{11}{4}$

2. $$\frac{2x}{3}+1=5$$

$\times$ **3:** $2x+3=15$

$2x=15-3$

$2x=12$

$x=6$

3. $6x-4+x-9x=4x-5$

$6x+x-9x-4x=-5+4$

$-6x=-1$

$x=-\frac{1}{6}$

4. $2(x+2)-(x-3)=5$

$2x+4-x+3=5$

$2x-x=5-4-3$

$x=-2$

5. $$7-x=\frac{2x-7}{5}$$

$\times$ **5:** $35-5x=2x-7$

$-5x-2x=-7-35$

$-7x=-42$

$x=6$

6. $$x-\frac{x-1}{2}=3$$

$\times$ **2:** $2x-(x-1)=6$

$2x-x+1=6$

$2x-x=6-1$

$x=5$

7. $$\frac{2x+1}{2} - 3 = \frac{x-1}{4}$$
$$\times 4: \quad 2(2x+1) - 12 = x - 1$$
$$4x + 2 - 12 = x - 1$$
$$4x - x = -1 + 12 - 2$$
$$3x = 9$$
$$x = 3$$

8. $$\frac{2x+1}{5} = \frac{x+1}{2} + 1$$
$$\times 10: \quad 2(2x+1) = 5(x+1) + 10$$
$$4x + 2 = 5x + 5 + 10$$
$$4x - 5x = 5 + 10 - 2$$
$$-x = 13$$
$$x = -13$$

9. $$\frac{3(2x-1)}{4} = 1 + \frac{2(x+2)}{3}$$
$$\times 12: \quad 9(2x-1) = 12 + 8(x+2)$$
$$18x - 9 = 12 + 8x + 16$$
$$18x - 8x = 12 + 16 + 9$$
$$10x = 37$$
$$x = \frac{37}{10}$$

10. $$\frac{a+1}{2} - 4 = \frac{a-1}{4}$$
$$\times 4: \quad 2(a+1) - 16 = a - 1$$
$$2a + 2 - 16 = a - 1$$
$$2a - a = -1 - 2 + 16$$
$$a = 13$$

11. $$\frac{a}{3} - \frac{a+1}{2} = 2$$
$$\times 6: \quad 2a - 3(a+1) = 12$$
$$2a - 3a - 3 = 12$$
$$2a - 3a - 3 = 12 + 3$$
$$-a = 15$$
$$a = -15$$

12. $$\frac{3p-2}{4} + \frac{4p-1}{5} = \frac{p+1}{20}$$
$$\times 20: \quad 5(3p-2) + 4(4p-1) = p + 1$$
$$15p - 10 + 16p - 4 = p + 1$$
$$15p + 16p - p = 1 + 4 + 10$$
$$30p = 15$$
$$p = \frac{1}{2}$$

13. $$\frac{a+3}{4} - \frac{a+2}{8} = \frac{a}{2} - 1$$
$$\times 8: \quad 2(a+3) - (a+2) = 4a - 8$$
$$2a + 6 - a - 2 = 4a - 8$$
$$2a - a - 4a = -8 - 6 + 2$$
$$-3a = -12$$
$$a = 4$$

14. $$\frac{3(y+6)}{4} - \frac{y}{3} = 2 - \frac{2y-3}{6}$$
$$\times 12: \quad 9(y+6) - 4y = 24 - 2(2y-3)$$
$$9y + 54 - 4y = 24 - 4y + 6$$
$$9y - 4y + 4y = 24 + 6 - 54$$
$$9y = -24$$
$$y = \frac{-24}{9} = \frac{-8}{3}$$

15. $$\frac{1}{2}(2x+4) + \frac{9}{4} = \frac{3}{4}(3x-5)$$
$$\times 4: \quad 2(2x+4) + 9 = 3(3x-5)$$
$$4x + 8 + 9 = 9x - 15$$
$$4x - 9 = -15 - 8 - 9$$
$$-5x = -32$$
$$x = \frac{32}{5}$$

16. $$\frac{2}{a} + \frac{3}{2a} = 7 \quad (a \neq 0)$$
$$\times 2a: \quad 4 + 3 = 14a$$
$$7 = 14a$$
$$\frac{7}{14} = a$$
$$\therefore a = \frac{1}{2}$$

17. $$\frac{5}{4} + \frac{2}{3a} = 4 - \frac{a-3}{12a} \quad (a \neq 0)$$
$$\times 12a: \quad 15a + 8 = 48a - (a-3)$$
$$15a + 8 = 48a - a + 3$$
$$15a - 48a + a = 3 - 8$$
$$-32a = -5$$
$$a = \frac{-5}{-32} = \frac{5}{32}$$

18. $$5 + \frac{8}{x} = \frac{2(x+4)}{x}$$
$$\times x: \quad 5x + 8 = 2(x+4)$$
$$5x + 8 = 2x + 8$$
$$5x - 2x = 8 - 8$$
$$3x = 0$$
$$x = 0$$
No solution $x \neq 0$

19. $$\frac{x-1}{3x} - \frac{1}{9} = \frac{-3+x}{9x} \quad (a \neq 0)$$
$$\times 9x: \quad 3(x-1) - x = -3 + x$$
$$3x - 3 - x = -3 + x$$
$$3x - x - x = -3 + 3$$
$$x = 0$$
No solution $x \neq 0$

20. $$\frac{4}{3x} + \frac{1}{5} = \frac{4}{3} + \frac{1}{5x} \quad (a \neq 0)$$
$$20 + 3x = 20x + 3$$
$$3x - 20x = 3 - 20$$
$$-17x = -17$$
$$x = 1$$

Exercise 5: "Story" sums – Solutions

1. Let the number be x.

$$\therefore 5x - 3 = 17$$
$$5x = 17 + 3$$
$$5x = 20$$
$$x = 4$$

$\therefore$ Number is 4

2. Let one number be x.

$\therefore$ the other is $60 - x$

$$\therefore x - (60 - x) = 36$$
$$x - 60 + x = 36$$
$$2x = 36 + 60 = 96$$
$$x = 48$$

$\therefore$ Numbers are 48 and 12

3. Let the number be x.

$$\therefore x + 3x = 28$$
$$4x = 28$$
$$x = 7$$

$\therefore$ the number is 7

4. Let the breadth be x.

$\therefore$ the length is $x + 10$

$$\underbrace{2x}_{2\times\text{Breadth}} + \underbrace{2(x+10)}_{2\times\text{Length}} = \textit{perimeter}$$
$$\begin{aligned} 2x + 2x + 20 &= 48 \\ 2x + 2x + 20 &= 48 \\ 4x &= 28 \\ x &= 7 \end{aligned}$$

$\therefore$ Breadth: 7 cm

Length: 17 cm

$\therefore$ Area: $= l \times b$

$= 17 \times 7$

$= 119\ cm^2$

5. Let the numerator be x

$\therefore$ denominator is $x + 1$ } $\therefore$ fraction: $\dfrac{x}{x+1}$

new fraction: $\dfrac{x+4}{x+1+6}$

$$\therefore \frac{x+4}{x+7} = \frac{x}{x+1}$$
$$[\times (x+7)(x+1)]:\ (x+4)(x+1) = x(x+7)$$
$$x^2 + 5x + 4 = x^2 + 7x$$
$$x^2 - x^2 + 5x - 7x = -4$$
$$-2x = -4$$
$$x = 2$$

$\therefore$ Original fraction: $\dfrac{2}{3}$

6. Let the breadth be x

If perimeter = 30 m

Then AB + BC = 15 m

$\therefore$ Length $15 - x$

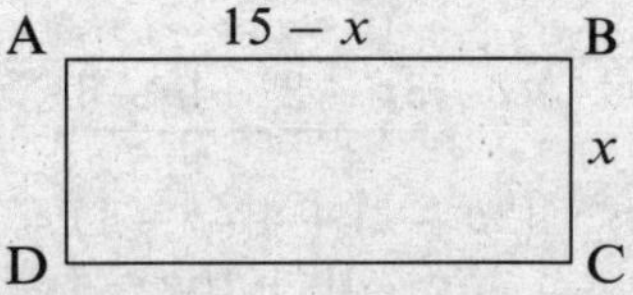

New rectangle: Breadth: $2x$
Length: $15 - x$ } square

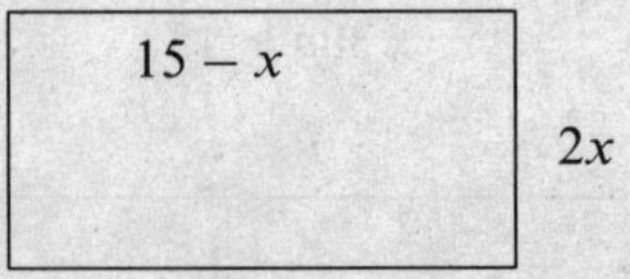

$$\therefore 15 - x = 2x$$
$$15 = 2x + x$$
$$15 = 3x$$
$$5 = x$$

$\therefore$ Original rectangle is $5m \times 10m$

7. Let there be x tables: 3 legs
$\therefore (36-x)$ tables: 4 legs

$$\therefore 3x + 4(36-x) = 124$$
$$3x + 144 - 4x = 124$$
$$-x = 124 - 144$$
$$-x = -20$$
$$x = 20$$

$\therefore$ 20 tables with 3 legs
16 tables with 4 legs

8. Now: Let A: x years old
B: $(82-x)$ years old
In 6 years time: A: $(x+6)$ years
4 years ago: B: $82 - x - 4$

$$x + 6 = 2(82 - x - 4)$$
$$x + 6 = 164 - 2x - 8$$
$$x + 2x = 164 - 8 - 6$$
$$3x = 150$$
$$\therefore x = 50$$

$\therefore$ A is now: 50 years
B is now: 32 years

9. Let sister be: x years
I am: $2x$ years
brother is: $(2x+2)$ years

$$\therefore x + 2x + 2x + 2 = 32$$
$$\therefore 5x = 32 - 2$$
$$5x = 30$$
$$x = 6$$

$\therefore$ I am 12 years old.

10. Let the tank hold: x liter water

$$\frac{2}{3}x - 35 = \frac{1}{6}x$$
$$\times 6:\ 4x - 210 = x$$
$$4x - x = 210$$
$$3x = 210$$
$$x = 70$$

$\therefore$ Tank holds 70 liters of water

11. Let square's side be x meter:

rectangle:
$\therefore$ length: $x+8$
breadth: $x-6$

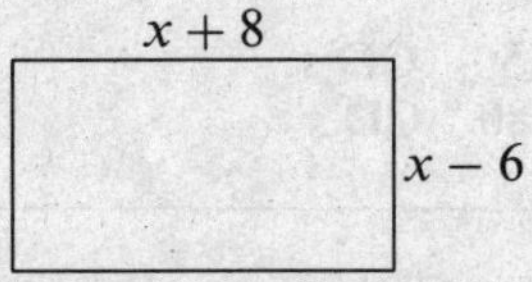

$\therefore$ 2 lengths + 2 breadths = perimeter

$$\therefore 2(x+8) + 2(x-6) = 100$$
$$2x + 16 + 2x - 12 = 100$$
$$4x = 100 - 16 + 12$$
$$4x = 96$$
$$x = 24$$

Area of square is: *side* $\times$ *side*
$= 24 \times 24$
$= 576\ \text{m}^2$

12. Let there be x: 5c coins
$\therefore$ $400 - x$: 10c coins

$$5x + 10(400 - x) = 2\,750$$
$$5x + 4\,000 - 10x = 2\,750$$
$$-5x = 2\,750 - 4\,000$$
$$-5x = -1\,250$$
$$x = 250$$

$\therefore$ 250 → 5c coins
$\therefore$ 150 → 10c coins

13. Let x be the number of hits
$\therefore$ $30 - x$ times he missed

$$\overbrace{25x}^{\text{hit}} - \overbrace{15(30-x)}^{\text{miss}} = 350$$
$$25x - 450 + 15x = 350$$
$$40x = 350 + 450$$
$$40x = 800$$
$$x = 20$$
$$\Rightarrow 20 \text{ hits}$$

14. Let the number of correct answers be: x
$\therefore$ wrong answers: $30 - x$

$$x - 2(30 - x) = 21$$
$$x - 60 + 2x = 21$$
$$3x = 21 + 60$$
$$3x = 81$$
$$x = 27$$

$\therefore$ 27 correct answers.

15. Let: $x \rightarrow$ R 50 CD's
$\therefore 25 - x \rightarrow$ R 80 CD's

$$\therefore \quad 50x + 80(25 - x) = 1\,430$$
$$\therefore \quad 50x + 2\,000 - 80x = 1\,430$$
$$-30x = -570$$
$$x = 19$$

$\therefore$ 19: R 50 CD's
6 R 80 CD's

16. Say last year CD's cost: R x each
$\therefore$ this year CD's cost: R$(x + 3)$

$$\underbrace{12(x+3)}_{\text{price}} - \underbrace{80}_{\text{discount}} = 496$$
$$\therefore 12x + 36 - 80 = 496$$
$$12x = 496 + 80 - 36$$
$$12x = 540$$
$$x = 45$$

$\therefore$ Price last year: R 45.

17. 1st Bottle: 120 ml
2nd Bottle: 240 ml
Let x ml be transferred

$$3(120 - x) = 240 + x$$
$$360 - 3x = 240 + x$$
$$360 - 240 = 3x + x$$
$$120 = 4x$$
$$30 = x$$

$\therefore$ 30 ml must be added.

18.

	1st part of journey		2nd part of journey	
D:	$4x$ km	(5)	$2(x + 20)$	(6)
S:	x km/h	(3)	$(x + 20)$	(4)
T:	4 hour	(1)	2 hour	(2)

Order in which completed: (arrows to the bracketed numbers)

$$4x + 2(x + 20) = 520$$
$$4x + 2x + 40 = 520$$
$$6x = 520 - 40$$
$$6x = 480$$
$$x = 80$$

$\therefore$ original speed: 80 km/h

19.

	Against the wind	With the wind
D:	36 km	36 km
S:	12 km/h	18 km/h
T:	3 hour	2 hour

$$\text{Average speed} = \frac{\text{total distance}}{\text{total time}}$$
$$= \frac{72}{5}$$
$$= 14{,}4 \text{ km/h}$$

20. Let there be x adults
$\therefore (120 - x)$ children
$\therefore 12x + 8(120 - x) = 1\,280$
$12x + 960 - 8x = 1\,280$
$4x = 1\,280 - 960$
$4x = 320$
$\therefore x = 80$
$\therefore$ 80 adults
40 children

7 *Graphs*

1. Coordinate plane:

- To represent $x = 2$ we use a number line.

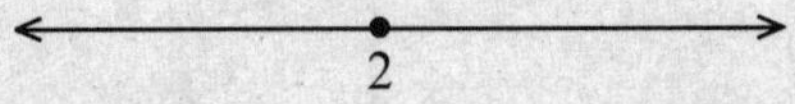

- But to represent $y = x + 2$ we need an ***X – Axis*** and a ***Y – Axis***. This coordinate plane is called the **Cartesian Plane**.

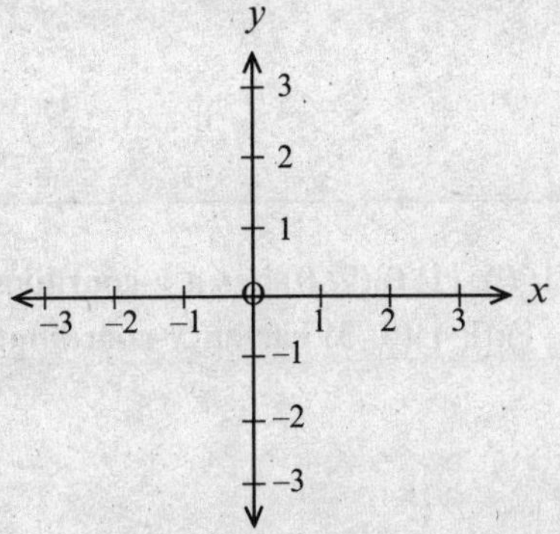

- **O** is the **origin** with coordinates (0; 0).
- **A** (2; 4) is an **ordered number pair**.

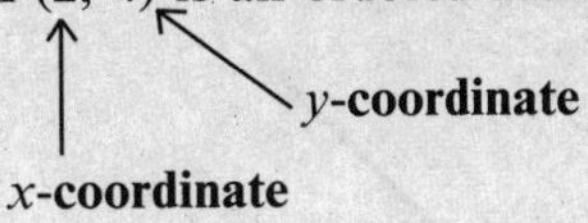

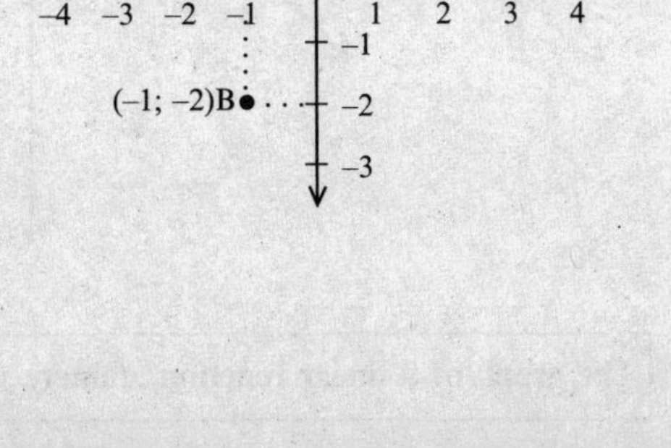

- Write down the coordinates of the points: A – E:

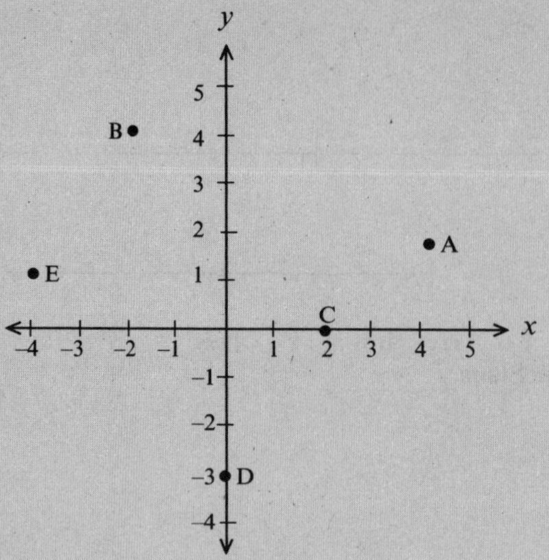

Solution: A (4; 2)

B (–2; 4)

C (2; 0)

D (0; –3)

E (–4; 1)

- **Note that:** **Every point on the X-axis** eg. $(-4; 0)$; $(-1; 0)$; $(1; 0)(2; 0)$ **has a *y*-coordinate of 0**
  similarly: **every point on the Y-axis** eg. $(0; -2)(0; -1)(0; 1)(0; 3)$ **has an *x*-coordinate of 0.**

2. Graph of $y = mx + c$ **using a table:**

Given: $y = x + 2$

To draw the graph, choose a few *x*-values and calculate their corresponding *y*-values.
Do this using a table.

x	-3	-2	0	1	2	4
$y = x + 2$	-1	0	2	3	4	6

The values in the table can be written as coordinates.
$(-3; -1)$; $(-2; 0)$; $(0; 2)$; $(1; 3)$; $(2; 4)$; $(4; 6)$

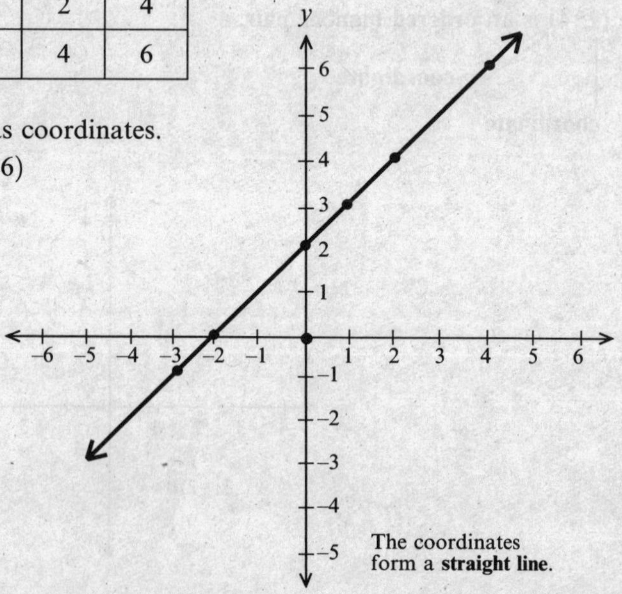

The coordinates form a **straight line.**

The graph of a linear function, namely $y = mx + c$, is a straight line.

3. Standard form: $y = mx + c$

$y = mx + c$

m : **gradient**/slope/inclination.

c : ***Y*-intercept, i.e. where the line cuts the *Y*-axis.**

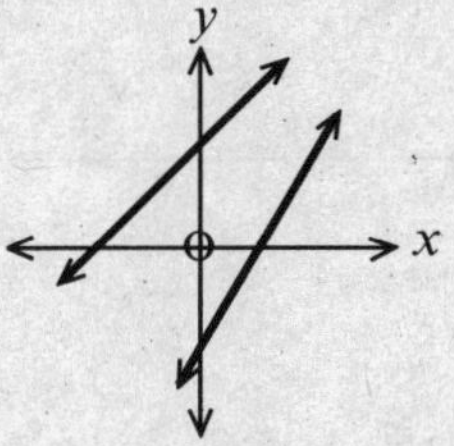

$m > o$

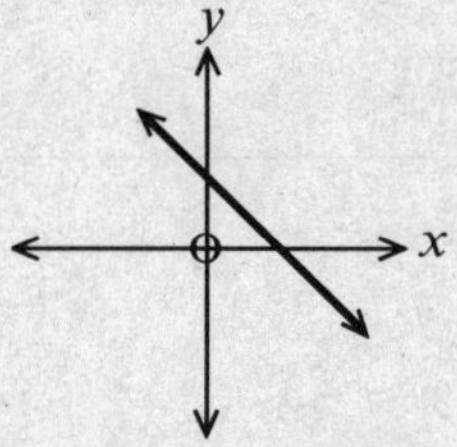

$m < o$

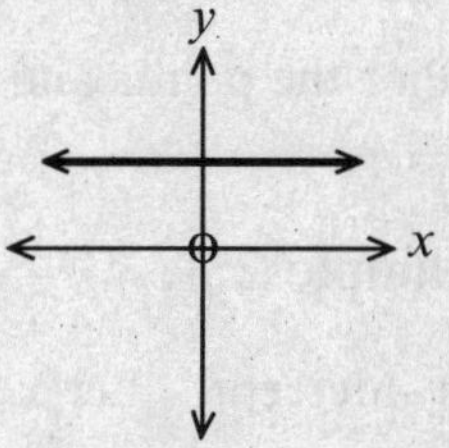

$m = o$

$$m = \frac{\Delta y}{\Delta x} = \frac{\textit{change in } y}{\textit{change in } x}$$

Examples:

1. $y = 3x - 2$ $\Rightarrow$ $m = 3; \quad c = -2$
2. $y = 4x + 1$ $\Rightarrow$ $m = -4; \quad c = 1$
3. $y = \frac{3}{4}x$ $\Rightarrow$ $m = \frac{3}{4}; \quad c = 0$
4. $y = -x - \frac{1}{2}$ $\Rightarrow$ $m = -1; \quad c = -\frac{1}{2}$
5. $y = \frac{x}{3} + 1$ $m = \frac{1}{3}; \quad c = 1$
6. $2y + x = 4$

   $2y = -x + 4$

   $y = -\frac{1}{2}x + 2$ $\Rightarrow$ $m = -\frac{1}{2}; \quad c = 2$

7. $2x - \frac{1}{3}y = 5$

   $\times 3$: $6x - y = 15$

   $-y = -6x + 15$

   $y = 6x - 15$ $\Rightarrow$ $m = 6; \quad c = -15$

4. Graph of $y = mx + c$ using the intercept method:

Method:

Find:

1. *Y*-intercept: let $x = 0$ (c-value)
2. *X*-intercept: let $y = 0$
3. Plot the 2 points;
   Draw the graph

Example 1: $y = -2x + 4$

1. *Y*-intercept: Let $x = 0$: $\mathbf{y = 4}$

2. *X*-intercept: Let $y = 0$: $0 = -2x + 4$

$-4 = -2x$

$\mathbf{2 = x}$

3. Plot the points and draw the graph.

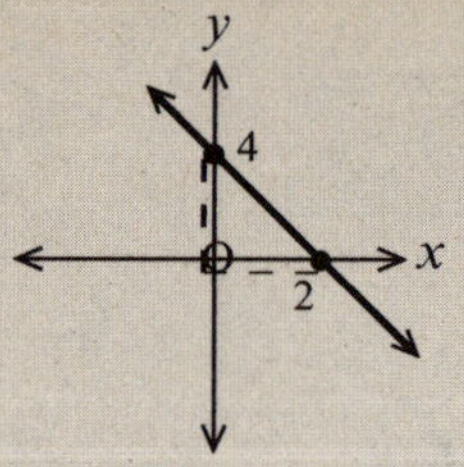

Example 2: $y = \frac{1}{2}x - 3$

1. *y*-intercept: Let $x = 0$ $\Rightarrow$ $\mathbf{y = -3}$

2. *x*-intercept: Let $y = 0$ $\Rightarrow$ $0 = \frac{1}{2}x - 3$

$3 = \frac{1}{2}x$

$\times 2$: $\mathbf{6 = x}$

3. Plot points and draw graph.

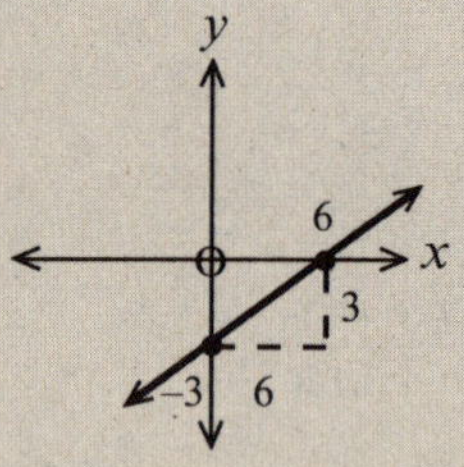

Example 3: Draw $y = 4x + 1$ and $y = 4x - 2$

<u>$y = 4x + 1$</u>

(1) *y*-intercept: $x = 0$: $y = 1$

(2) *x*-intercept: $y = 0$: $0 = 4x + 1$

$-1 = 4x$

$-\frac{1}{4} = x$

<u>$y = 4x - 2$</u>

(1) *y*-intercept: $x = 0$: $y = -2$

(2) *x*-intercept: $y = 0$: $0 = 4x - 2$

$2 = 4x$

$\frac{1}{2} = x$

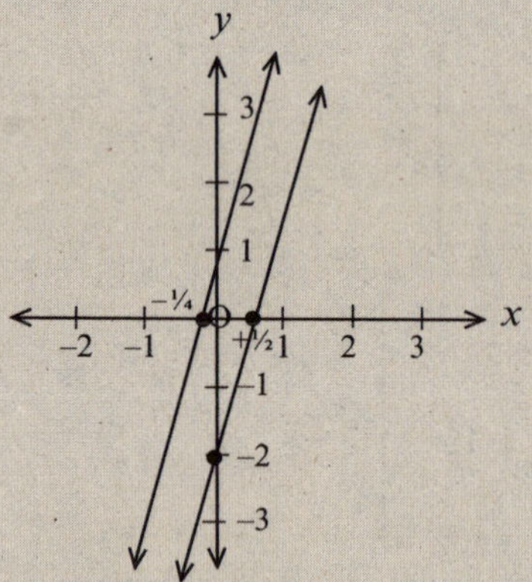

The lines are **parallel**, because **both** $m = 4$.

As mentioned before m is the gradient/slope/inclination.

m **the same $\Leftrightarrow$ lines parallel.**

5. Graph of $y = a$ **and** $x = b$:

Example 1: Draw $y = 3$

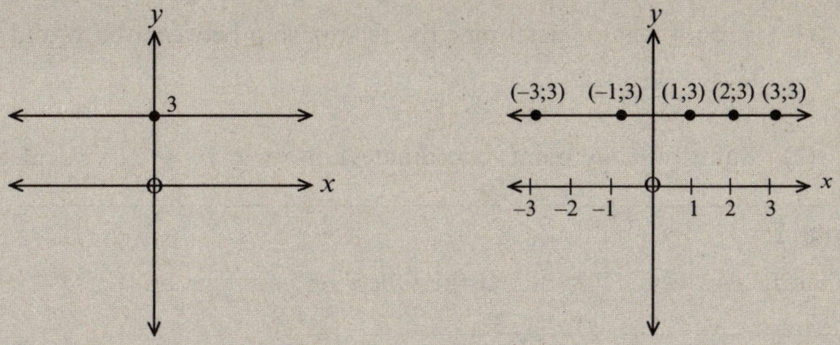

Line is parallel to X-axis.

Note that, in the example above, every point on the line has a y-coordinate of 3.

Example 2: Draw $x = -2$

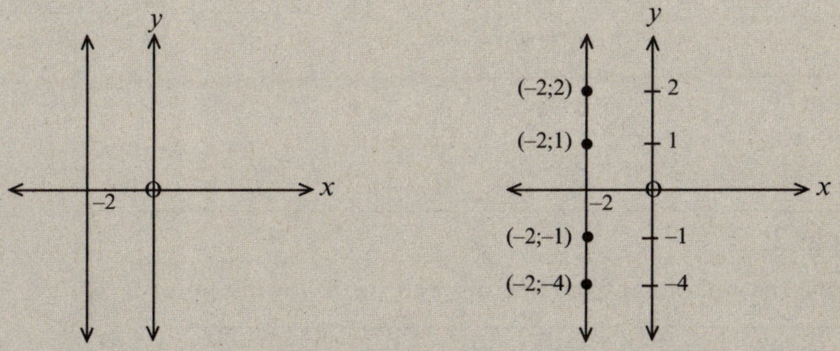

Line is parallel to Y-axis.

Note that, in the example above, every point on the line has an x-coordinate of -2

6. Graph of $ax + by = c$ **using intercept method** (as in 4):

Example: $3x + 2y = -6$

(1) ***Y*-intercept:** Let $x = 0$: $0 + 2y = -6$

$y = -3$

(2) ***X*-intercept:** Let $y = 0$: $3x + 0 = -6$

$x = -2$

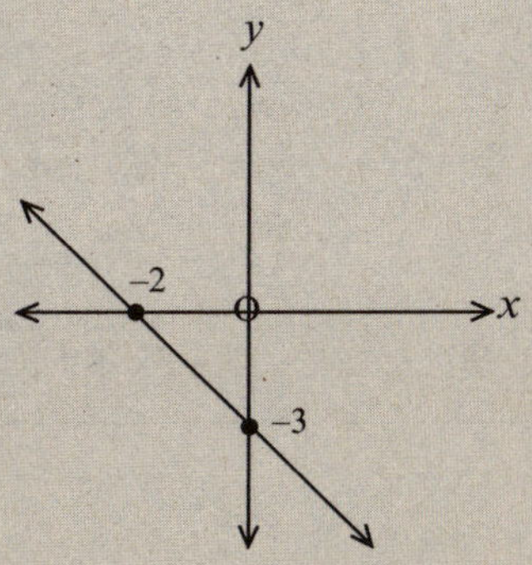

7. Finding the equation from the table:

A table with values represents a few coordinates on the graph.

Remember: Every point (in coordinate form) on the graph *satisfies* the *equation* of the graph.

Method: (1) From the table determine the relationship between the x- and y-values, namely m.

$$m = \frac{\Delta y}{\Delta x} = \frac{y_2 - y_1}{x_2 - x_1} = \frac{y_4 - y_3}{x_4 - x_3} = \frac{y_6 - y_5}{x_6 - x_5}$$

(2) Substitute **any** point (coordinates) into $y = mx + c$ to calculate c.

Example 1:

Determine the equation of the line if the following points lie on it.

	x_1	x_2	x_3	x_4	x_5	x_6
x	−3	−2	−1	1	2	3
y	−9	−7	−5	−1	1	3
	y_1	y_2	y_3	y_4	y_5	y_6

Solution: $m = \frac{\Delta y}{\Delta x} = \frac{3-1}{3-2} = \frac{2}{1} = 2$

Note: The same as eg. $\left[\frac{-5-(-7)}{-1-(-2)} = \frac{-5+7}{-1+2} = \frac{2}{1}\right]$

$y = mx + c$

$\therefore\ y = 2x + c$

Substitute (1;−1) into: $-1 = 2(1) + c$

$-1 = 2 + c$

$-3 = c \quad \Rightarrow \quad y = 2x - 3$

Example 2:

Determine the equation of the line through the following points:

x	−3	−2	−1	0	1	2	3
y	10	7	4	1	−2	−5	−8

Solution: $m = \frac{\Delta y}{\Delta x} = \frac{4-7}{-1-(-2)}$ (Choose any 2 points)

$= \frac{-3}{-1+2}$

$= \frac{-3}{1} = -3$

$\therefore\ y = -3x + c$

(2;−5) $\quad -5 = -3(2) + c$ (Substitute any point)

$-5 = -6 + c$

$1 = c$

$\therefore\ y = -3x + 1$

8. Determining the equation from the graph:

$$y = mx + c$$

Method: (1) Find the **c-value** namely the <u>y-intercept.</u>

(2) Determine whether m is pos. 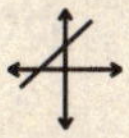 or neg. 

(3) Find m by drawing a right-angle Δ using the Y-intercept and the X-intercept.

$$m = \frac{\Delta y}{\Delta x}.$$

Examples: Find the equations of the following graphs:

1.

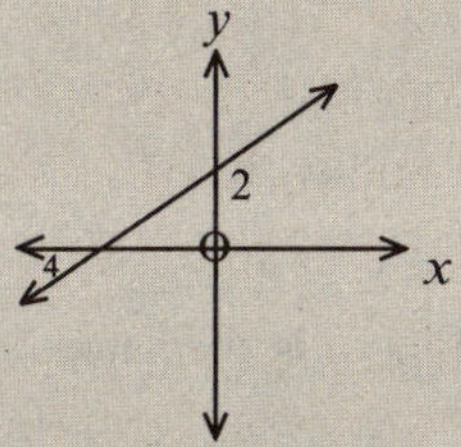

m pos

$y = \frac{2}{4}x + 2$

$y = \frac{1}{2}x - 2$

2.

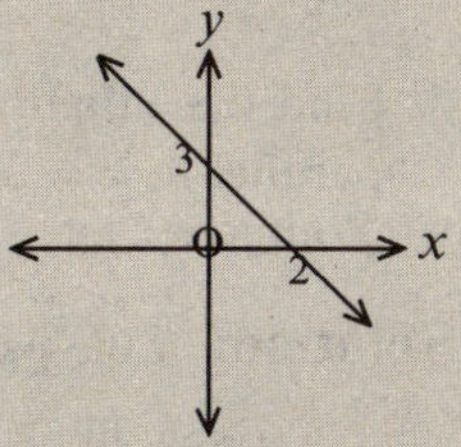

m neg

$y = -\frac{3}{2}x + 3$

3.

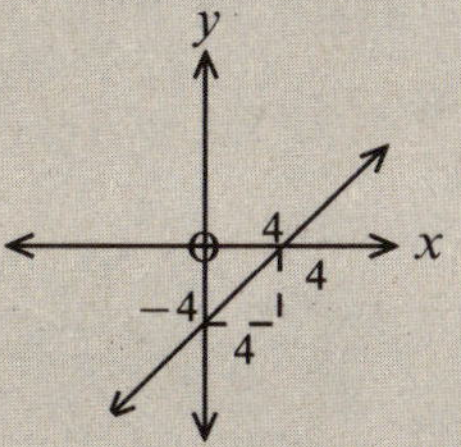

m pos

$y = \frac{4}{4}x - 4$

$y = x - 4$

4.

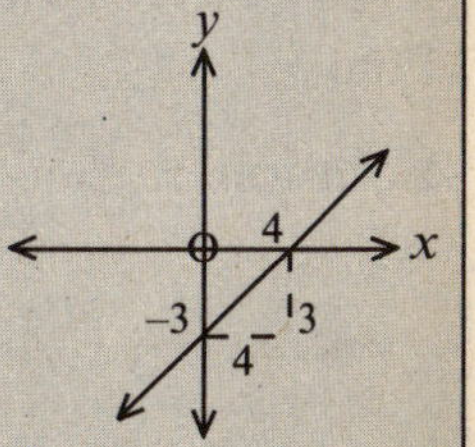

m pos

$y = \frac{3}{4}x - 3$

5.

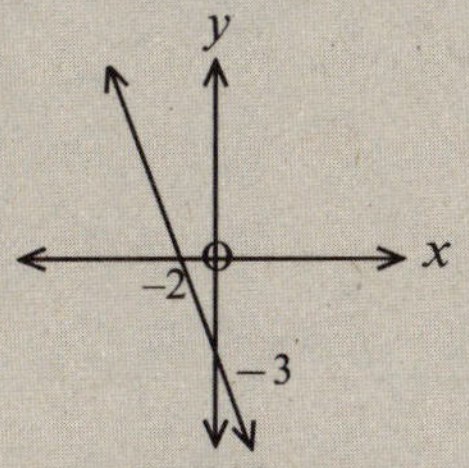

m neg

$y = -\frac{3}{2}x - 3$

6.

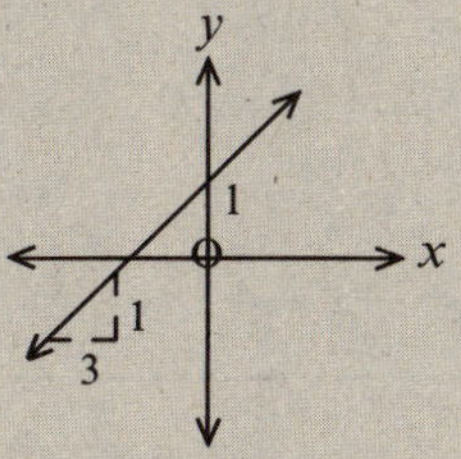

m pos

$y = \frac{1}{3}x + 1$

7.

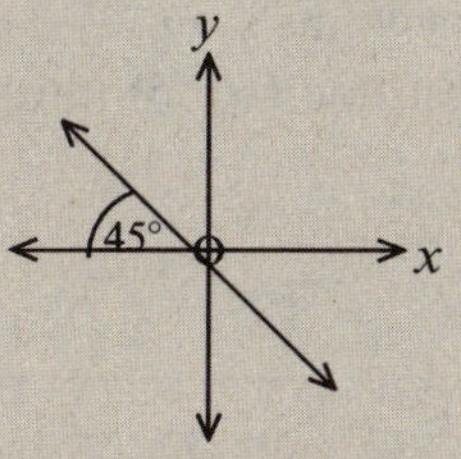

8.

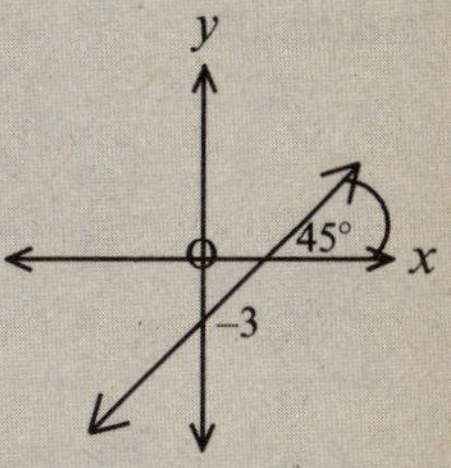

45°/135°: $m = 1$ *or* -1

7: m neg, $y = -x$

8: m pos, $y = x - 3$

9.

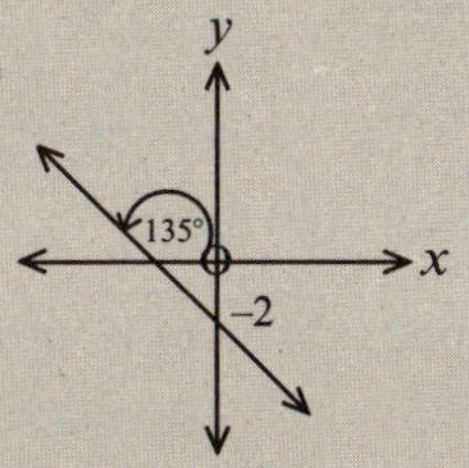

m neg

$y = -x - 2$

10.

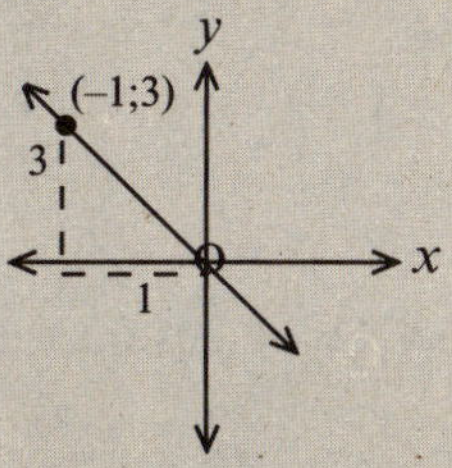

m neg

$y = -\frac{3}{1}x$

$y = -3x$

11.

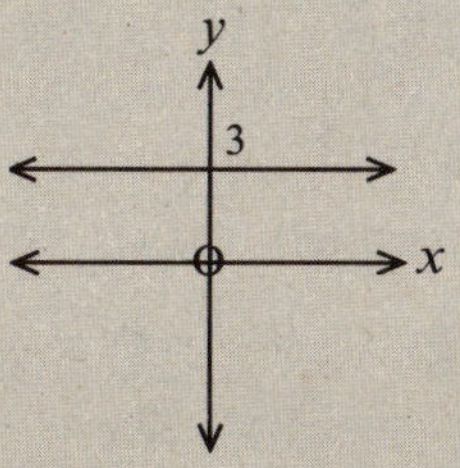

$y = 3$

12.

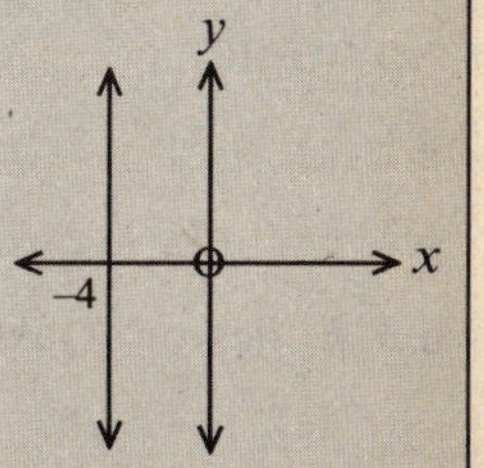

$x = -4$

More Examples:

1. (3; −2) lies on the graph of $y = 4x + k$.
Determine k:

Solution:
$$y = 4x + k$$
$$-2 = 4(3) + k$$
$$-2 = 12 + k$$
$$-14 = k$$

2. (5; a) lies on the graph of $y = \frac{1}{2}x - 1$.
Find a:

Solution:
$$y = \frac{1}{2}x - 1$$
$$a = \frac{1}{2} . 5 - 1$$
$$a = 2\frac{1}{2} - 1$$
$$a = 1\frac{1}{2}$$

3. Determine whether (−1; 7) lies on the graph of $y + 2x = 5$.

Solution: Substitute (−1; 7) into the LHS of the equation.

LHS. $= y + 2x$
$= 7 + 2(-1) = 7 - 2 = 5$

$\therefore$ (−1; 7) satifies the equation,
$\therefore$ (−1; 7) lies on the line.

4. Find the equation of the line parallel to $y = 2x + 3$ and with a Y-intercept of −5.

Solution: **Line parallel $\Rightarrow m$ is the same.**

$\therefore$ (New line) $y = 2x + c$
(Y-intercept is −5) $y = 2x - 5$

5. Determine the equation of the line parallel to $y = -x + 3$ and which passes through the point (−1; 5).

Solution:
(The new line's $m = -1$): $y = -x + c$
(Substitute (−1; 5) into): $5 = -(-1) + c$
$5 = 1 + c$
$4 = c$
$y = -x + 4$

6. Find **6.1** the equation of AB
6.2 the equation of CD
6.3 the coordinates of E
6.4 the value of a

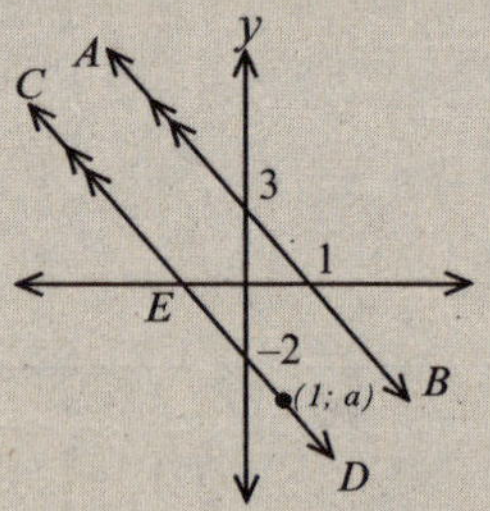

Solution:

6.1 $y = -3x + 3$ (m neg.)

6.2 $y = -3x - 2$ (lines parallel)

6.3 E lies on X-axis: $y = 0$: $0 = -3x - 2$
$$2 = -3x$$
$$\frac{-2}{3} = x$$
$$\therefore E\left(-\frac{2}{3}; 0\right)$$

6.4 (1; a) lies on the graph, substitute it into:
$$y = -3x - 2$$
$$a = -3(1) - 2$$
$$a = -3 - 2$$
$$a = -5$$

Exercise:

1. Standard form of a straight line graph:
$y - mx + c$.

1.1 What does m represent?
1.2 What does c represent?
1.3 Draw a graph if $c < 0$ *and* $m < 0$.
1.4 Draw a graph if $c = 0$ *and* $m > 0$.

2. Draw sketch graphs of each of the following:

2.1 $y = 2x - 2$
2.2 $y = 2x + 4$
2.3 $y = x + 5$
2.4 $y = -3x + 3$
2.5 $y = -\frac{1}{3}x - 1$
2.6 $y = -\frac{1}{3}x + 2$
2.7 $4y - x + 8 = 0$
2.8 $y = x$
2.9 $4y + 5x + 20 = 0$
2.10 $x = 2$
2.11 $x = 2y$
2.12 $y = -\frac{1}{2}x + 2$
2.13 $y = 4x$
2.14 $y = 4x - 2$
2.15 $y = -3x + 3$
2.16 $y = -3x + 5$
2.17 $y = 4$
2.18 $2y + x = 4$
2.19 $3x - 2y = 6$
2.20 $x = -2$

3. The following table shows the linear relationship between x and y.

x	−3	−2	−1	1	3
y	−11	−9	−7	−3	1

Determine the equation of the line passing through the points above.

4. Determine the equation from the following table:

x	−10	5	20
y	50	5	−40

5. Determine whether the points **5.1** (−2; 1) **5.2** (4; 6)
lie on the graph of: $3y - 2x = 7$

6. Test whether the following points lie on the graph of $5x - y = -3$?

6.1 $(-2; -7)$ **6.2** $\left(\frac{1}{2}; -\frac{1}{2}\right)$

7. If the point $(k; -5)$ lies on the line $2x + 4y = 8$, find k.

8. If $(3; a)$ and $(b; 2)$ lie on the line $y = 2x + 6$, find a en b.

9. If $y = 4x + 2$ is parallel to $y = ax + 7$, find a.

10. Determine b if $y = bx - 1$ is parallel to $3x - 2y = 4$.

11. Determine the equations of each of the following lines:

11.1

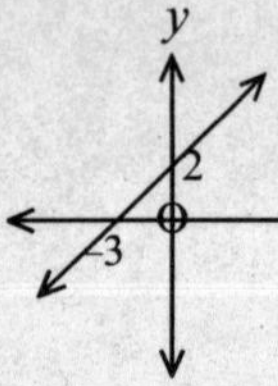

11.2

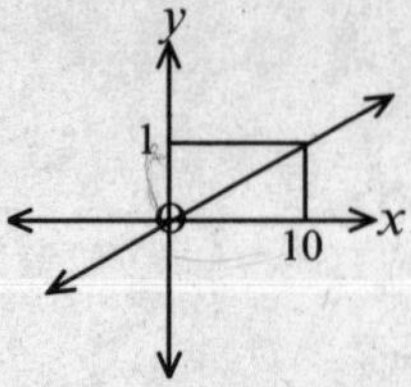

11.3

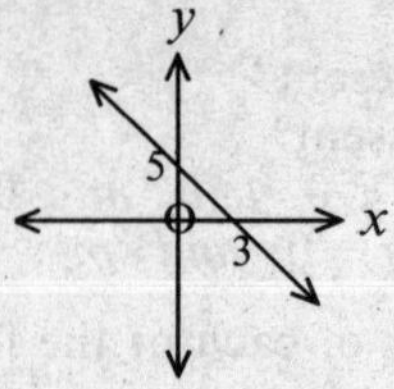

11.4

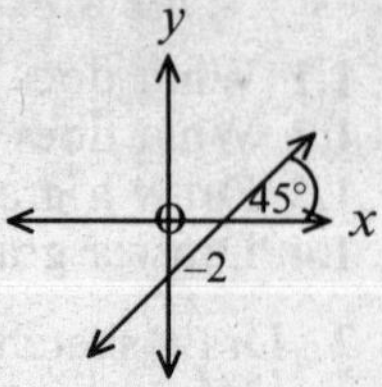

11.5

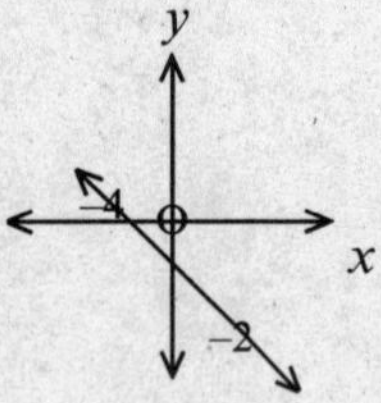

11.6

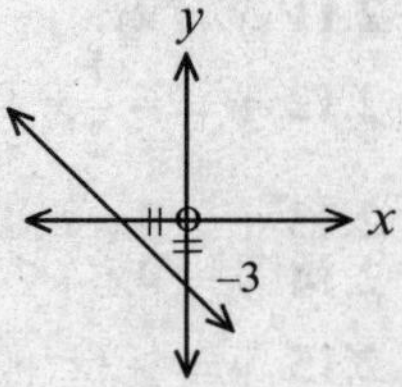

11.7

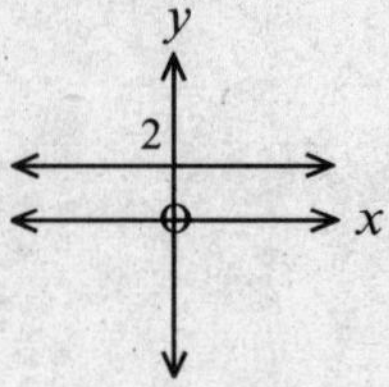

11.8

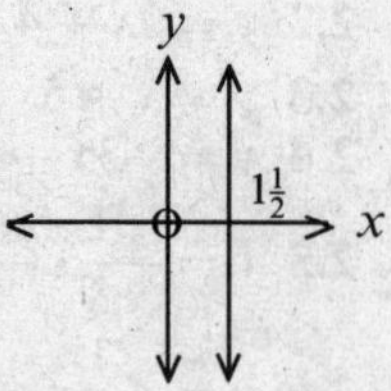

11.9

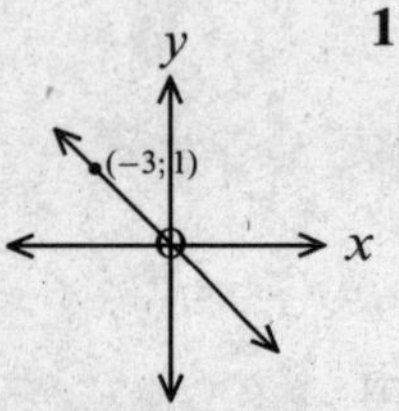

11.10

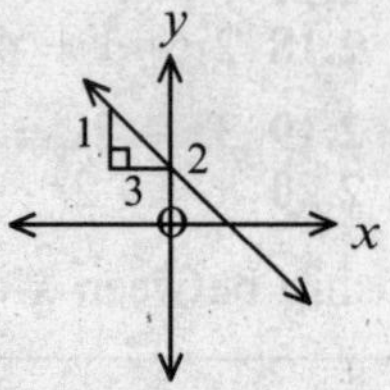

11.11

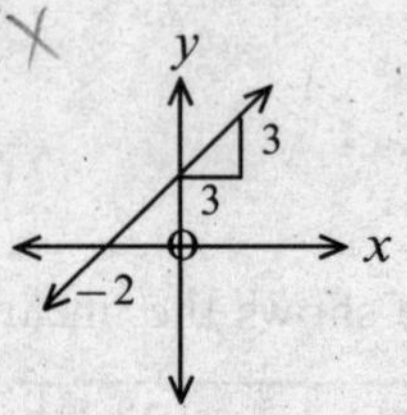

11.12

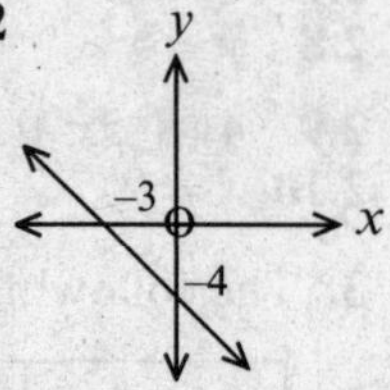

12.

12.1 Determine the equations of f and g.

12.2 Determine the coordinates of A.

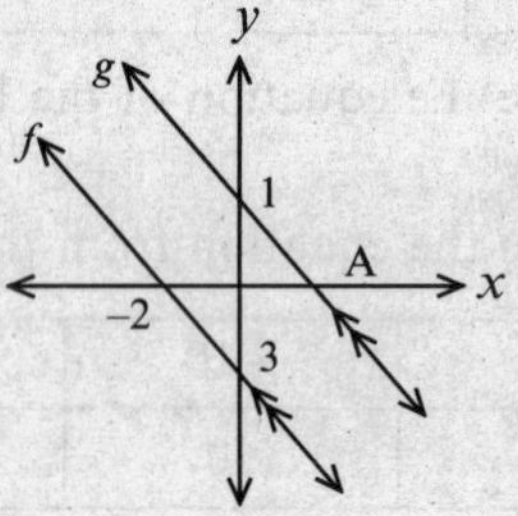

13.

13.1 Determine the equations of AB and CD.

13.2 Determine the coordinates of P and Q.

13.3 What is the equation of a line perpendicular to CD and passing through the point $(1; 5)$?

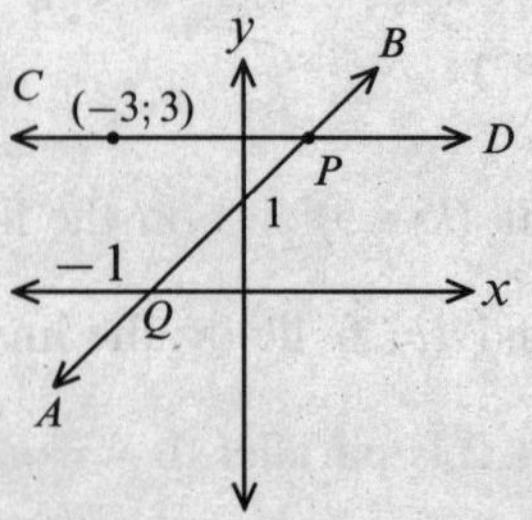

14. Determine the equation of the line which

14.1 cuts the Y-axis at 2 and has a gradient of -1.
14.2 passes through the point $(3; 2)$ and cuts the Y-axis at -2.
14.3 has a gradient of 3 and passes through the point $(-1; 2)$.
14.4 passes through the points $(5; -1)$ and $(4; -3)$.

(hint: remember $\dfrac{\Delta y}{\Delta x} = m$)

14.5 is parallel to $y = \frac{1}{2}x + 3$ and has a y-intercept of -2.

15. Complete the table if it is given that
$-4x + y = -2$.

x	0		-1		$-\frac{1}{2}$
y		2		0	

16. Determine the value of k, if $(2; 1)$ lies on the line with equation $kx + y = 2$.

17.1 Determine the equations of line f, g and h.

17.2 Determine the coordinates of A and B.

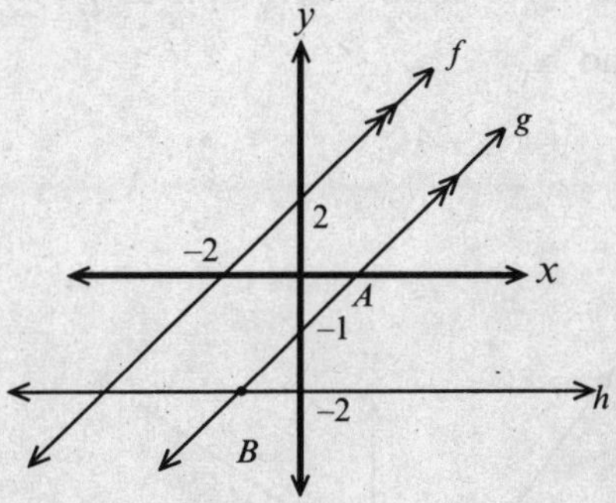

18.

18.1 Determine the equation of line AB.

18.2 Determine the values of a and b.

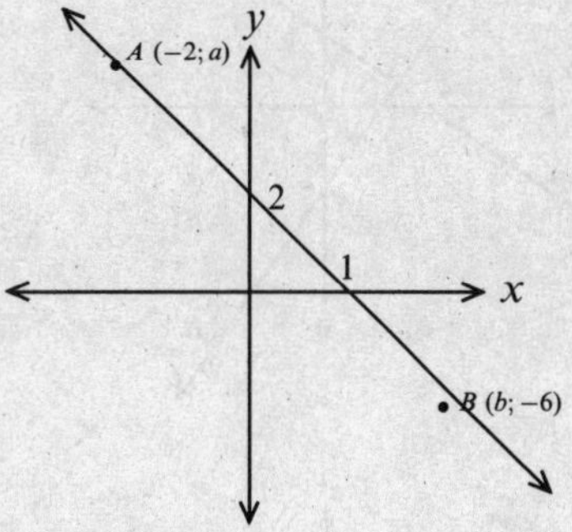

19.

19.1 Determine the equations of f, g and k.

19.2 Determine the coordinates of A and B.

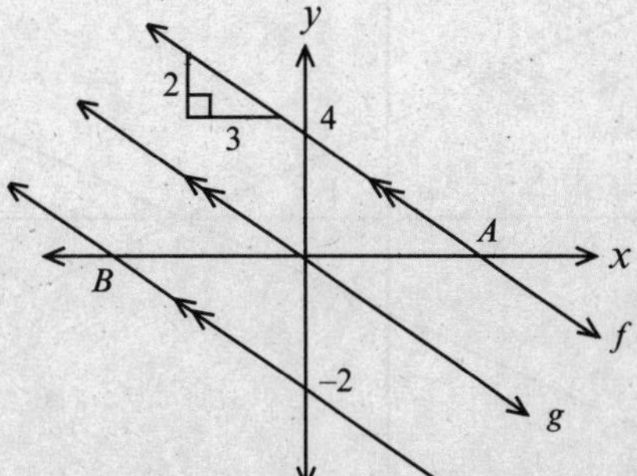

20. Determine

20.1 the equation of AM.

20.2 the equation of CM.

20.3 the coordinates of M.

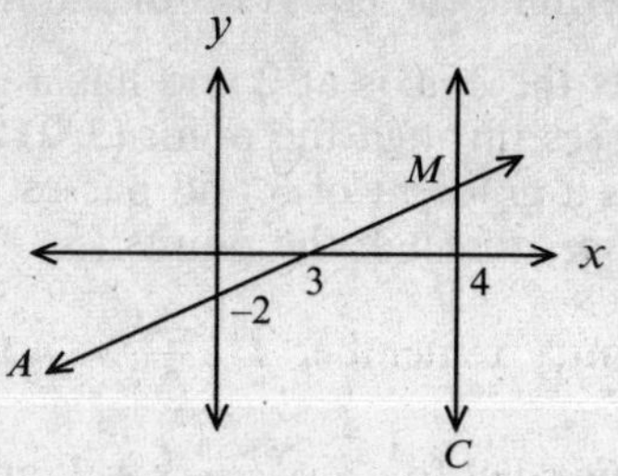

21. Given in the diagram: $y = -2x + 3$.

If $OB = 1$ unit, find the length of AB.

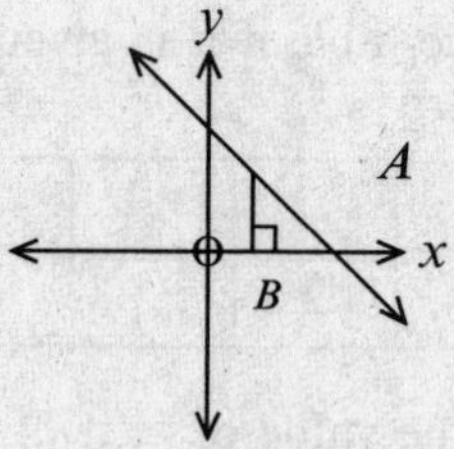

22. The figure shows the graph of $y = -3x + 9$.

Find x.

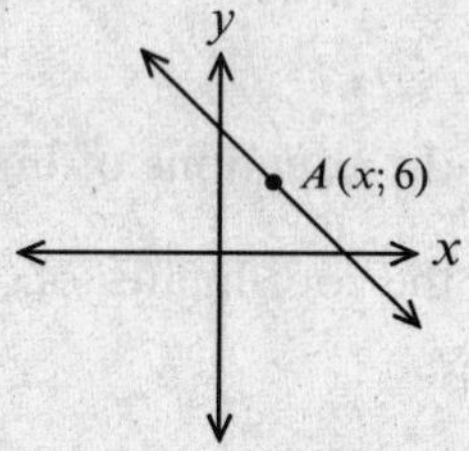

23.

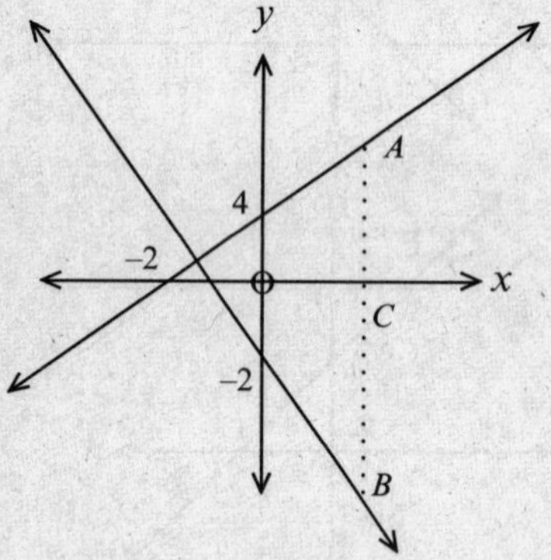

The diagram represents the graphs of:

$y = 2x + 4$

and $y = -3x - 2$.

Determine the length of AB if $OC = 2$ units.

24.

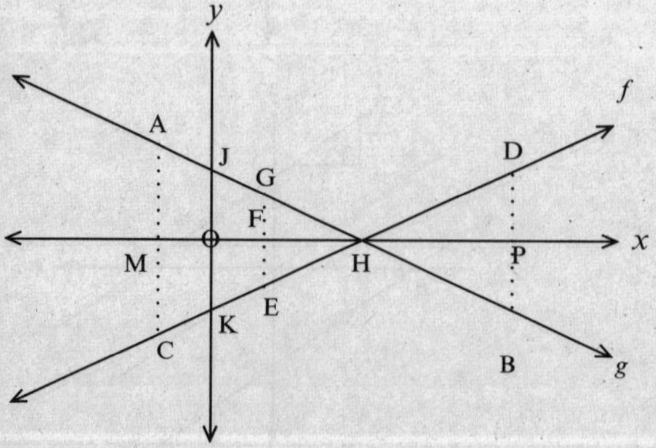

$y = x - 4 \quad \ldots (f)$

$y = -\frac{1}{2}x + 2 \ldots (g)$

Determine the lengths of:

24.1 OH

24.2 JK

24.3 AC if OM = 2 units

24.4 GE if OF = 1 unit

24.5 BD if OP = 8 units

SOLUTIONS

1.

1.1 m: gradient
1.2 c: y-intercept
1.3 $y = mx + c : c < 0;\ m < 0$

1.3

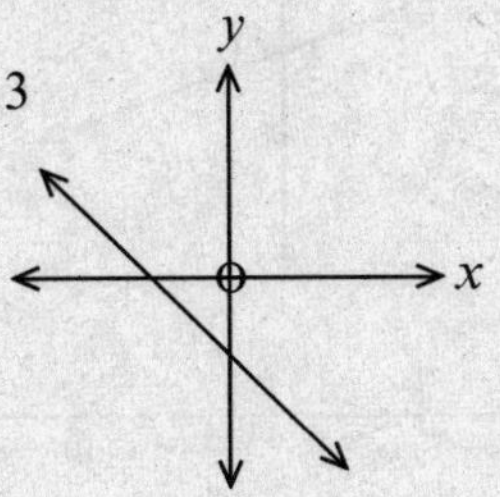

1.4 $y = mx + c : c = 0;\ m > 0.$

1.4

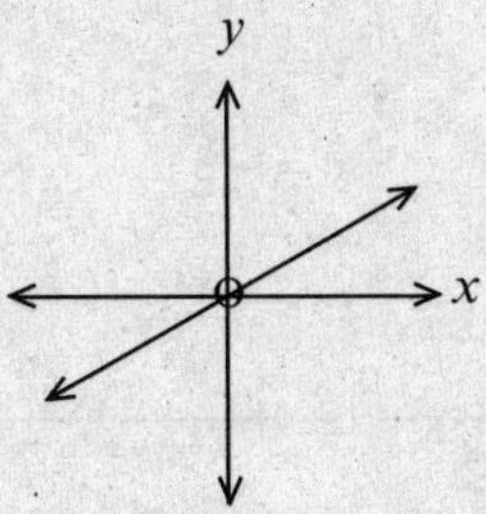

2.

2.1 $\mathbf{y = 2x - 2}$

Y-intercept: -2
X-intercept: $y = 0$
$0 = 2x - 2$
$2 = 2x$
$1 = x$

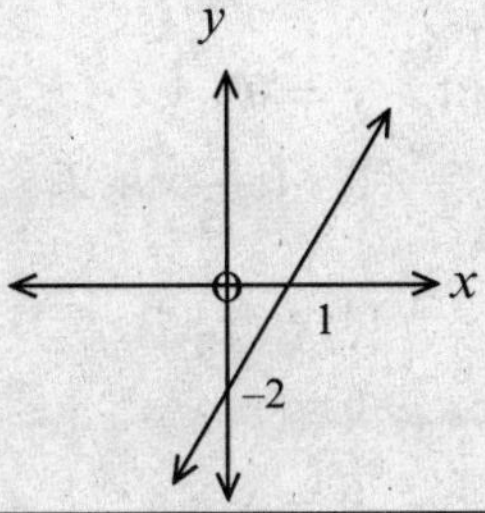

2.2 $\mathbf{y = 2x + 4}$

Y-intercept: 4
X-intercept: $y = 0$
$0 = 2x + 4$
$-4 = 2x$
$-2 = x$

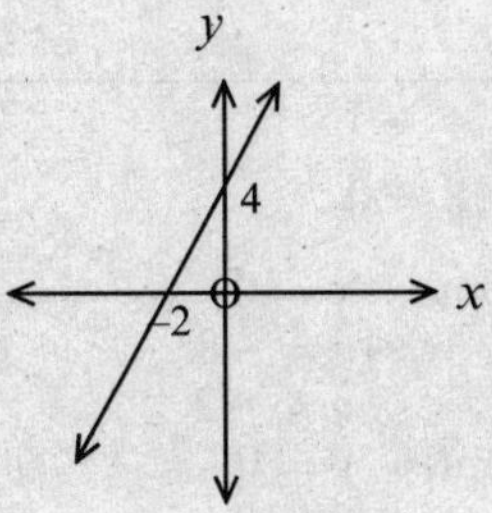

2.3 $\mathbf{y = x + 5}$

Y-intercept: 5
X-intercept: $y = 0$
$x + 5 = 0$
$x = -5$

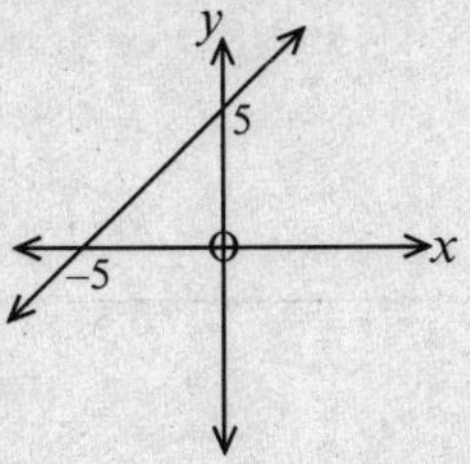

2.4 $\mathbf{y = -3x + 3}$

Y-intercept: 3
X-intercept: $y = 0$
$0 = -3x + 3$
$-3 = -3x$
$1 = x$

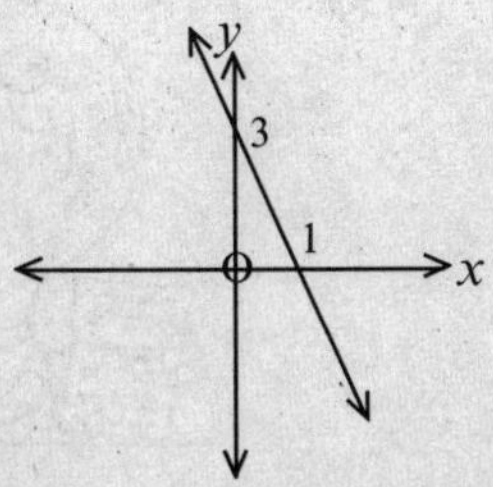

2.5 $\boldsymbol{y = -\frac{1}{3}x - 1}$

Y-intercept: -1

X-intercept: $y = 0$

$$-\frac{1}{3}x - 1 = 0$$

$$-\frac{1}{3}x = 1$$

$$x = -3$$

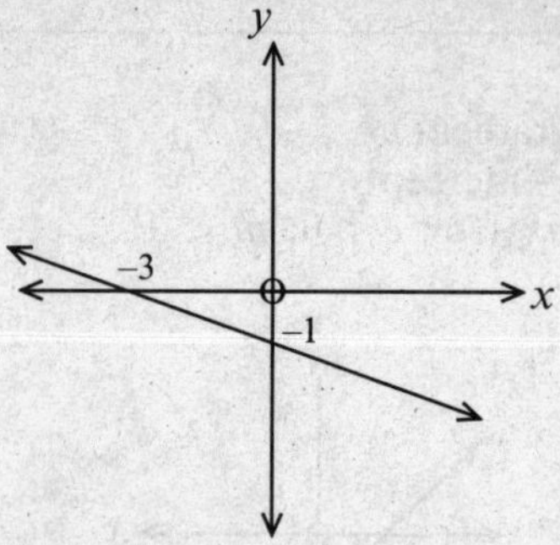

2.6 $\boldsymbol{y = -\frac{1}{3}x + 2}$

Y-intercept: 2

X-intercept: $y = 0$

$$0 = -\frac{1}{3}x + 2$$

$$-2 = -\frac{1}{3}x$$

$$\times 3: \quad -6 = -x$$

$$6 = x$$

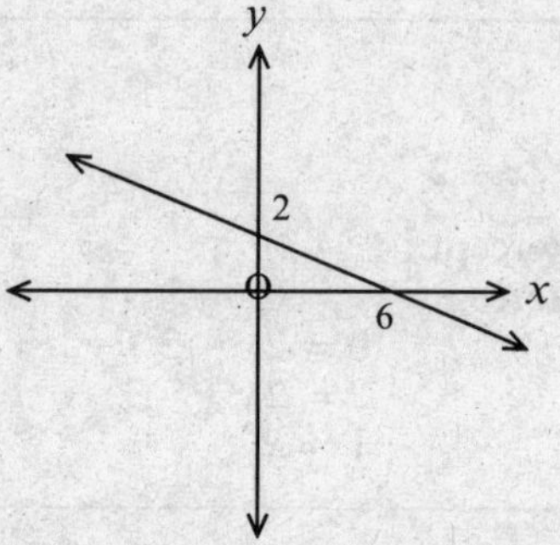

2.7 $\boldsymbol{4y - x + 8 = 0}$

Y-intercept: $x = 0$

$$4y = -8$$

$$y = -2$$

X-intercept: $y = 0$

$$-x + 8 = 0$$

$$-x = -8$$

$$x = 8$$

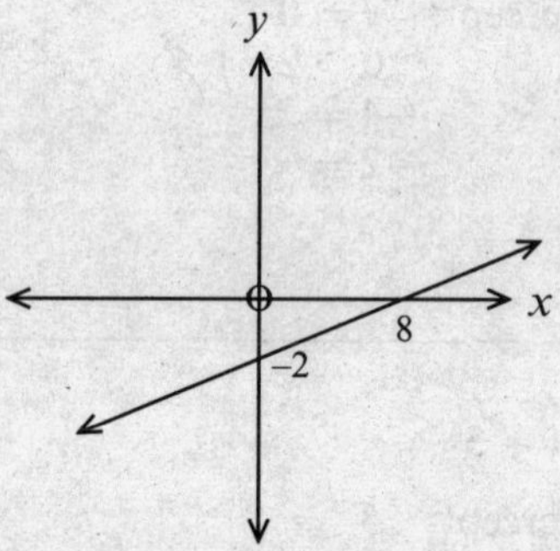

2.8 $\boldsymbol{y = x}$

Y-intercept: 0

Through (1; 1)

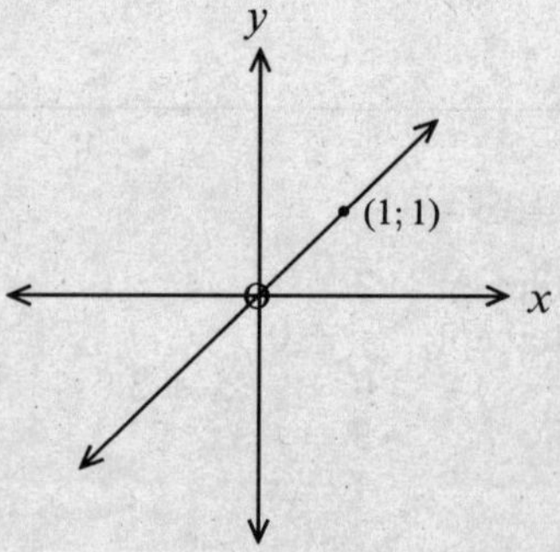

2.9 $\mathbf{4y + 5x + 20 = 0}$

Y-intercept: $x = 0$

$4y = -20$

$y = -5$

X-intercept: $y = 0$

$5x = -20$

$x = -4$

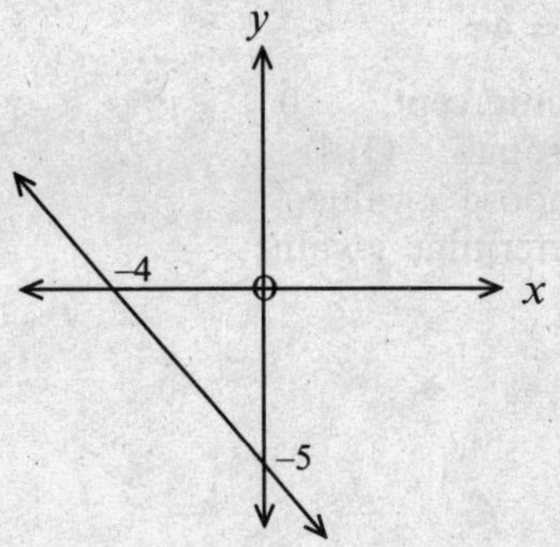

2.10 $\mathbf{x = 2}$

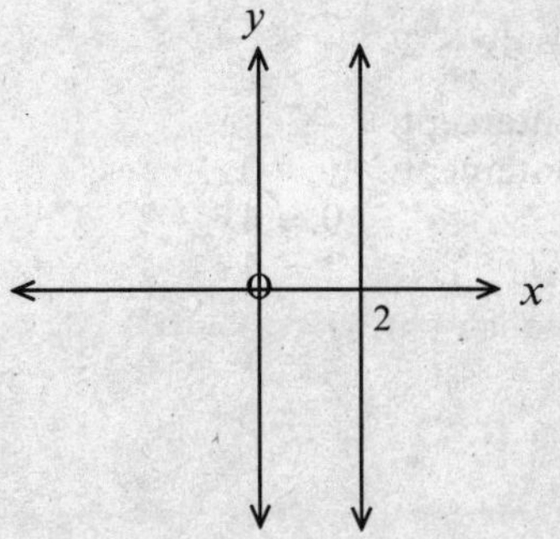

2.11 $\mathbf{x = 2y}$

through: (0; 0)
and (2; 1)
↑
choose x-value, eg. $x = 2$
determine y-value, $\therefore\ y = 1$

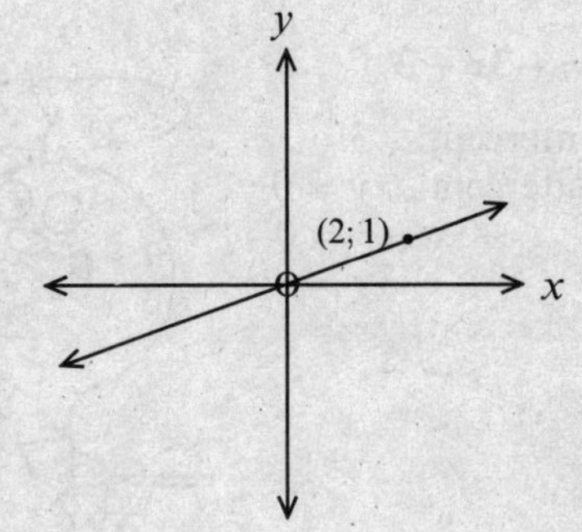

2.12 $\mathbf{y = -\frac{1}{2}x + 2}$

Y-intercept: 2

X-intercept: $y = 0$

$-\frac{1}{2}x + 2 = 0$

$-\frac{1}{2}x = -2$

$\times 2$: $-x = -4$

$x = 4$

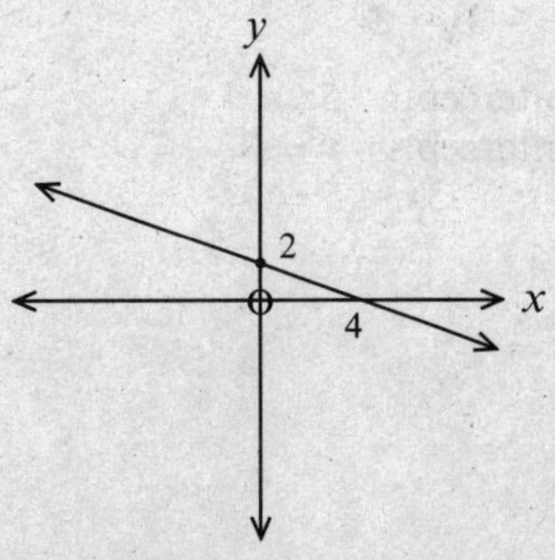

2.13 $\boldsymbol{y = 4x}$

Y-intercept: 0
through: $(1; 4)$
Choose x-value.
Determine y-value.

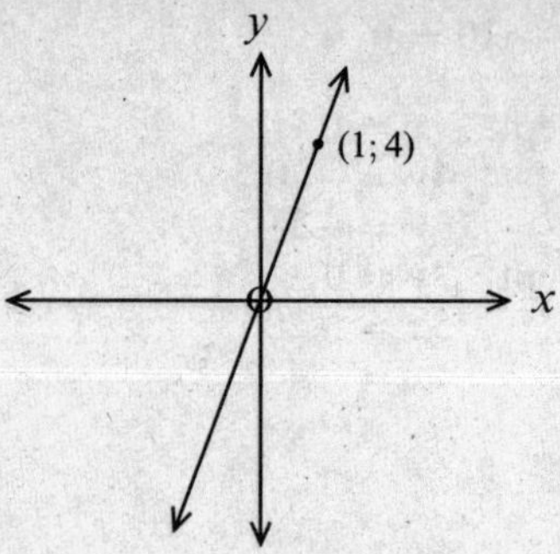

2.14 $\boldsymbol{y = 4x - 2}$

Y-intercept: -2
X-intercept: $y = 0$

$$0 = 4x - 2$$
$$2 = 4x$$
$$\frac{1}{2} = x$$

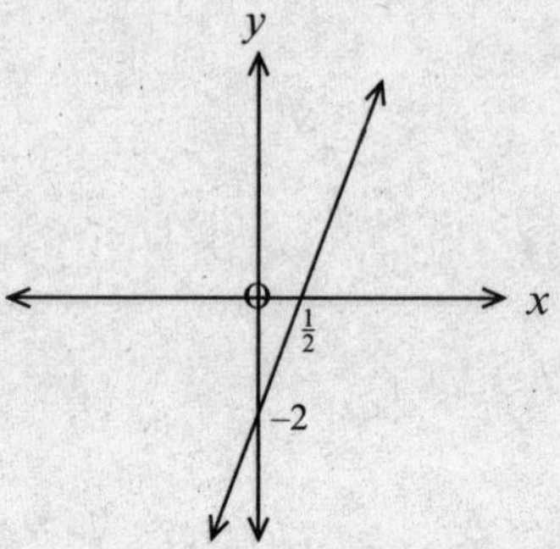

2.15 $\boldsymbol{y = -3x + 3}$

Y-intercept: 3
X-intercept: $y = 0$

$$-3x + 3 = 0$$
$$-3x = -3$$
$$x = 1$$

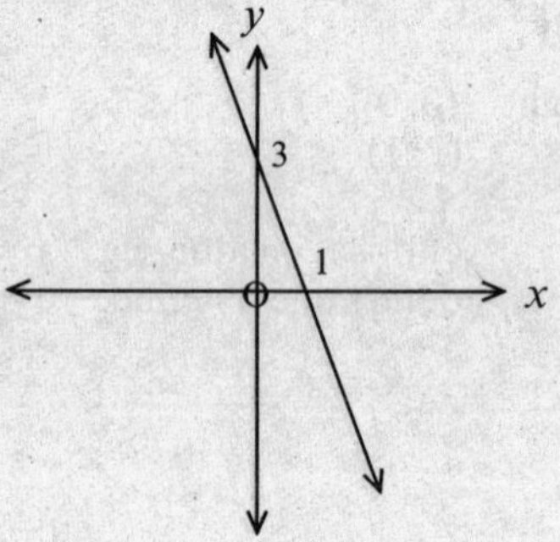

2.16 $\boldsymbol{y = -3x + 5}$

Y-intercept: 5
X-intercept: $y = 0$

$$-3x + 5 = 0$$
$$-3x = -5$$
$$x = \frac{5}{3}$$

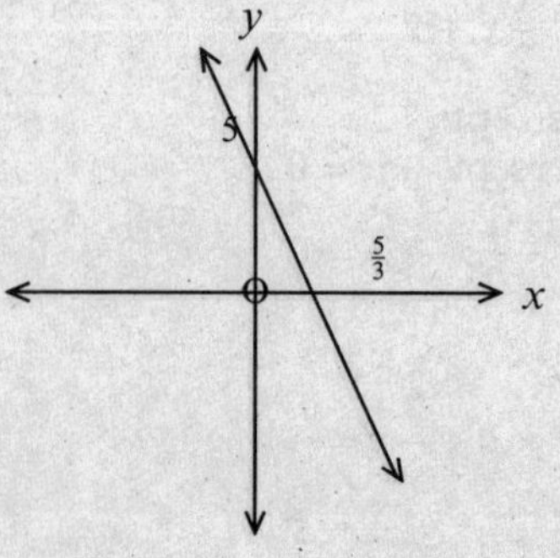

2.17 $y = 4$

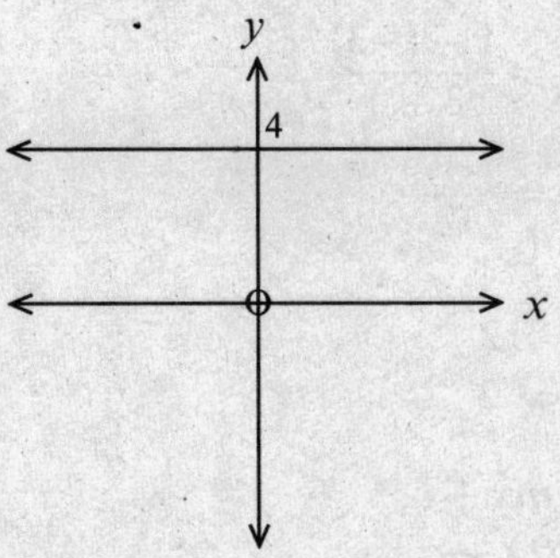

2.18 $2y + x = 4$

Y-intercept: $x = 0$

$2y = 4$

$y = 2$

X-intercept: $y = 0$

$x = 4$

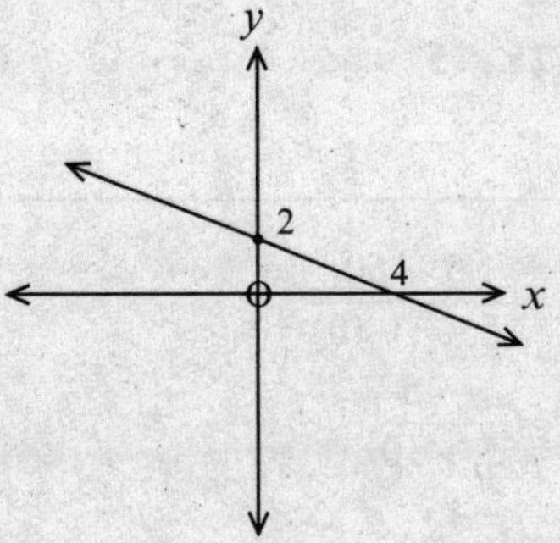

2.19 $3x - 2y = 6$

Y-intercept: $x = 0$

$-2y = 6$

$y = -3$

X-intercept: $y = 0$

$3x = 6$

$x = 2$

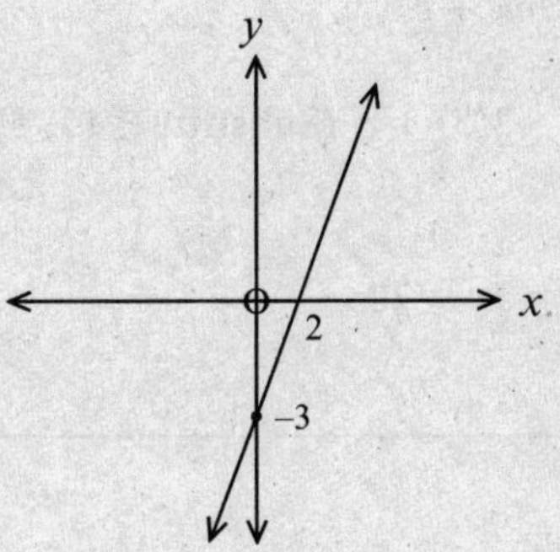

2.20 $x = -2$

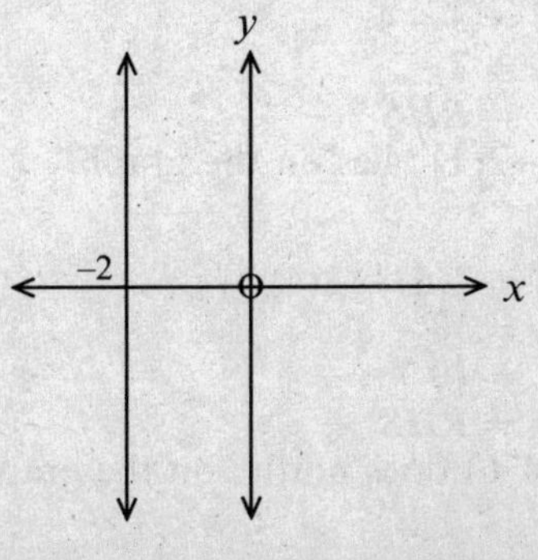

3. $m = \frac{\Delta y}{\Delta x} = \frac{1-(-3)}{3-1}$

$= \frac{1+3}{2}$

$= \frac{4}{2}$

$= 2$

$y = mx + c$

$y = 2x + c$

$1 = 2(3) + c$ (Substitute (3; 1) into)

$1 - 6 = c$

$-5 = c$

$\therefore$ $\boldsymbol{y = 2x - 5}$

4. $m = \frac{\Delta y}{\Delta x} = \frac{5-50}{5-(-10)}$

$= \frac{-45}{5+10}$

$= \frac{-45}{15}$

$= -3$

$y = mx + c$

$y = -3x + c$

$5 = -3(5) + c$ (Substitute (5; 5) into)

$5 + 15 = c$

$20 = c$

$\therefore$ $\boldsymbol{y = -3x + 20}$

5. $\boldsymbol{3y - 2x = 7}$

5.1 $(-2; 1)$

LHS $= 3(1) - 2(-2)$

$= 3 + 4$

$= 7$

$= RHS$

$\therefore$ $(-2; 1)$: lies on the graph.

5.2 $(4; 6)$

LHS $= 3(6) - 2(4)$

$= 18 - 8$

$= 10$

$\neq RHS$

$\therefore$ $(4; 6)$ does not lie on the graph.

6. $\boldsymbol{5x - y = -3}$

6.1 $(-2; 7)$

$LHS = 5(-2) - (-7)$

$= -10 + 7$

$= -3$

$= RHS$

$\therefore$ $(-2; -7)$ lies on the graph.

6.2 $(\frac{1}{2}; -\frac{1}{2})$

$LHS = 5(\frac{1}{2}) - (-\frac{1}{2})$

$= 2\frac{1}{2} + \frac{1}{2}$

$= 3$

$\neq RHS$

$\therefore$ $(\frac{1}{2}; -\frac{1}{2})$ does not lie on the graph.

7. $\boldsymbol{2x + 4y = 8}$

Substitute $(k; -5)$ into:

$2(k) + 4(-5) = 8$

$2k - 20 = 8$

$2k = 28$

$k = 14$

8. $\boldsymbol{y = 2x + 6}$

Substitute $(3; a)$ into:

$a = 2(3) + 6$

$= 6 + 6$

$a = 12$

Substitute $(b; 2)$ into:

$2 = 2(b) + 6$

$2 = 2b + 6$

$-4 = 2b$

$-2 = b$

9. $\left.\begin{array}{l} y = 4x + 2 \\ y = ax + 7 \end{array}\right\}$ parallel

$\boldsymbol{a = 4}$

10. $3x - 2y = 4$

$-2y = -3x + 4$

$\left.\begin{array}{l} y = \frac{3}{2}x - 2 \\ y = bx - 1 \end{array}\right\}$ parallel

$\therefore$ $\boldsymbol{b = \frac{3}{2}}$

11.

11.1 $y = \frac{2}{3}x + 2$

11.2 $y = \frac{1}{10}x$

11.3 $y = -\frac{5}{3}x + 5$

11.4 $y = x - 2$

11.5 $y = -\frac{2}{4}x - 2$
$y = -\frac{1}{2}x - 2$

11.6 $y = -x - 3$

11.7 $y = 2$

11.8 $x = 1\frac{1}{2}$

11.9
$$\left.\begin{aligned} y &= mx \\ 1 &= (-3)m \\ -\tfrac{1}{3} &= m \end{aligned}\right\} \; y = -\tfrac{1}{3}x$$

11.10 $y = -\frac{1}{3}x + 2$

11.11 $y = x + 2$

11.12 $y = -\frac{4}{3}x - 4$

12. $f: \; y = -\frac{3}{2}x - 3$
$g: \; y = -\frac{3}{2}x + 1$

At A: Substitute $y = 0$ (X-intercept)
$0 = -\frac{3}{2}x + 1$
$-1 = -\frac{3}{2}x$
$-2 = -3x$
$\frac{2}{3} = x$
$\therefore$ $\boldsymbol{A(\frac{2}{3}; 0)}$

13.

13.1 ***AB*:** $\boldsymbol{y = x + 1}$
***CD*:** $\boldsymbol{y = 3}$

13.2 Substitute into $y = 3$ in $y = x + 1$
$\therefore\; 3 = x + 1$
$2 = x$
$\therefore$ ***P*(2; 3)**
***Q*(-1; 0)**

13.3 Vertical line through (1; 5)
$\boldsymbol{x = 1}$

14.

14.1 $\boldsymbol{y = -x + 2}$

14.2 $y = mx - 2$
Substitute (3; 2) into $2 = 3m - 2$
$4 = 3m$
$\frac{4}{3} = m$
$\boldsymbol{y = \frac{4}{3}x - 2}$

14.3 $y = 3x + c$
Substitute (–1; 2) into: $2 = 3(-1) + c$
$2 = -3 + c$
$5 = c$
$\boldsymbol{y = 3x + 5}$

14.4 $m = \dfrac{\Delta y}{\Delta x} = \dfrac{-3-(-1)}{4-5} = \dfrac{-3+1}{4-5}$
$= \dfrac{-2}{-1} = 2$
$y = 2x + c$

Substitute (5; –1) $-1 = 2(5) + c$
$-1 - 10 = c$
$-11 = c$
$\boldsymbol{y = 2x - 11}$

14.5 $y = \frac{1}{2}x + 3$
$\boldsymbol{y = \frac{1}{2}x - 2}$

15. $-4x + y = -2$

x	0	**+1**	−1	$\boldsymbol{\frac{1}{2}}$	$-\frac{1}{2}$
y	**−2**	2	**−6**	0	**−4**

16. $kx + y = 2$

Substitute (2; 1) $k(2) + 1 = 2$
$2k = 2 - 1$
$2k = 1$
$k = \frac{1}{2}$

17. $f: \; y = x + 2$
$g: \; y = x - 1$
$h: \; y = -2$

***A* (1; 0)**
***B* Substitute $\boldsymbol{y = -2}$ into:**

$\boldsymbol{y = x - 1}$
$-2 = x - 1$
$-1 = x$
$\therefore$ ***B* (−1; −2)**

18. $AB: \ y = -2x + 2$

$(-2; a): \ a = -2(-2) + 2$

$a = 6$

$(b; -6): \ -6 = -2b + 2$

$-6 - 2 = -2b$

$-8 = -2b$

$b = 4$

19.

19.1 $f: \ y = -\frac{2}{3}x + 4$

$g: \ y = -\frac{2}{3}x$

$h: \ y = -\frac{2}{3}x - 2$

19.2 $y = -\frac{2}{3}x + 4 \quad (f)$

X-intercept: $y = 0$

$0 = -\frac{2}{3}x + 4$

$-4 = -\frac{2}{3}x$

$-12 = -2x$

$6 = x$

$\therefore \ A(6; 0)$

$y = -\frac{2}{3}x - 2 \quad (h)$

X-intercept: $y = 0$

$0 = -\frac{2}{3}x - 2$

$6 = -2x$

$-3 = x$

$\therefore \ B(-3; 0)$

20.1 $AM: \ y = \frac{2}{3}x - 2$

20.2 $CM: \ x = 4$

20.3 Substitute $x = 4$ into $y = \frac{2}{3}x - 2$

$y = \frac{2}{3}(4) - 2$

$= \frac{8}{3} - 2$

$= \frac{8}{2} - 2$

$= 2\frac{2}{3} - 2$

$= \frac{2}{3} \Rightarrow M(4; \frac{2}{3})$

21. $y = -2x + 3$

$OB = 1 \Rightarrow x = 1$ at A

$\therefore \ y = -2(1) + 3$

$= 1 \Rightarrow \ A(1; 1)$

$\therefore \ AB = 1$

22. $y = -3x + 9$

Substitute $(x; 6)$ in:

$6 = -3(x) + 9$

$-3 = -3x$

$1 = x$

23. $y = 2x + 4$

Substitute $x = 2: \quad y = 2(2) + 4 = 8 \Rightarrow A(2; 8)$

$y - 3x - 2$

Substitute $x = 2$

$y - 3(2) - 2$

$= -8 \Rightarrow B(2; -8)$

$\therefore \ AB = 16$

24.1 $CD: \ y = x - 4$

X-intercept: $y = 0$

$0 = x - 4$

$4 = x \Rightarrow \boldsymbol{OH = 4}$

24.2 $\boldsymbol{JK = 6}$ Difference between Y-intercepts

24.3 $y = x - 4$

Substitute $x = -2: \quad y = -2 - 4 = -6$

$\Rightarrow \boldsymbol{C(-2; -6)}$

$y = -\frac{1}{2}x + 2$

Substitute $x = -2: \quad y = -\frac{1}{2}(-2) + 2$

$= 3 \Rightarrow A(-2; 3)$

$\therefore \ \boldsymbol{AC = 9}$

24.4 $y = x - 4$

Substitute $x = 1: \quad y = 1 - 4$

$= -3 \Rightarrow \boldsymbol{E(1; -3)}$

$y = -\frac{1}{2}x + 2$

Substitute $x = 1: \quad y = -\frac{1}{2}(1) + 2$

$= 1\frac{1}{2} \Rightarrow \boldsymbol{G(1; 1\frac{1}{2})}$

$\therefore \ \boldsymbol{EG = 4\frac{1}{2}}$

24.5 $y = x - 4$

Substitute $x = 8: \quad y = 8 - 4 = 4 \Rightarrow \boldsymbol{D(8; 4)}$

$y = -\frac{1}{2}x + 2$

Substitute $x = 8: \quad y = -\frac{1}{2}(8) + 2$

$= -2 \Rightarrow \boldsymbol{B(8; -2)}$

$\therefore \ \boldsymbol{BD = 6}$

8 *Simultaneous solution of 2 equations*

The **simultaneous solution** of x and y in 2 equations, **means**:
Find the $(x; y)$ which satisfies **both** equations.
It is the point where the graphs of the 2 equations cut one another.

N.B.: The first equation has 1 unknown, eg. x. . . . (1)
The second equation has 2 unknowns, eg. x and y. . . . (2)

Method: 1. Use the 1st equation and make the variable (x/y) the subject.
2. Now substitute this x/y-value into the 2nd equation.

Example 1:

Solve for x and y if: $x + 2 = 0$ and $x - y = 3$

Solutions: $x + 2 = 0$ .... (1)
$x - y = 3$ .... (2)

From (1): $x = -2$

Substitute $x = -2$ **into (2):**

$$\begin{aligned} x - y &= 3 \\ (-2) - y &= 3 \\ \therefore -y &= 3 + 2 \\ -y &= 5 \\ y &= -5 \end{aligned}$$

$\therefore$**(−2; −5)** satifies both equations.

A graphic representation of the example above:

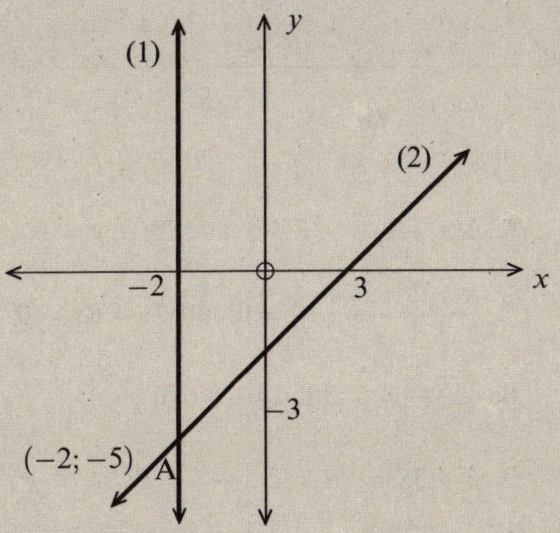

$x = -2$ .... (1)
and
$x - y = 3$ .... (2)
(0; −3) and (3; 0)

Solution obtained: **A (−2; −5)**

Example 2:

Solve for x and y: $2x - 5y = 4$ and $2y - 4 = 2$

Solution: $2y - 4 = 2$ .... (1)
$2x - 5y = 4$ .... (2)

From (1):

$$2y = 2 + 4$$
$$2y = 6$$
$$y = 3$$

Substitute $y = 3$ into (2):

$$\begin{aligned} 2x - 5y &= 4 \\ 2x - 5(3) &= 4 \\ 2x &= 4 + 15 \\ 2x &= 19 \\ x &= 9\tfrac{1}{2} \end{aligned}$$

$(9\frac{1}{2}; 3)$ is the solution.

Example 3:

Solve simultaneously: $2(x - 4) = x - 2$
$x - 2y + 4 = 0$

Solution: $2(x - 4) = x - 2$ .... (1)
$x - 2y + 4 = 0$ .... (2)

From (1):

$$2x - 8 = x - 2$$
$$2x - x = 8 - 2$$
$$x = 6$$

Substitute $x = 6$ into (2):

$$\begin{aligned} x - 2y + 4 &= 0 \\ 6 - 2y + 4 &= 0 \\ -2y &= -4 - 6 \\ -2y &= -10 \\ y &= 5 \end{aligned}$$

(6;5) is the solution.

Exercise:

Solve the following equations simultaneously:

1. $x + 2 = 6$ and $2x + y = 6$
2. $y = -6$ and $y = x + 4$
3. $3x - 9 = 0$ and $6x - 9y = 15$
4. $2x + 1 = -5$ and $y - x = 2$
5. $3p + 4q = 12$ and $3p - 9 = 0$
6. $3(x + 1) = -3 + x$ and $2x - y = 1$
7. $\dfrac{x+2}{3} - \dfrac{2x-1}{2} = 0$ and $7y + 8x = 0$
8. $\dfrac{x+y}{2} = 5$ and $2y + 8 = 0$
9. $3 + 3x = 9x$ and $x - y = 0$
10. $2(x - 4) = x - 2$ and $x - 2y + 4 = 0$

SOLUTIONS

1. $x + 2 = 6$ .... (1)
   $2x + y = 6$ .... (2)

From (1): $x + 2 = 6$
$x = 6 - 2$
$x = 4$

Substitute $x = 4$ into (2):

$2x + y = 6$
$2(4) + y = 6$
$y = 6 - 8$
$y = -2$

Solution: (4;–2)

2. $y = -6$ .... (1)
   $y = x + 4$ .... (2)

Substitute $y = -6$ into (2):

$-6 = x + 4$
$-6 - 4 = x$
$-10 = x$

Solution: (–10;–6)

3. $3x - 9 = 0$ .... (1)
   $6x - 9y = 15$ .... (2)

From (1): $3x = 9$
$x = 3$

Substitute $x = 3$ into (2):

$6(3) - 9y = 15$
$-9y = 15 - 18$
$-9y = -3$
$y = \frac{-3}{-9} = \frac{1}{3}$

Solution: $(3; \frac{1}{3})$

4. $2x + 1 = -5$ .... (1)
   $y - x = 2$ .... (2)

From (1): $2x + 1 = -5$
$2x = -5 - 1$
$2x = -6$
$x = -3$

Substitute $x = -3$ into (2):

$y - x = 2$
$y - (-3) = 2$
$y + 3 = 2$
$y = 2 - 3$
$y = -1$

Solution: (–3;–1)

5. $3p - 9 = 0$ .... (2)
   $3p + 4q = 12$ .... (1)

From (1): $3p = 9$
$p = 3$

Substitute $p = 3$ into (2):

$3(3) + 4q = 12$
$4q = 12 - 9$
$4q = 3$
$q = \frac{3}{4}$

Solution: $p = 3$; $q = \frac{3}{4}$

6. $3(x + 1) = -3 + x$ .... (1)
   $2x - y = 1$ .... (2)

From (1): $3x + 3 = -3 + x$
$2x = -6$
$x = -3$

Substitute $x = -3$ into (2):

$2(-3) - y = 1$
$-y = 1 + 6 = 7$
$y = -7$

Solution: (–3;–7)

7. $\frac{x+2}{3} - \frac{2x-1}{2} = 0$ .... (1)

$7y + 8x = 0$ .... (2)

From (1): $\times$ **LCM 6:**

$$2(x+2) - 3(2x-1) = 0$$
$$2x + 4 - 6x + 3 = 0$$
$$-4x = -7$$
$$x = -\tfrac{7}{4}$$

Substitute $x = -\frac{7}{4}$ **into (2):**

$$7y + 8\left(-\tfrac{7}{4}\right) = 0$$
$$7y - 14 = 0$$
$$7y = 14$$
$$y = 2$$

Solution: $\left(-\frac{7}{4}; 2\right)$

8. $2y + 8 = 0$ .... (1)

$\frac{x+y}{2} = 5$ .... (2)

From (1): $2y + 8 = 0$

$$2y = -8$$
$$y = -4$$

Substitute $y = -4$ **into (2):**

$$\frac{x + (-4)}{2} = 5$$

$\times$ **LCM 2:** $x - 4 = 10$

$$x = 14$$

Solution: $(14; -4)$

9. $3 + 3x = 9x$ .... (1)

$x - y = 0$ .... (2)

From (1): $3 + 3x = 9x$

$$3x - 9x = -3$$
$$-6x = -3$$
$$x = \tfrac{1}{2}$$

Substitute $x = \frac{1}{2}$ **into (2):**

$$\tfrac{1}{2} - y = 0$$
$$-y = -\tfrac{1}{2}$$
$$y = \tfrac{1}{2}$$

Solution: $\left(\frac{1}{2}; \frac{1}{2}\right)$

10. $2(x-4) = x - 2$ .... (1)

$x - 2y + 4 = 0$ .... (2)

From (1): $2x - 8 = x - 2$

$$x = 8 - 2$$
$$x = 6$$

Substitute $x = 6$ **into (2):**

$$x - 2y + 4 = 0$$
$$6 - 2y + 4 = 0$$
$$-2y = -10$$
$$y = 5$$

Solution: $(6; 5)$

9 *Interest*

Interest may be **Income** on an **amount** invested.
Interest may be **Expenditure** on an **amount** borrowed (on a loan).

Simple Interest: (SI)

The formula:

$$\textbf{SI} = \frac{prt}{100} \text{ i.e. } \frac{Principal \times rate \times time}{100}$$

p: Principal is the amount of money deposited, invested or borrowed. The interest is calculated on this amount.

r: Interest rate given as a percentage, is the rate used to calculate the interest per annum (p.a.) on the principal.

t: Time in years for which the principal has been invested or borrowed.

SI: Simple Interest

Final amount: Principal + Interest

Examples:

1. Sophia wants to buy a colour television set. She is granted a loan of R1 200 at 16% p.a. for 2 years. Calculate
(*a*) the simple interest.
(*b*) the final amount she has to repay.

(*a*) $$\textbf{SI} = \frac{prt}{100} = \frac{1\,200 \times 16 \times 2}{100} = \text{R } 384$$

∴ Simple Interest: R 384

(*b*) Final amount: R 1 200 + 384 = R 1 584

2. Mr Jafta invests R750 at $10\frac{1}{2}$ % p.a. simple interest for 4 years and 3 months. Calculate the amount he will receive at the end of the period.

$$\textbf{SI} = \frac{prt}{100} = \frac{750 \times 10\frac{1}{2} \times 4\frac{3}{12}}{100} = \frac{750 \times 10,5 \times 4,25}{100} = \text{R } 334,69$$

Final amount: R 750 + 334,69 = R 1084,69

Compound Interest:

Interest is calculated after each time period (yearly/quarterly/monthly) and then **added to the principle** to become the **new principal** which is then used to calculate the next term's interest. **Compound interest** is used more often in real life.

Example:

Calculate the final amount which Mr. Myendeki will receive if he invests R12 000 at 6% **compound interest** for 4 years.

Principal:	1st Year:		R 12 000,00
Interest:	1st Year:	$\frac{12\,000 \times 6}{100}$	720,00
Principal:	2nd Year:		R 12 720,00
Interest:	2nd Year:	$\frac{12\,720 \times 6}{100}$	763,20
Principal:	3rd Year:		R 13 483,20
Interest:	3rd Year:	$\frac{13\,483,20 \times 6}{100}$	808,99
Principal:	4th Year:		R 14 292,19
Interest:	4th Year:	$\frac{14\,292,19 \times 6}{100}$	857,53
FINAL AMOUNT:	after 4 years:		R 15 149,72

With **compound interest** the interest is calculated as follows:

Interest = R 15 149,72 − R 12 000

= R 31 49,72

The following example compares **simple interest** to **compound interest**.

Example:

Calculate the final amount after 3 years on R5 000 invested at:

(*a*) 12% p.a.: Simple interest

(*b*) 12% p.a.: Compound interest

(*a*) **Simple Interest** $= \frac{prt}{100}$

$$= \frac{5\,000 \times 12 \times 3}{100}$$

= R 1 800

Final amount = R 6 800

(*b*) **Compound interest**

Principal:	1st Year:		R 5 000,00
Interest:	1st Year:	$\frac{5\,000 \times 12}{100}$	600,00
Principal:	2nd Year:		R 5 600,00
Interest:	2nd Year:	$\frac{5\,600 \times 12}{100}$	672,00
Principal:	3rd Year:		R 6 272,00
Interest:	3rd Year:	$\frac{6\,272 \times 12}{100}$	752,64
Final amount:			R 7 024,84

Thus: **Final amount (SI): R 6 800,00**
Final amount (CI): R 7 024,84

In the following example the compound interest is added half-yearly: this **12% p.a. now becomes 6%;** while **2 years** the interest is calculated **4 times** (4 periods).

Example:

Calculate the **compound interest** on an investment of R12 500 for 2 years if interest at 12% p.a. is compounded half-yearly:

Solution:

Principal:	1st Period:		R 12 500,00
Interest:	1st Period:	$12\,500 \times \frac{6}{100}$	750,00
Principal:	2nd Period:		R 13 250,00
Interest:	2nd Period:	$13\,250 \times \frac{6}{100}$	795,00
Principal:	3rd Period:		R 14 045,00
Interest:	3rd Period:	$14\,045 \times \frac{6}{100}$	842,70
Principal:	4th Period:		R 14 887,70
Interest:	4th Period:	$14\,887,70 \times \frac{6}{100}$	893,26
Final amount after 2 years:			R 15 780,96

Compound interest for 2 years: R 15 780,96 − R 12 500,00 = **R 3 280,96**

Exercise:

1. Calculate the final amount which I shall have available if I invest R 8 000 for 5 years at 11% p.a. simple interest.

2. Mimi has to borrow R 1 800 from a Building Society to buy a new washing machine. The loan is for 3 years at an interest rate of $16\frac{1}{2}$% p.a. simple interest.
What is the final amount that she will have to pay back?

3. Calculate the final amount if R 1 350 is invested at $15\frac{1}{4}$% p.a. simple interest for 3 years and 2 months.

4. Mr. Mabope invests R 8 000 in his bank account at a simple interest rate of 13,5% p.a. What final amount will be available to him after 8 months?

5. If I borrow R 1 500 from a bank at 8% p.a. compound interest, what amount will I have to pay back after 2 years?

6. Ms Dunn invests R 10 000 in her bank account at 11% p.a. compound interest. How much interest has she earned after 2 years?

7. Calculate the compound interest on an investment of R 12 000, invested for $1\frac{1}{2}$ years at 14% p.a., compounded half-yearly.

8. George invests R 800 at 10% p.a. simple interest. After 2 years the interest rate changes to 12% p.a. Determine the final amount which he will receive after 5 years.

9. Edson borrows R 12 000 from Jane at the beginning of 1993 at 10% p.a. compound interest. At the end of each year he pays back R500 on the loan. How much does he owe her at the end of 1996?

10. A customer buys a motor-car for R 45 000. He has to deposit a quarter of the price immediately. On the remainder he has to pay 15% simple interest p.a. over a period of 4 years. This amount has to be paid back in 48 equal monthly instalments. What is the monthly instalment?

11. An amount of R12 500 is invested at 15% p.a. compound interest. After the first year the interest rate drops to 11% p.a. Determine the interest earned (in total) on this investment after 4 years.

12. Calculate the book value of a truck after 4 years, if it costs R 180 000 initially and it depreciates (that is, its value is reduced) by 12% p.a. (compounded annually).

SOLUTIONS

1. Simple interest: $\frac{prt}{100}$

$= \frac{8\,000 \times 11 \times 5}{100}$

$= R\,4\,400$

Final amount: R 8 000 + R 4 400 = R 12 400

2. Simple interest: $\frac{1\,800 \times 16{,}5 \times 3}{100}$

$= R\,891$

Final amount to be paid back:

R 1 800 + R 891 = R 2 691

3. Simple interest: $\frac{1\,350 \times 15{,}25 \times 3\frac{2}{12}}{100}$

$= R\,651{,}94$

Final amount = R 2 001,94

4. Simple interest: $\frac{R\,8\,000 \times 13{,}5 \times \frac{8}{12}}{100}$

$= R\,720$

Final amount = R 8 720

5. Principal:	1st Year:		R 1 500,00
Interest:	1st Year:	$1\,500 \times \frac{8}{100}$	120,00
Principal:	2nd Year:		R 1 620,00
Interest:	2nd Year:	$1\,620 \times \frac{8}{100}$	129,60
Final amount after 2 years:			R 1 749,60

6. Principal:	1st Year:		R 10 000,00
Interest:	1st Year:	$10\,000 \times \frac{11}{100}$	1 100,00
Principal:	2nd Year:		R 11 100,00
Interest:	2nd Year:	$11\,100 \times \frac{11}{100}$	1 221,00
Final amount:			R 12 321,00

Interest: R 12 321 – R 10 000 = R 2 321

7. Principal:	1st Period:		R 12 000,00
Interest:	1st Period:	$12\,000 \times \frac{7}{100}$	840,00
Principal:	2nd Period:		R 12 840,00
Interest:	2nd Period:	$12\,840 \times \frac{7}{100}$	898,80
Principal:	3rd Period:		R 13 738,80
Interest:	3rd Period:	$13\,738,80 \times \frac{7}{100}$	961,72
Final amount after $1\frac{1}{2}$ years:			R 14 700,52
Compound interest:	= R 14 700,52 − R 12 000,00 =		R 2 700,52

8. Simple interest: $\frac{R\,800 \times 10 \times 2}{100} = R\,160$

Simple interest: $\frac{R\,800 \times 12 \times 3}{100} = R\,288$

Final amount after 5 years: R 800 + R 160 + R 288 = R 1 248

9. Principal:	1993:		R 12 000,00
Interest:	1993:	$12\,000 \times \frac{10}{100}$	1 200,00
End:	1993:		R 13 200,00
Instalment:			500,00
Principal:	1994:		R 12 700,00
Interest:	1994:	$12\,700 \times \frac{10}{100}$	1 270,00
End:	1994:		R 13 970,00
Instalment:			500,00
Principal:	1995:		R 13 470,00
Interest:	1995:	$13\,470 \times \frac{10}{100}$	1 347,00
End:	1995:		R 14 817,00
Instalment:			500,00
Principal:	1996:		R 14 317,00
Interest:	1996:	$14\,317 \times \frac{10}{100}$	1 431,70
Total debt end 1996			R 1 5748,70

10. Immediate payment: $\frac{1}{4} \times 45\,000 : 11\,250$

Balance: R 33 750

SI: $33\,750 \times \frac{15}{100} \times 4 =$ R 20 250

Final amount: R 33 750 + 20 250 = R 54 000

Instalment: R 54 000 ÷ 48 = R1 125 per month.

11. Principle: 1st Year: .... R 12 500,00

Interest: 1st Year: $12\,500 \times \frac{15}{100}$ 1 875,00

Principle: 2nd Year: .... R 14 375,00

Interest: 2nd Year: $14\,375 \times \frac{11}{100}$ 1 581,25

Principle: 3rd Year: .... R 15 956,25

Interest: 3rd Year: $15\,956,25 \times \frac{11}{100}$ 1 755,19

Principle: 4th Year: .... R 17 711,44

Interest: 4th Year: $17\,711,44 \times \frac{11}{100}$ 1 948,26

Final amount after 4 years: R 19 659,70

Interest in 4 years: R 7 159,70

12. Purchase price: 1st Year: .... R 180 000,00

Depreciation: 1st Year: $180\,000 \times \frac{12}{100}$ 21 600,00

Book value: Beginning 2nd Year: .... R 158 400,00

Depreciation: 2nd Year: $158\,400 \times \frac{12}{100}$ 19 008,00

Book value: Beginning 3rd Year: .... R 139 392,00

Depreciation: 3rd Year: $139\,392 \times \frac{12}{100}$ 16 727,04

Book value: Beginning 4th Year: .... R 122 664,96

Depreciation: 4th Year: $122\,664,96 \times \frac{12}{100}$ 14 719,80

BOOK VALUE END OF 4th YEAR R 107 945,16

10 *Statistics*

Statistics is the collection, classification and interpretation of numerical data.

For the **classification** and **presentation** of data, we shall use:

1. Frequency table
2. Pictogram
3. Bar graph and histogram
4. Pie Chart
5. Jagged-line graph

For the **interpretation** of data we shall use:

1. Frequency
2. Average
3. Modus
4. Median
5. Range

1. Frequency Table:

The frequency is the figure which represents the number of **times** a specific incident occurred.

Example:

The following are the marks (out of 10) which a group of 30 pupils obtained in a test.

5	6	4	8	5	1	2	9	5	5
2	3	3	4	8	7	9	2	3	1
5	4	6	10	9	6	7	6	5	6

Set up a frequency table and answer the following questions:

(*a*) Which test mark appears most frequently?

(*b*) Which test mark appears least frequently?

(*c*) How many pupils scored 9 out of 10?

(*d*) How many pupils scored 50% or more?

Solution:

Points gained	Tally	Frequency
1	11	2
2	111	3
3	111	3
4	111	3
5	~~1111~~1	6
6	~~1111~~	5
7	11	2
8	11	2
9	111	3
10	1	1
		30

(*a*) Most frequently: $\frac{5}{10}$
(highest frequency)

(*b*) Least frequently $\frac{10}{10}$

(*c*) $\frac{9}{10}$: 3 pupils

(*d*) 50% or more: 19 pupils

2. Pictogram:

A pictogram uses identical pictures to represent the data.

Example:

Draw a pictogram to represent the following:

The number of trees planted on the school grounds during the last 5 years.

1990 : 20
1991 : 30
1992 : 15
1993 : 20
1994 : 10

Use the scale:  = 5 trees

Solution:

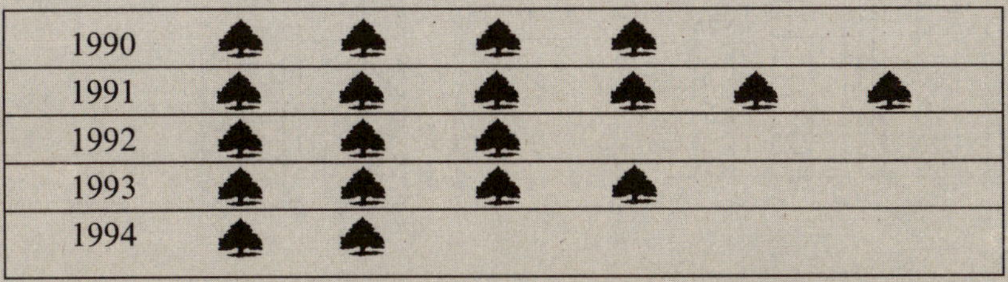

3. Bar graph and histogram:

A bar graph and a histogram are used in cases where the data is grouped into intervals and the frequency of each interval is then determined.

Example:

The age of 60 teachers in a certain school is as follows:

23	20	24	26	29	34	35	37	39	40
25	24	26	28	30	34	39	45	54	58
60	65	64	63	62	69	20	23	21	24
27	28	39	31	32	35	40	42	43	45
52	57	58	46	64	67	48	51	56	59
42	45	27	56	53	52	45	46	32	33

Draw up a frequency table and then use a bar graph and a histogram to represent the data.

Solution:

Age-Intervals	Tally	Frequency
0 - 29	𝍸 𝍸 𝍸 |	16
30 - 39	𝍸 𝍸 |||	13
40 - 49	𝍸 𝍸 ||	12
50 - 59	𝍸 𝍸 |	11
60 - 69	𝍸 |||	8
		60

Bar graph:

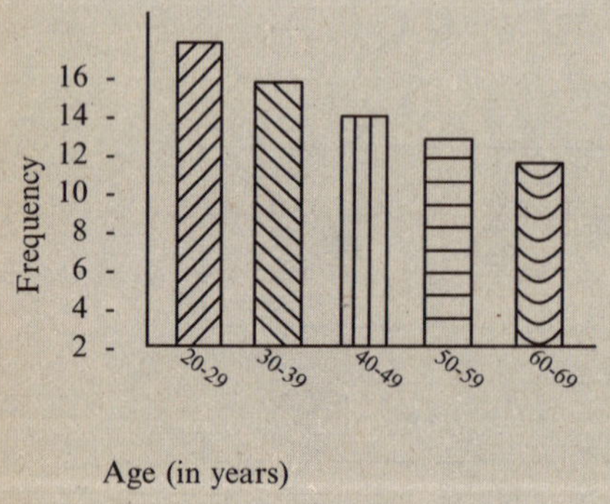

Histogram:

In a histogram the columns are displayed next to one another.

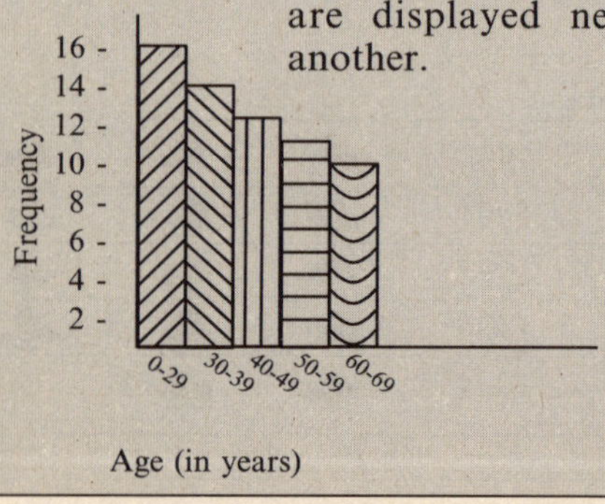

4. Pie Chart

In a pie chart a circle (360°) is divided into sectors.

The following background knowledge is necessary to enable you to express a sector as degrees and as a %.

Example:

1. Given:

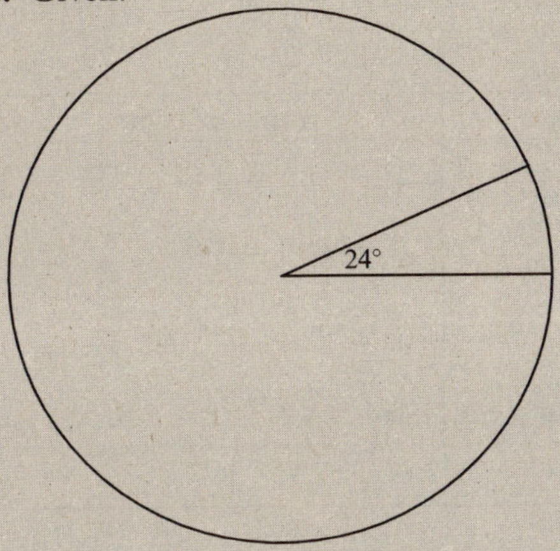

This sector represents:

(i) As a fraction: $\frac{24}{360} = \frac{1}{15}$ of the total.

(ii) As %: $\frac{1}{15} \times 100\%$

$= 6{,}67\%$

2. Given: $\frac{1}{5}$ of the class has blond hair.

Therefore (i) $\frac{1}{5} \times 360° = 72°$

(ii) $\frac{1}{5} \times 100\% = 20\%$ is blond

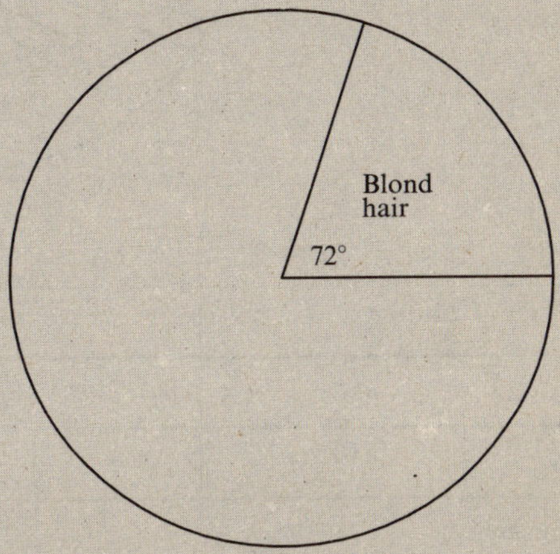

Example:

In a class with 30 pupils there are 12 with blue eyes; 9 with brown eyes; 4 with dark brown eyes and 5 with green eyes.

(*a*) Draw a pie chart showing this information.

(*b*) Calculate the fraction and percentage represented by each eye colour.

Solution:

Colour of eyes	Number of pupils	Angle at centre
Blue	12	$\frac{12}{30} \times 360° = 144°$
Brown	9	$\frac{9}{30} \times 360° = 108°$
Dark brown	4	$\frac{4}{30} \times 360° = 48°$
Green	5	$\frac{5}{30} \times 360° = 60°$
	30	360°

(*a*)

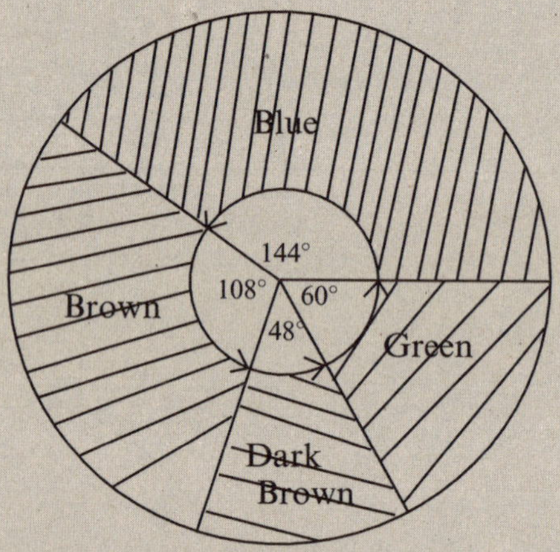

(*b*)

Colour	Fraction	%
Blue	$\frac{12}{30} = \frac{2}{5}$	$\frac{12}{30} \times 100 = 40\%$
Brown	$\frac{9}{30} = \frac{3}{10}$	$\frac{9}{30} \times 100 = 30\%$
Dark brown	$\frac{4}{30} = \frac{2}{15}$	$\frac{4}{30} \times 100 = 13\frac{1}{3}\%$
Green	$\frac{5}{30} = \frac{1}{6}$	$\frac{5}{30} \times 100 = 16\frac{2}{3}\%$

5. Broken-line graphs/Jagged-line graphs:

In this case the points are joined using straight lines, thus forming a broken-line graph.

Example:

The following table shows the temperature (in °C) as measured at certain times during the day.

Time (in hours)	7h00	8h00	11h00	12h00	13h00	15h00	17h00	19h00
Temp (°C)	8	12	16	20	22	18	10	6

Draw a broken-line graph in order to estimate the following values:

(*a*) the temperature at 10h00

(*b*) the temperature at 14h00

(*c*) the am. time when the temperature was 13 °C

(*d*) the pm. time when the temperature was 12 °C

Solution:

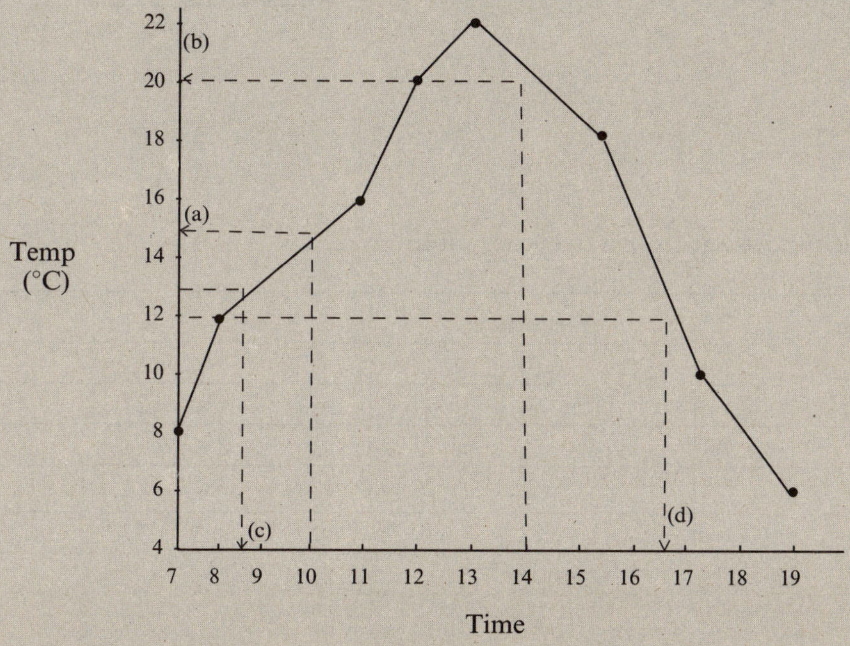

(*a*) 14,6 °C (*b*) 20°C (*c*) between 08h00 and 09h00 (*d*) between 16h00 and 17h00

Interpretation of data:

1. Frequency
2. Average
3. Modus
4. Median
5. Range

Example: 3; 5; 7; 6; 2; 5; 4; 6; 5; 1; 8

1. Frequency:

It means the number of times a certain value appears.

From the example: Frequency of 5 is 3.

2. Average:

$$\text{Average} = \frac{\textit{sum of all the values}}{\textit{number of items}}$$

$$= \frac{3+5+7+6+2+5+4+6+5+1+8}{11}$$

$$= \frac{52}{11} = 4{,}73$$

How to determine the average from a frequency table:

Point	Frequency	fx
5	3	$3 \times 5 = 15$
6	4	$4 \times 6 = 24$
7	2	$2 \times 7 = 14$
8	3	$3 \times 8 = 24$
	12	77

$$\therefore \quad \text{Average: } \bar{x} = \frac{77}{12}$$

$$= 6{,}42$$

3. Mode:

It is the **number** occurring most often, in other words with the highest frequency.

From: 1; 2; 3; 4; 5; 5; 5; 6; 6; 7; 8

Mode: 5

4. Median:

It is the number exactly in the middle, or halfway between the two numbers in the middle. Arrange the numbers in chronological order:

1; 2; 3; 4; 5; 5; 5; 6; 6; 7; 8
Median is : 5

5. Range:

Range = highest number – lowest number
= 8 – 1
= 7

An approximate average can be calculated from a frequency table with intervals.

Example:

The following table shows the exam marks of a class of 40 pupils. Calculate the approximate average of the class.

Symbol	Interval	Frequency	$\bar{x}$ of interval	*fx*
G	20-29	2	**25**	2 × 25 = 50
F	30-39	6	**35**	6 × 35 = 210
E	40-49	4	**45**	4 × 45 = 180
D	50-59	8	**55**	8 × 55 = 440
C	60-69	12	**65**	12 × 65 = 780
B	70-79	4	**75**	4 × 75 = 300
A	80-89	3	**85**	3 × 85 = 255
A^+	90-99	1	**95**	1 × 95 = 95
		40		2 310

$$\text{Approximate average: } (\bar{x}) = \frac{2310}{40}$$

$$= 57,75\%$$

A **stem-and-leaf table** is a useful aid when a **large number of values** must be **interpreted**.

Example:

Represent the data below in a stem-and-leaf table:

The length (in cm) of 30 pupils in a Std 7 class is as follows:

152;	165;	143;	139;	138;	144;	150;	158;	161;	157;
143;	156;	162;	165;	139;	158;	151;	167;	166;	160;
170;	132;	145;	151;	148;	147;	171;	155;	146;	141;

Determine: **1.** the frequency of 143.
2. the median of the data.
3. the range

Solution:

The first two digits of each number form the stem and the last digit forms the leaf.

13	9;	8;	9;	2;						(4)
14	3;	4;	3;	5;	8;	7;	6;	1;		(8)
15	2;	0;	8;	7;	6;	8;	1;	1;	5;	(9)
16	5;	1;	2;	5;	7;	6;	0;			(7)
17	0;	1;								(2)
										30

1. Frequency of 143 is 2.

2. The median of the 30 pupils lies between the 15th and 16th values,

namely between 151 and 152

$151\frac{1}{2}$ cm.

3. Range: 171 − 132 = 39 cm.

Exercise: Statistics

1. In a Maths exam 40 pupils scored the following marks (%):

25;	40;	61;	43;	29;	87;	94;	66;	63;	59;	29	28;
63;	54;	39;	86;	89;	95;	96;	54;	48;	49;	53;	54;
63	67	69	84	83	85	74	73	83	95	74	72
83;	84;	85;	63;								

(*a*) Complete the frequency table below.

(*b*) How many pupils obtained more than 90%?

(*c*) 40% is a pass. How many pupils failed Maths?

(*d*) Represent the data on a stem-and-leaf table, and then answer the following questions:

(*e*) What is the median of the class?

(*f*) What is the modus of the class?

(*g*) What is the class average for Maths?

(*h*) Draw a histogram of the Maths marks of the class.

%-Intervals	**Tally**	**Frequency**
20-29		
30-39		
40-49		
50-59		
60-69		
70-79		
80-89		
90-99		

2. Draw a pictogram to show the population increase in a certain town.

Year	**Population**
1983	1200
1984	1500
1985	2000
1986	2200
1987	2500

Use scale:  = 200 people and 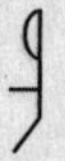 = 100 people

3. Draw a pictogram showing the favourite subjects of a std 7 class.

Subject	English	Maths	Science	History	Art
Frequency	5	10	8	3	4

(Use your imagination and draw an interesting diagram.)

4. The bar graph shows the number of pupils per standard in a school.

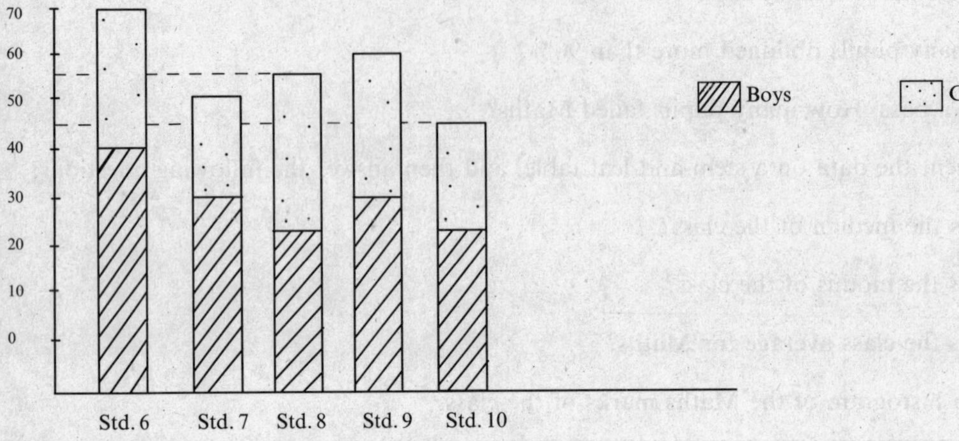

(*a*) How many pupils are there in Std 8?

(*b*) How many boys are there in Std 10?

(*c*) How many more boys than girls are there in Std 6?

(*d*) Which standard has an equal number of boys and girls?

(*e*) How many pupils are there in the school?

(*f*) Which standards have the same number of girls?

5. The following table shows a boy's test marks in Maths through the year.

	Real Numbers	Indices	Geometry	Factorization	Equations	Graphs	Statistics
Test no.	1	2	3	4	5	6	7
% obtained	52	67	74	60	78	85	94

(*a*) Draw a broken-line graph of the information.

(*b*) What is his average mark for the year?

(*c*) In which test did he achieve a mark higher than his average?

(*d*) Which sections need more attention?

6. The pupils in a class have been numbered from 1 to 36 in order to determine their participation in wintersport.

R: Rugby	C: Cross country
H: Hockey	S: Soccer
N: Netball	O: No sport

1 R	2 R	3 R	4 H	5 N	6 N
7 N	8 H	9 S	10 H	11 O	12 C
13 H	14 R	15 R	16 N	17 O	18 S
19 R	20 O	21 R	22 C	23 H	24 R
25 S	26 R	27 N	28 R	29 C	30 R
31 R	32 S	32 C	34 O	35 N	36 S

6.1 Set up a frequency table.

6.2 Which sport is the most popular?

6.3 Which percentage of the pupils play netball?

6.4 What fraction of the class does not take part in any sport?

6.5 How many pupils may kick a ball?

6.6 Comment on the popularity of cross country.

6.7 Draw an interesting pictograph to portray the data.

7. Determine the (*a*) Average; (*b*) Mode; (*c*) Median; (*d*) range of each of the following groups of numbers.

7.1 5; 10; 12; 5; 5; 14; 6; 10; 12; 5;

7.2 3; 6; 1; 8; 9; 1; 4;

7.3 5; 6; 8; 9; 6; 5; 6; 9; 5; 6; 8; 9;

8. Determine the average from each of the following frequency tables:

(*a*)

Data	Frequency	*fx*
5	3	
7	8	
9	2	
10	3	
12	4	

(*b*)

Data	Frequency	*fx*
10	3	
15	5	
20	2	
25		
30	1	
	15	

(*c*)

Interval	Frequency	$\bar{x}$ of Interval	*fx*
10-19	2		
20-29	3		
30-39	7		
40-49	1	45	
50-59	2		

9. The lengths of 40 pupils are given below. Set up a frequency table using 6 intervals.

(*a*) Calculate the approximate average

(*b*) Determine the mode

(*c*) Determine the median
(A stem-and-leaf table may be a help for (*b*) and (*c*).)

(*d*) How many pupils are shorter than your approximate average in (*a*)?

(*e*) Compare the median and the average in (*a*). Comment on this.

142	170	162	131	145	146	147	160
159	150	141	132	169	173	139	46
152	154	140	145	161	163	156	157
171	168	166	151	152	132	142	150
161	138	170	132	149	150	138	152

SOLUTION

Solution: Statistics

1. (*a*)

% Intervals	Tally	Frequency
20-29	𝍷𝍷𝍷𝍷	4
30-39	𝍷	1
40-49	𝍷𝍷𝍷𝍷	4
50-59	𝍸	5
60-69	𝍸 𝍷𝍷𝍷	8
70-79	𝍷𝍷𝍷𝍷	4
80-89	𝍸 𝍸	10
90-99	𝍷𝍷𝍷𝍷	4
		40

(*b*) 4 pupils

(*c*) 5 pupils

(*d*) Stem-and-leaf:

(*e*) Median: $66\frac{1}{2}$
(between 66 and 67)

(*f*) Mode: 63
(Appears 4 times)

(*g*) Average:

$$\frac{2641}{40} = 66{,}0\%$$

Stem-and-leaf table:

2	5 9 9 8	(4)
3	9	(1)
4	0 3 8 9	(4)
5	9 4 4 3 4	(5)
6	1 6 3 3 3 7 9 3 . .	(8)
7	3 4 4 2	(4)
8	7 6 9 3 4 5 3 3 4 5	(10)
9	4 5 6 5	(4)
		40

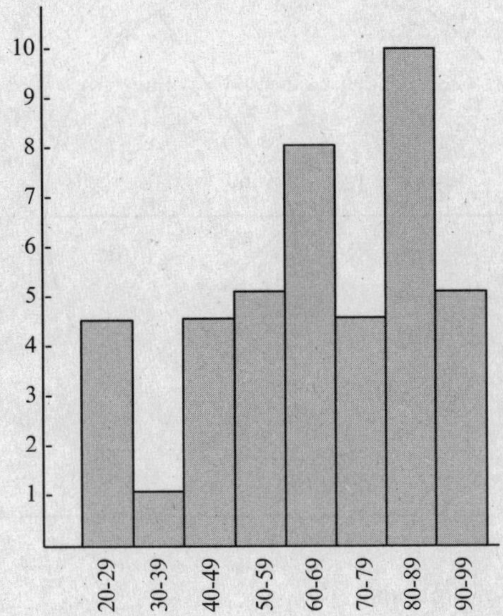

2.

1983	
1984	
1985	
1986	
1987	

3.

Subject	Freq.	Degrees
Eng.	5	$\frac{5}{30} \times 360° = 60°$
Maths.	10	$\frac{10}{30} \times 360° = 120°$
Science	8	$\frac{8}{30} \times 360° = 96°$
History	3	$\frac{3}{30} \times 360° = 36°$
Art	4	$\frac{4}{30} \times 360° = 48°$
	30	360°

↑ Test

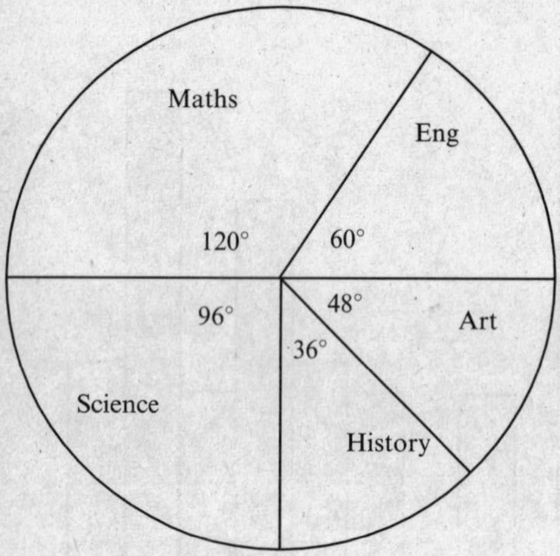

4. (*a*) 55 pupils

(*b*) 20 boys

(*c*) 70 – 40 = 30 girls
Difference: 10 more boys

(*d*) Std 9

(*e*) 280 pupils

(*f*) In Std 6 and Std 9.

5. (*a*)

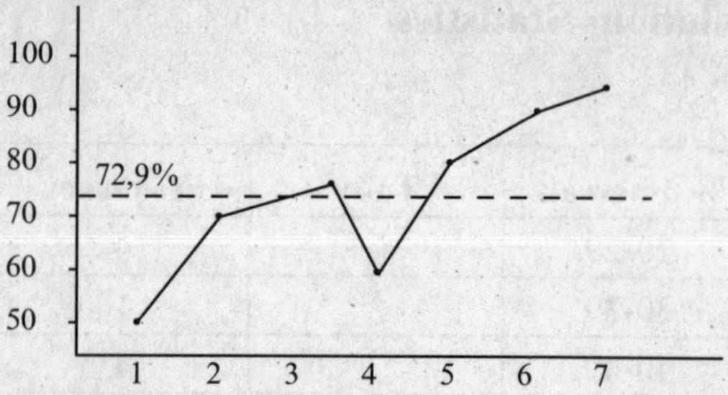

(*b*) average: $(\bar{x}) = \frac{510}{7} = 72{,}9\%$

(*c*) Tests 3; 5; 6; 7.

(*d*) Real numbers and factorization.

6.1

Sport	Tally	Frequency	Degree
Rugby (R)	𝍸 𝍸 ||	12	$\frac{12}{36} \times 360° = 120°$
Hockey (H)	𝍸	5	$\frac{5}{36} \times 360° = 50°$
Netball (N)	𝍸 |	6	$\frac{6}{36} \times 360° = 60°$
Cross Country (C)	||||	4	$\frac{4}{36} \times 360° = 40°$
Soccer (S)	𝍸	5	$\frac{5}{36} \times 360° = 50°$
No Sport (O)	||||	4	$\frac{4}{36} \times 360° = 40°$
		36	360°

6.2 Rugby

6.3 Netball: $\frac{6}{36} \times 100\% = 16,7\%$

6.4 No Sport: $\frac{4}{36} = \frac{1}{9}$

6.5 17 pupils

6.6 Because the frequency of crosscountry is so low we may assume that pupils prefer a ball sport.

6.7

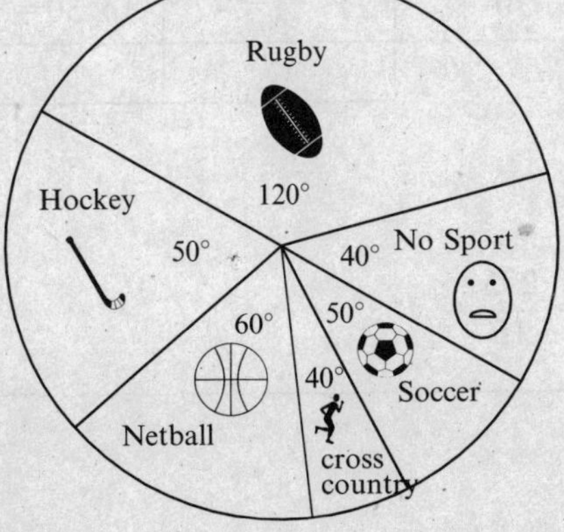

7.1 5 5 5 5 6 10 10 12 12 14

(*a*) Average: $\frac{84}{10} = 8{,}4$

(*b*) Mode: 5 (highest frequency of 4)

(*c*) Median: 8 (middle value, halfway between 6 and 10)

7.2 1 1 3 4 6 8 9

(*a*) Average: $\frac{32}{7} = 4{,}57$

(*b*) Mode: 1 (highest frequency of 2)

(*c*) Median: 4 (middle value)

7.3 5 5 5 6 6 6 6 8 8 9 9 9

(*a*) Average: $\frac{82}{12} = 6{,}83$

(*b*) Mode: 6 (highest frequency of 4)

(*c*) Median: 6 (middle value)

8. (*a*)

Data	Frequency	*fx*
5	3	$3 \times 5 = 15$
7	8	$8 \times 7 = 56$
9	2	$2 \times 9 = 18$
10	3	$3 \times 10 = 30$
12	4	$4 \times 12 = 48$
	20	167

$(\bar{x}) = \frac{167}{20} = 8{,}35$

8. (*b*)

Data	Frequency	*fx*
10	3	$3 \times 10 = 30$
15	5	$5 \times 15 = 75$
20	2	$2 \times 20 = 40$
25	4	$4 \times 25 = 100$
30	1	$1 \times 30 = 30$
	15	275

$(\bar{x}) = \frac{275}{15} = 18{,}3$

8. (*c*)

Interval	Frequency	$\bar{x}$ in Int.	*fx*
10-19	2	15	$2 \times 15 = 30$
20-29	3	25	$3 \times 25 = 75$
30-39	7	35	$7 \times 35 = 245$
40-49	1	45	$1 \times 45 = 45$
50-59	2	55	$2 \times 55 = 110$
	15		505

Approximate Average $(\bar{x}) = \frac{510}{7} = \frac{505}{15} = 33{,}7$ cm

9. Range: $173 - 131 = 42$; 6 Intervals: each: $\frac{42}{6} = 7$

Intervals	Tally	Frequency	$\bar{x}$ in Int.	*fx*
131-137	||||	4	134	$4 \times 134 = 536$
138-144	~~||||~~ ||	7	141	$7 \times 141 = 987$
145-151	~~||||~~ ~~||||~~	10	148	$10 \times 148 = 1480$
152-158	~~||||~~ |	6	155	$6 \times 155 = 930$
159-165	~~||||~~ |	6	162	$6 \times 162 = 972$
166-172	~~||||~~ ||	7	169	$7 \times 169 = 1183$
		40		6088

(*a*) Approximate Ave: $(\bar{x}) = \frac{510}{7} = 152{,}2$ cm

(*b*) Mode:132 and 152 and 150

(*c*) Median: between 150 and 151

$\therefore$ 150,5 cm

(*d*) 24 Pupils

13	8 1 2 2 2 9 8	(7)
14	2 1 0 5 5 9 6 7 2 6	(10)
15	9 2 0 4 1 2 0 6 7 0 2 ..	(11)
16	1 8 2 6 9 1 3 0	(8)
17	1 0 0 3	(4)
		40

(*e*) Because the average (152,2) > median (150,5) and according to the answer in (*d*) namely that 24 of the 40 pupils are shorter than the average, we may conclude that there are a few exceptionally tall pupils in the class.

11 *Probability*

Probability (P) $= \dfrac{\text{number of successful outcomes}}{\text{total number of possible outcomes}}$

Example 1:

A card is drawn from a pack of playing cards (52). What is the probability that the card is:

(*a*) a Queen?
(*b*) a Club?
(*c*) a Jack of Hearts?

Solution:

(*a*) p *(Queen)* $= \dfrac{4}{52} = \dfrac{1}{13}$

(*b*) p *(Club)* $= \dfrac{13}{52} = \dfrac{1}{4}$

(*c*) p *(Jack of Hearts)* $= \dfrac{1}{52}$

Example 2:

There are 5 green balls, 2 red balls and 4 yellow balls. One ball is drawn from the bag.

What is the probability that the ball:
(*a*) will be green?
(*b*) will be red?
(*c*) will be yellow?

Solution:

There are 11 balls in the bag.

(*a*) p *(green)* $= \dfrac{5}{11}$

(*b*) p *(red)* $= \dfrac{2}{11}$

(*c*) p *(yellow)* $= \dfrac{4}{11}$

Example 3:

The King of Hearts and the Jack of Spades are drawn from a pack of playing cards.

What is the probability that the next card drawn will be:

(*a*) another King?

(*b*) another Heart?

(*c*) the Ace of Spades?

(*d*) not a Heart?

Solution:

(*a*) p *(King)* $= \dfrac{3}{50}$

(*b*) p *(Heart)* $= \dfrac{12}{50} = \dfrac{6}{25}$

(*c*) p *(Ace of Spades)* $= \dfrac{4}{50}$

(*d*) p *(not a Heart)* $= \dfrac{38}{50} = \dfrac{19}{25}$

Tree diagram:

It is a useful diagram if two actions are carried out, one after the other

Example 4:

There are 5 yellow balls and 4 red balls in a bag. A ball is drawn from the bag, replaced and a second ball is drawn. Determine the probability of drawing:

(*a*) 2 yellow balls.
(*b*) 1 ball of each colour.

Solution:

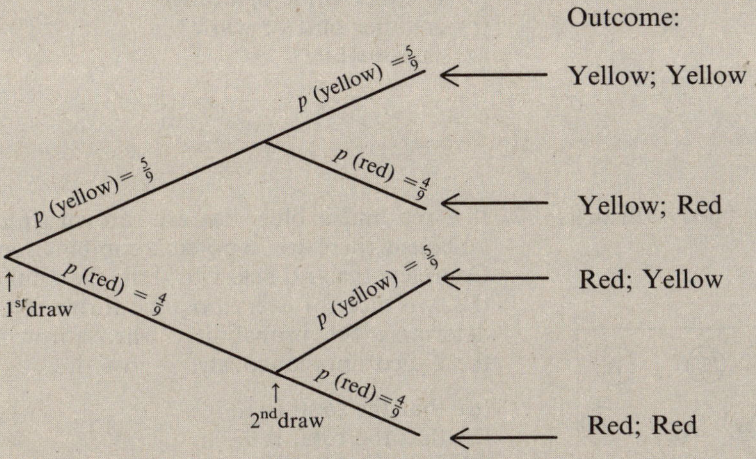

(*a*) p (2 yellow) $= \frac{5}{9} \times \frac{5}{9} = \frac{25}{81}$

Yellow; Red   Red; Yellow

(*b*) p (1 of each colour) $= \left(\frac{5}{9} \times \frac{4}{9}\right) + \left(\frac{4}{9} \times \frac{5}{9}\right)$

$= \frac{20}{81} + \frac{20}{81}$

$= \frac{40}{81}$

Exercise: Probability

1. A dice is thrown. Determine the probability of
   (*a*) throwing a six.
   (*b*) throwing a four.
   (*c*) throwing an even number.
   (*d*) throwing a number greater than 4.

2. A pack of playing cards (52) are shuffled well, and 1 card is drawn. What is the probability that the card will be
   (*a*) a Jack?
   (*b*) a Queen or a Jack?
   (*c*) the ten of Hearts?
   (*d*) a Spade?
   (*e*) a Club greater than 9?
   (*f*) a card greater than 9?

3. A ball is drawn from a box filled with balls as in the diagram below.

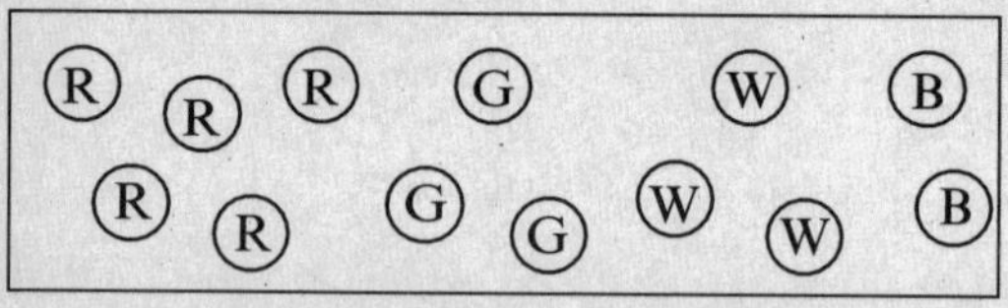

R: Red G: Green
W: White B: Blue

Determine the probability of drawing:
(*a*) a white ball.
(*b*) a green or a blue ball.
(*c*) a ball which is not red.

4. The number of matches in 10 boxes are:

48; 46; 45; 49; 44;
46; 47; 48; 45; 46

One box is drawn. What is the probability of drawing a box with
(*a*) 49 matches?
(*b*) 46 matches?
(*c*) more than 47 matches?

5. The Ace, King and Queen of Diamonds are drawn from a pack of playing cards and are not replaced. Determine the probability of drawing
   (*a*) another Ace.
   (*b*) another Diamond.
   (*c*) the Ace of Spades.
   (*d*) a card greater than 10.

6. A bag holds 5 blue discs, 4 orange discs and 2 white discs. What is the probability of drawing a disc which
   (*a*) is blue?
   (*b*) is either white or orange?
   (*c*) is either blue or white?
   (*d*) is not white?

7. If a red and a blue dice are thrown simultaneously there are 36 possible combinations. Complete the grid below to show the total of the two dice for every possible throw. Now determine the probability, when throwing the 2 dice simultaneously:
   (*a*) that the total is 12
   (*b*) that the total is 9.
   (*c*) that the total is at least 10.
   (*d*) of getting a double.
   (*e*) that the total is more than 6.
   (*f*) of getting a double and a total of more than 10.

6						
5						
4					9	
3	4					
2	3			6		
1	2					
	1	2	3	4	5	6

8. There are 4 green balls and 3 white balls in a bag. A ball is drawn at random and replaced. A second ball is drawn. Complete the following tree diagram:

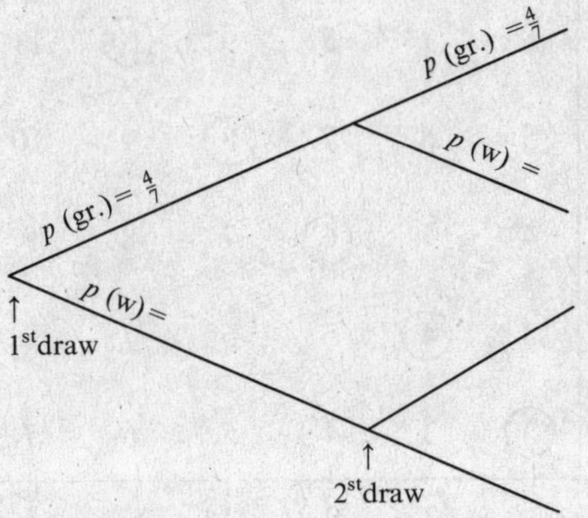

Now determine the probability that:
(*a*) both balls are green.
(*b*) both balls are white.
(*c*) the first ball is green and the second ball is white.

9. A bag contains 4 white balls, 2 black balls and 1 pink ball. A ball is drawn at random and replaced. What is the probability that:
(*a*) both balls are pink?
(*b*) both balls are white?
(*c*) the first ball is white, and the second ball is black?
(*d*) the two balls are pink and white in any order.

10. A team consisting of 2 members has to be chosen. Only Ann, John, Nina, Sipho and Thandi are available. The team is chosen by drawing 2 names simultaneously from a hat.

(*a*) How many possible combinations are there?
Determine the probability that:
(*b*) the team will consist of 2 girls
(*c*) 2 boys will be chosen.
(*d*) John will be a member of the team.
(*e*) the team will consist of one boy and one girl.

11. Look at question 10 again. This time the names are drawn one after the other: the first draw for the A-member of the team and the second draw for the B-member. Therefore (A; B)
Now determine the probability of:
(*a*) the team consisting of 2 girls
(*b*) the A-member being a boy
(*c*) the A-member being a boy, and the B - member a girl.
(*d*) the team consisting of one boy and one girl.
(*e*) John being in the team.

SOLUTIONS

Solution: Probability

1. (*a*) $\frac{1}{6}$

(*b*) $\frac{1}{6}$

(*c*) $\frac{3}{6} = \frac{1}{2}$ (There are 3 even numbers: 2; 4; 6)

(*d*) $\frac{2}{6} = \frac{1}{3}$ (2 numbers greater than 4: 5 and 6)

2 (*a*) $\frac{4}{52} = \frac{1}{13}$ (4 Jacks in die pack)

(*b*) $\frac{8}{52} = \frac{2}{13}$

(*c*) $\frac{1}{52}$

(*d*) $\frac{13}{52} = \frac{1}{4}$

(*e*) $\frac{5}{52}$ (possible cards are: 10; J; Q; K; A of Clubs)

(*f*) $\frac{20}{52} = \frac{5}{13}$ (10; J; Q; K; A of all 4 suits)

3. (*a*) $\frac{3}{13}$

(*b*) $\frac{5}{13}$

(*c*) $\frac{8}{13}$

4. (*a*) $\frac{1}{10}$

(*b*) $\frac{3}{10}$

(*c*) $\frac{3}{10}$ (48; 48; 49)

5. There are only 49 cards left in the pack.

(*a*) $\frac{3}{49}$

(*b*) $\frac{10}{49}$ (10 Diamonds left)

(*c*) $\frac{1}{49}$

(*d*) $\frac{13}{49}$ (J; Q; K; A of 3 other suits + J of Diamonds)

6. (*a*) $\frac{5}{11}$

(*b*) $\frac{6}{11}$

(*c*) $\frac{7}{11}$

(*d*) $\frac{9}{11}$

7.

	1	2	3	4	5	6
6	7	8	9	10	11	(12)
5	6	7	8	9	(10)	11
4	5	6	7	(8)	9	10
3	4	5	(6)	7	8	9
2	3	(4)	5	6	7	8
1	(2)	3	4	5	6	7

Here are the 36 possible combinations.

(*a*) $\frac{1}{36}$

(*b*) $\frac{4}{36} = \frac{1}{9}$

(*c*) $\frac{6}{36} = \frac{1}{6}$ (of getting a 10; 11 or 12)

(*d*) $\frac{6}{36} = \frac{1}{6}$

(*e*) $\frac{21}{36} = \frac{7}{12}$

(*f*) $\frac{1}{36}$

8.

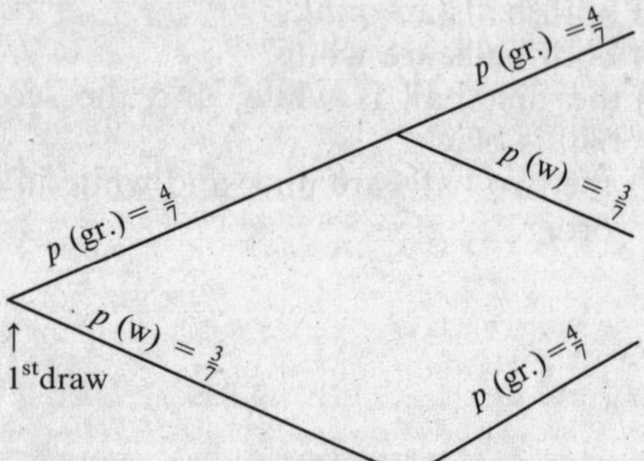

(*a*) $\frac{4}{7} \times \frac{4}{7} = \frac{16}{49}$

(*b*) $\frac{3}{7} \times \frac{3}{7} = \frac{9}{49}$

(*c*) $\frac{4}{4} \times \frac{3}{7} = \frac{12}{49}$

9. (*a*) $\frac{1}{7} \times \frac{1}{7} = \frac{1}{49}$

(*b*) $\frac{4}{7} \times \frac{4}{7} = \frac{16}{49}$

(*c*) $\frac{4}{7} \times \frac{2}{7} = \frac{8}{49}$

(*d*) $\left(\frac{1}{7} \times \frac{4}{7}\right) + \left(\frac{4}{7} \times \frac{1}{7}\right) = \frac{4}{49} + \frac{4}{49} = \frac{8}{49}$

$(p; w)$ $(w; p)$

10.

(*a*) 10 possible combinations

(*b*) $\frac{3}{10}$

(*c*) $\frac{1}{10}$

(*d*) $\frac{4}{10} = \frac{2}{5}$

(*e*) $\frac{6}{10} = \frac{3}{5}$

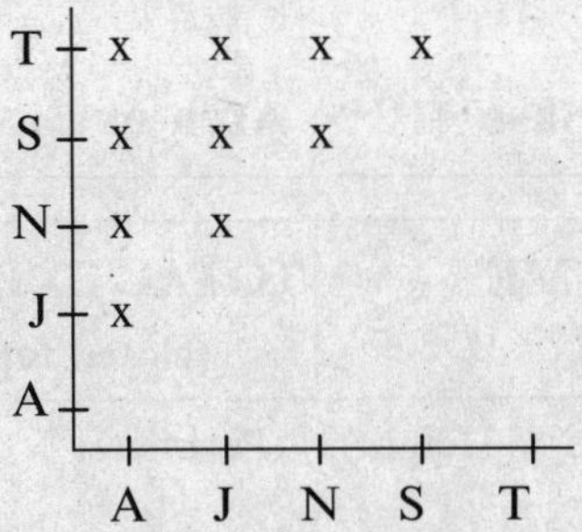

11.

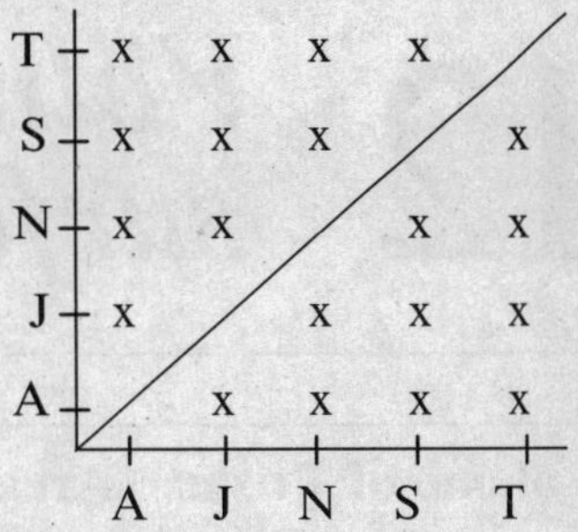

(*a*) $\frac{6}{20} = \frac{3}{10}$ (Count on table)

or

$\left(\frac{3}{5} \times \frac{2}{4} = \frac{6}{20} = \frac{3}{10}\right)$

(1st girl) (only 2 girls left)

(*b*) $\frac{8}{20} = \frac{2}{5}$ (From table)

or

$\left(\frac{2}{5} \times \frac{4}{4} = \frac{2}{5}\right)$

(*c*) $\frac{6}{20} = \frac{3}{10}$ (From table)

or

$\left(\frac{2}{5} \times \frac{3}{4} = \frac{6}{20} = \frac{3}{10}\right)$

(*d*) $\frac{12}{20} = \frac{3}{5}$ (From table)

or

$\left(\frac{2}{5} \times \frac{3}{4} + \frac{3}{5} \times \frac{2}{4} = \frac{6}{20} + \frac{6}{20} = \frac{12}{20} = \frac{3}{5}\right)$

(*e*) $\frac{8}{20} = \frac{2}{5}$ (From table)

or

$\left(\frac{1}{5} \times \frac{4}{4} + \frac{4}{5} \times \frac{1}{4} = \frac{8}{20} = \frac{2}{5}\right)$

12 *Volume and external surface area of prisms*

Volume of Prism: Area of Base × Height

External Surface Area of Prism:

1. Only the sides (eg. pipe): **Perimeter of Base × Height**
2. Prism "without lid": **Perimeter of Base + 1 × Area of Base**
3. Prism "with lid" i.e. closed top and bottom: **Perimeter of Base + 2 × Area of Base**

		VOLUME	TOTAL EXTERNAL AREA (closed top and bottom)
1.	**Cylinder:** Base ⊙ 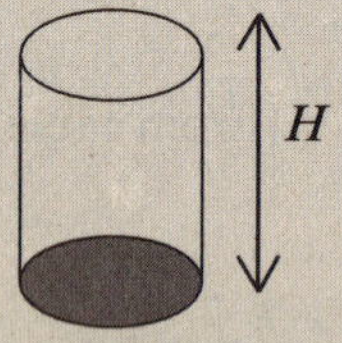	Area of Base × Height ⊙ Vol $= \pi r^2 H$	Perimeter × H + 2 Area. ⊙ ⊙ $= 2\pi r H + 2\pi r^2$
2.	**Rectangular Prism:** 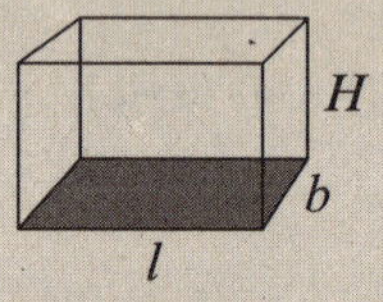	Area Base × Height ▭ $= l \times b \times H$	Perimeter × H + 2 Area. ▭ ▭ $= 2(l+b)H + 2(l \times b)$
3.	**Cube:** (Square) 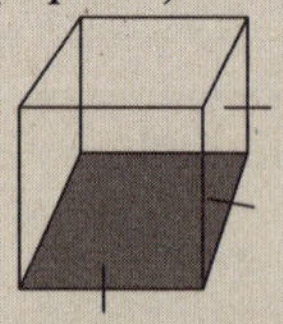	Area Base × Height □ $= side \times side \times side$ $= side^3$	Perimeter × H + 2 Area. □ □ $= 4 \times side \times side(H) + 2side^2$ $= 4side^2 + 2side^2 = 6side^2$
4.	**Prism with Base:** Triangle 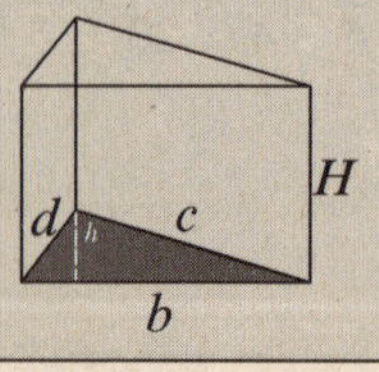	Area Base × Height △ $= \frac{1}{2}bh \times H$	Perimeter × H + 2 Area. △ △ $= (b+c+d)H + 2 \times \frac{1}{2}bh$ $= (b+c+d)H + bh$

Remember: **(1)** Volume – cm^3

Area – cm^2

(2) $1000mm^3 = 1cm^3$

$100mm^2 = 1cm^2$

(3) $1l = 1000cm^3$

Examples:

1. Calculate the **volume** and **total external area** of the cylinder.

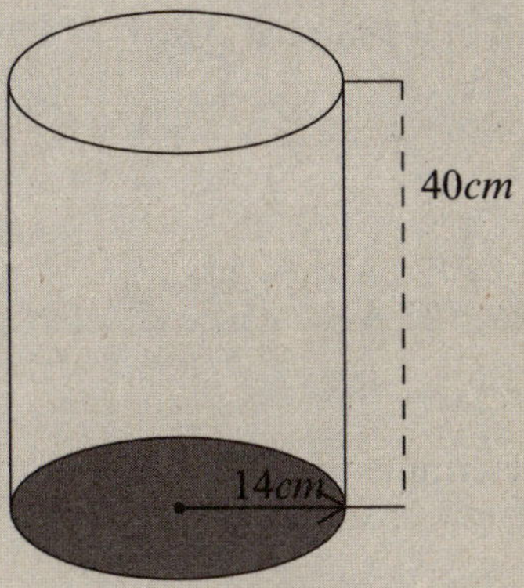

Solution:

Volume $= \text{Area} \times H$
⊙

$= \pi r^2 H$

$= \pi\, 14^2\, 40$

$= 24630,09\ cm^3$

Total External area $= \text{Perimeter}\ H + 2 \times \text{Area}$
⊙ ⊙

$= 2\pi rH + 2\pi r^2$

$= 2\pi.14.40 + 2\pi.14^2$

$= 4750,09\ cm^2$

2. Calculate: 2.1 The volume
2.2 Total external area

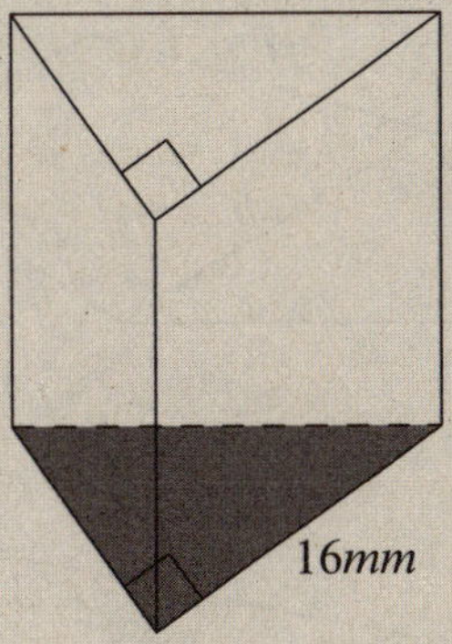

Solution:

2.1 Volume $= \text{Area} \times \text{Height}$
Δ

$= \frac{1}{2}bh \times H$

$= \frac{1}{2}16 \times 12^* \times 18$

$= 1\,728\ mm^3$

2.2 Total external area

$= \text{Perimeter} \times H + 2\ \text{Area}$
Δ Δ

$= (12+16+20)18 + 2\frac{1}{2}b.h$

$= (48 \times 18) + (16 \times 12)$

$= 1\,056\ mm^2$

*AC: In Δ ABC is $\hat{C} = 90°$

$AC^2 = 20^2 - 16^2$ (Pyth.)

$= 400 - 256$

$= 144$

$\therefore AC = 12\ mm$

3. The accompanying prism's measurements are in cm.

Calculate: 3.1 The volume
3.2 The external area

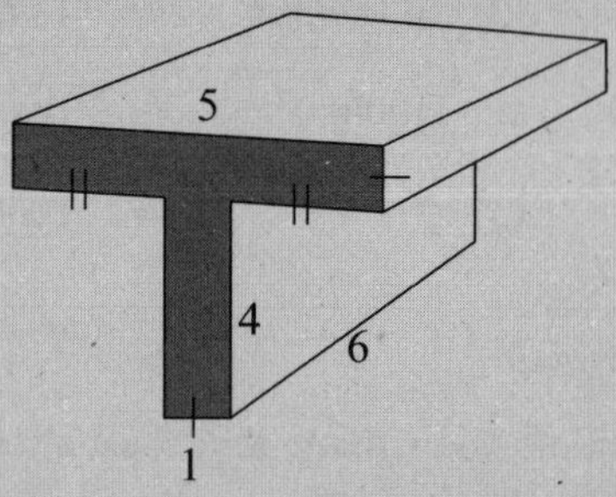

Solution:

3.1 Vol. $= \text{Area}_{\text{Base}} \times H$

$= (5 \times 1 + 4 \times 1)6$

$= (5 + 4)6$

$= 54\ cm^3$

3.2 Total External Area $= \text{Perimeter}_{\text{Base}} \times H + \text{Area}_{\text{Base}}$

$= (5 + 1 + 2 + 4 + 1 + 4 + 2 + 1)6 + 2(5 \times 1 + 4 \times 1)$

$= 20 \times 6 + 2 \times 9$

$= 120 + 18 = 138\ cm^2$

Exercise:

1. Determine the (*a*) volume and (*b*) total external area of each of the following prisms in terms of: *a*; *b*; *c*; *x*; *y*; *H* etc.

1.1

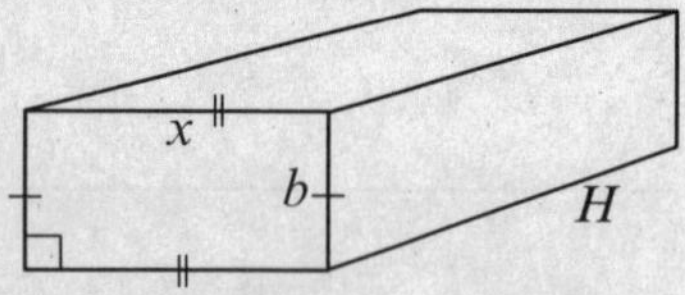

1.2

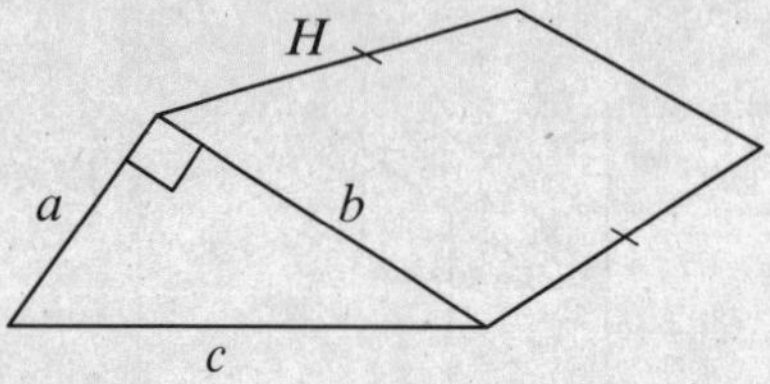

1.3

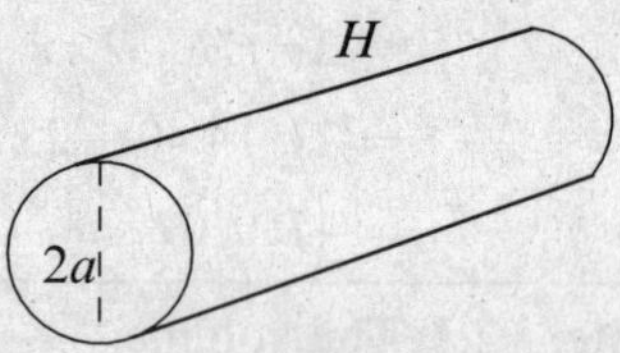

1.4

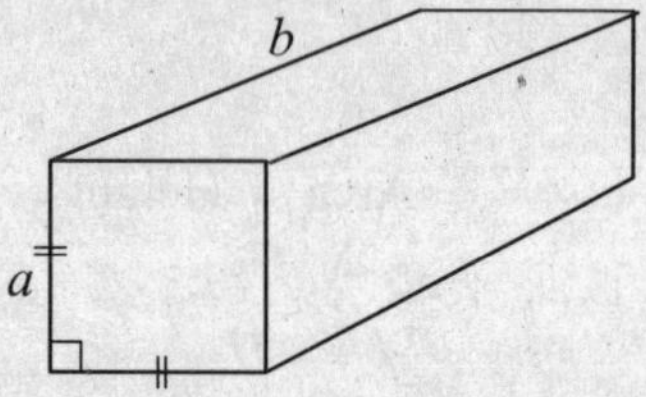

2. Calculate (*a*) the volume and (*b*) the total external area of each of the following figures:

2.1

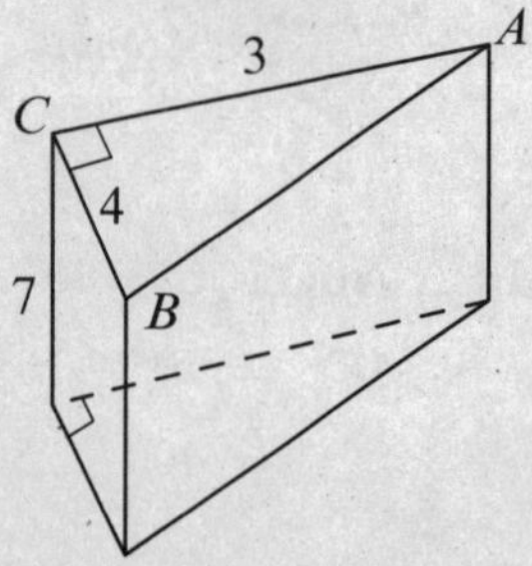

2.2

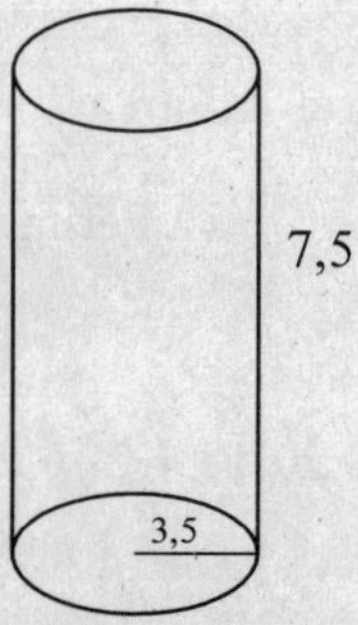

2.3

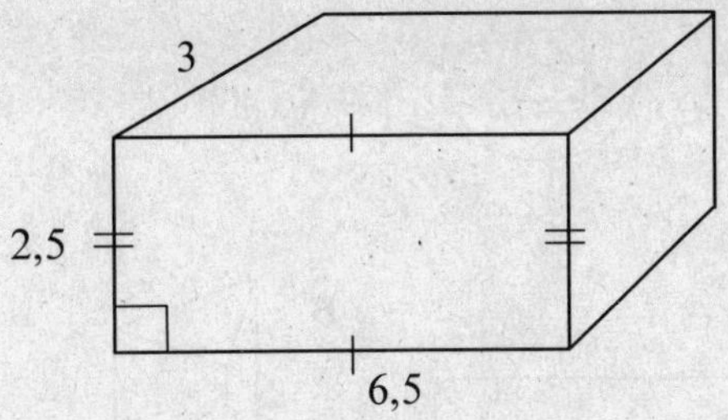

2.4

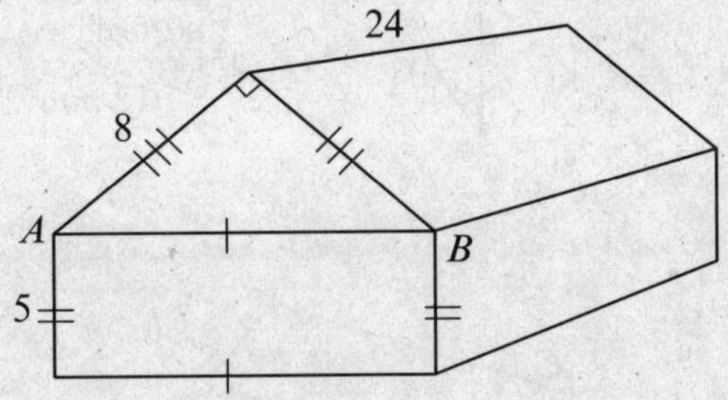

2.5

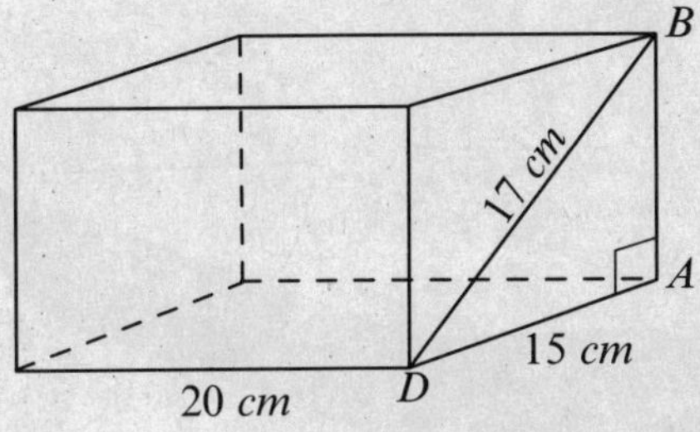

2.6

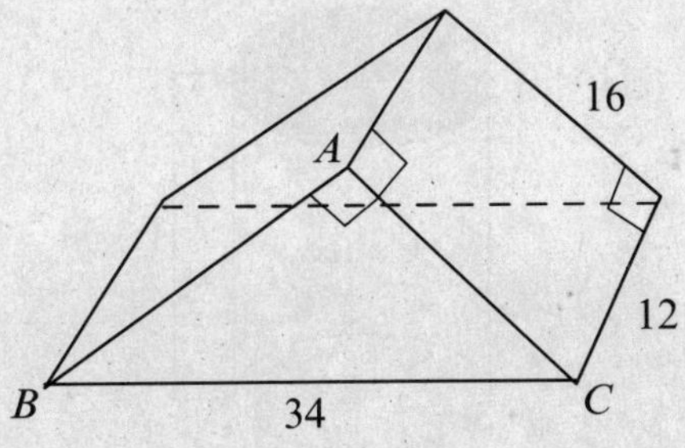

2.7

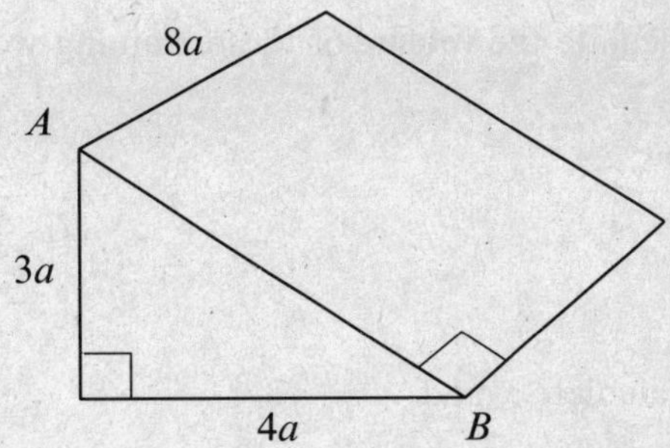

2.8

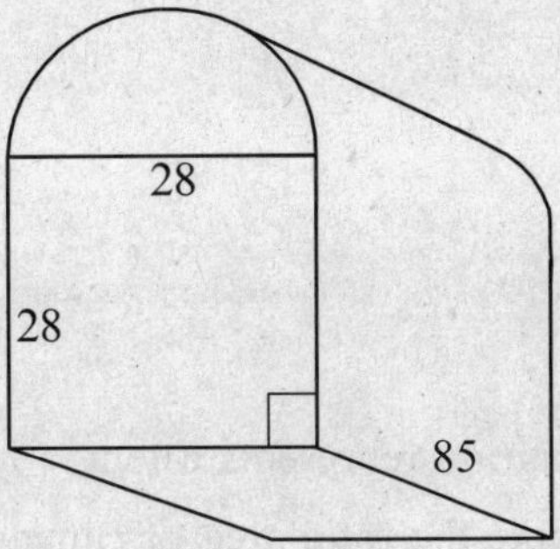

2.9

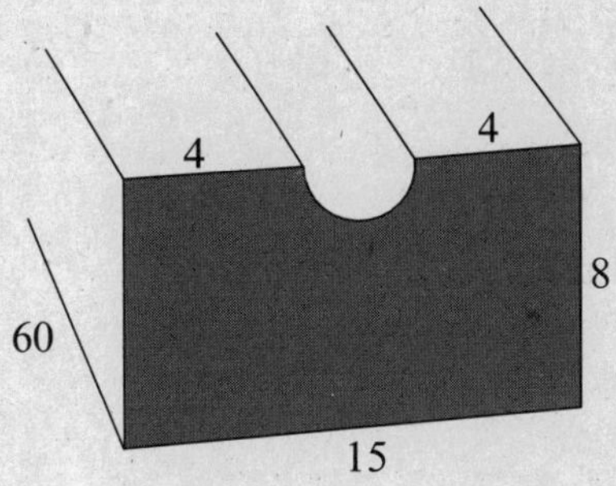

2.10

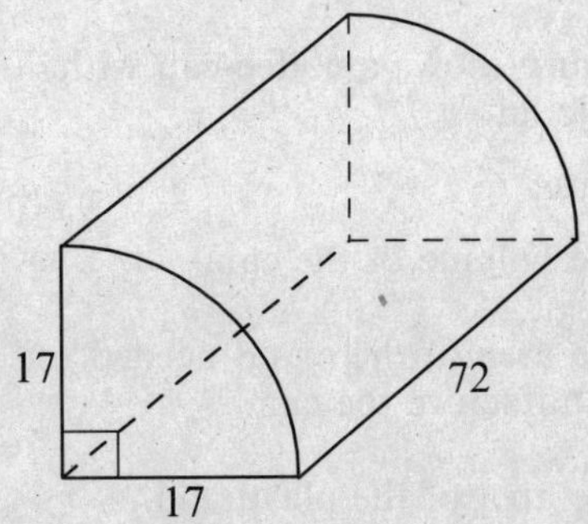

3. A wooden cylinder has a square hole bored out of it as shown in the diagram.

Calculate the **volume** of the remaining wood.

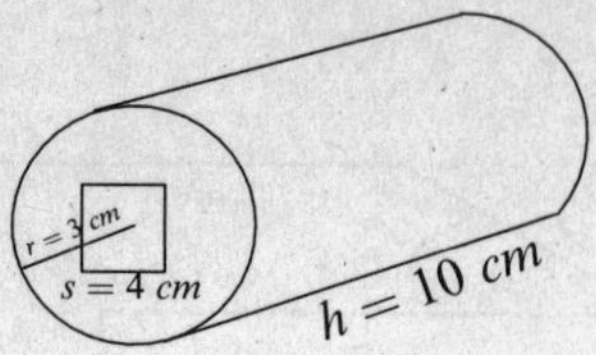

4.1 Calculate x and y.

4.2 Which of the containers A or B has the greater volume?

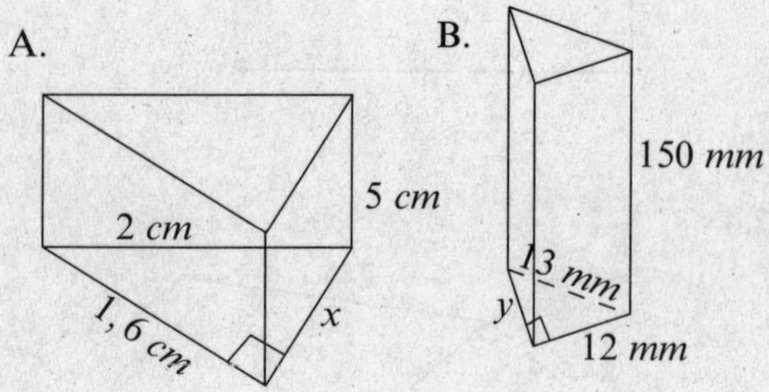

5. The diagram represents a tent.

Determine the total area of canvass (material) used to make the tent.

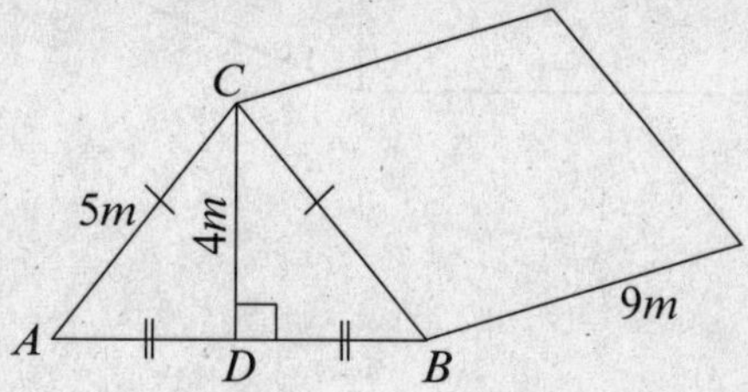

6 Find the volume of the trough in the diagram.

7. The figure shows a coffee-can with a plastic lid.

Calculate:

7.1 The volume of the can.

7.2 The area of the metal needed to manufacture the can.

7.3 The area of the plastic lid.

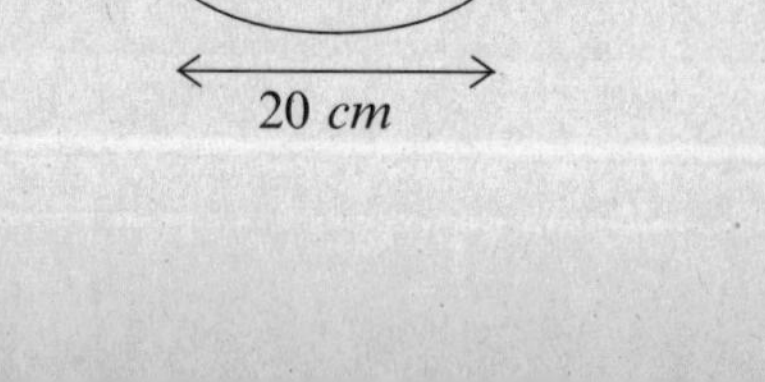

8. Find the total external area of the following figure:

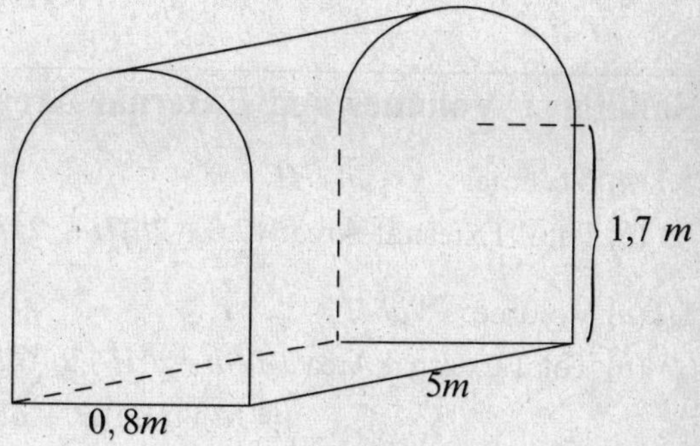

9. Determine the area of the carpet on the stairs:

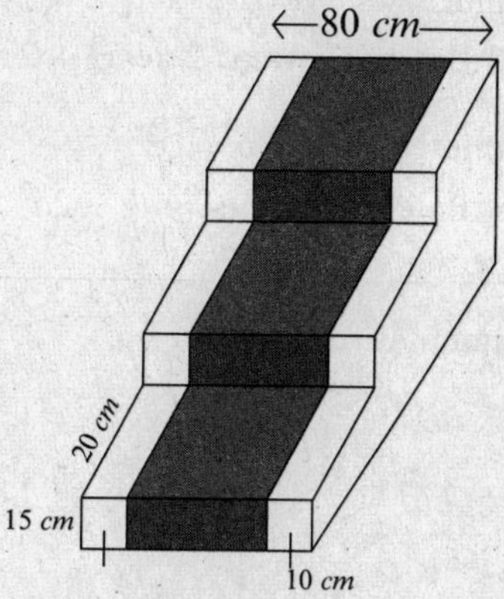

10. Twelve condensed milk cans are packed into a box. Each can has a diameter of 4 cm it also has a height of 4 cm. Find the volume of the remaining space in the box.

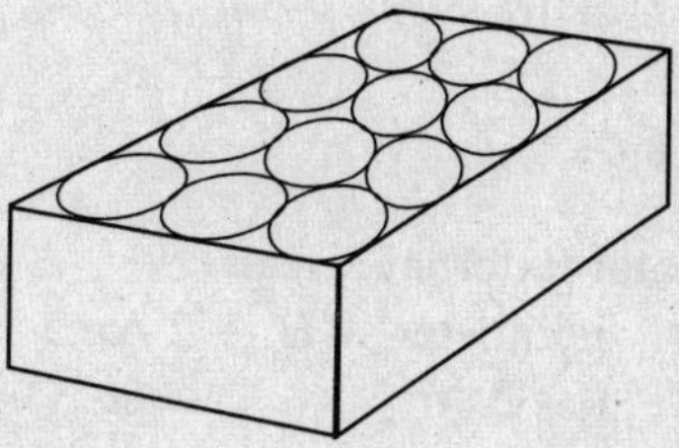

SOLUTIONS

Solutions: Volumes and External Areas

1.1 (*a*) Volume: $x \times b \times H$

(*b*) Total External Area: $(2x + 2b)H + 2xb$

1.2 (*a*) Volume: $\frac{1}{2}abH$

(*b*) Total External Area: $(a+b+c)H + \cancel{2}.\frac{1}{\cancel{2}}ab$

$= (a+b+c)H + ab$

1.3 (*a*) Volume: $\pi a^2 H$

(*b*) Total External Area: $2\pi a H + 2\pi a^2$

1.4 (*a*) Volume: $a \times a \times b = a^2$

(*b*) Total External Area: $4ab + 2a^2$

2.1 (*a*) Volume: Area × H
base: Δ

$= (\frac{1}{2}b.h) \times H$

$= \frac{1}{2} \times 4 \times 3 \times 7$

$= 42\ cm^3$

(*b*) $AB^2 = 4^2 + 3^2$ (Pyth.)

$= 16 + 9$

$= 25$

$AB = 5$

∴ Total External Area:

= Perimeter × H + 2 Area
base Δ base Δ

$= (3 + 4 + 5)7 + 2.\frac{1}{2}.4.3.$

$= 84 + 12$

$= 96\ cm^2$

2.2 Volume: Area × H $= \pi 3,5^2 \times 7,5$
⊙ $= 288,6\ cm^3$

Total External Area:

Perimeter × H + 2 Area
⊙ ⊙

$= 2\pi 3,5 \times 7,5 + 2\pi 3,5^2$

$= 241,9\ cm^2$

2.3 Volume: Area × H $= 2,5 \times 6,5 \times 3$
□ $= 48,75\ cm^3$

Total External Area:

Perimeter × H + 2 Area
□ □

$=(2,5+6,5)2\times3+2\times2,5\times6,5$

$= 54 + 32,5 = 86,5\ cm^2$

2.4 $AB^2 = 8^2 + 8^2 = 128$ (Pyth.)

$\therefore AB = 11,3$

Volume: Area × H
⌂

= (Area: □ + Area Δ) × H

$= (5 \times 11,3 + \frac{1}{2} \times 8 \times 8) \times 24$

$= 88,5 \times 24$

$= 2124\ cm^3$

Total External Area:

Perimeter × H + 2 Area
⌂ ⌂

$= (5 + 11,3 + 5 + 8 + 8)24 +$
$2(5 \times 11,3 + \frac{1}{2} \times 8 \times 8)$

$= 1\,072,2\ cm^2$

2.5 $AB^2 = 17^2 - 15^2$ (Pyth.)

$= 64$

$AB = 8$

Volume: Area × H $= 15 \times 8 \times 20$
□ $= 2\,400\ cm^3$

Total External Area:

Perimeter × H + 2 Area
□ □

$= 2(15 + 8) \times 20 + 2 \times 15 \times 8$

$= 1160\ cm^2$

2.6 $AB^2 = 34^2 - 16^2$
$= 900$
$AB = 30$

Volume: Area $\Delta \times H$
$= \frac{1}{2}.30 \times 16 \times 12$
$= 2\,880\ cm^3$

Total External Area:
Perimeter $\times H + 2$ Area
$\Delta \qquad \Delta$
$= (34 + 30 + 16) \times 12 + 2\frac{1}{2}.30 \times 16$
$= 960 + 480$
$= 1440\ cm^2$

2.7 Volume: Area $\times H$
$= \frac{1}{2}.4a.3a \times 8a$
$= 48a^3\ cm^3$

$AB = 5a$ (using. Pyth.)

Total External Area:
Perimeter $\times H + 2$ Area
$= (3a + 4a + 5a)8a +$
$2 \times \frac{1}{2} \times 4a \times 3a$
$= 12a \times 8a + 12a^2$
$= 96a^2 + 12a^2$
$= 108a^2 cm^2$

2.8 Volume: Area $\times H$
$=^*(28 \times 28 + \frac{1}{2}\,\pi\,14^2)85$
$= 92\,809,5\ cm^3$

Total External Area:
Perimeter $\times H + 2$ Area
$= (28 + 28 + 28 + \frac{1}{2}2.\,\pi\,14)85$
$+2^*(28 \times 28 + \frac{1}{2}\,\pi\,14^2)$
$= 13\,062,2\ cm^2$

2.9 Radius Semi-circle is: $\frac{7}{2} = 3,5$

Volume: Area $\times H$
$= (15 \times 8 - \frac{1}{2}.\pi\,3,5^2) \times 60$
$= 6\,045,5\ cm^3$

Total External Area:
Perimeter $\times H + 2$ Area
$= (4+8+15+8+4+\frac{1}{2}.2\,\pi\,3,5)60.$
$+2(15 \times 8 - \frac{1}{2}\,\pi\,3,5^2)$
$= 3\,201,2\ cm^2$

2.10 Volume: Area $\times H$
$= \frac{1}{4}.\,\pi\,17^2 \times 72$
$= 16\,342,6\ cm^3$

Total External Area:
Perimeter $\times H + 2$ Area
$= (17 + 17 + \frac{1}{4}.2\,\pi\,17)72+$
$2(\frac{1}{4}.\,\pi\,17^2)$
$= 4\,824,6\ cm^2$

3. Volume of Cylinder: $\pi r^2 H$
$= \pi\ 3^2 \times 10$
$= 282,74\ cm^3$

Volume of square hole:
$4 \times 4 \times 10$
$= 160\ cm^3$

Volume of remaining wood:
$122,74\ cm^3$

4.1 $x^2 = 2^2 - 1,6^2$
$= 4 - 2,56$
$= 1,44$
$x = 1,2$

$y^2 = 13^2 - 12^2$
$= 169 - 144$
$= 25$
$y = 5$

Pythagoras:

4.2 Volume A: $\frac{1}{2}b.h. \times H$

$= \frac{1}{2}1,6 \times 1,2 \times 5$

$= 4,8\ cm^3$

Volume B: $= \frac{1}{2}b.h.H$

$= \frac{1}{2}.5 \times 12 \times 150$

$= 4\ 500\ mm^3$

$= 4,5\ cm^3$

$\therefore$ A's volume is greater.

5. $AD^2 = 5^2 - 4^2$ (Pyth.)

$= 9$

$AD = 3$

Total Area:

Perimeter $\times H + 2 \times$ Area

$\triangle$ $\triangle$

$= (5+5+6)9 + 2\frac{1}{2}.6 \times 4$

$= 168\ m^2$

6. Volume: Area $\times H$

⌓

$= \frac{1}{2}.\pi r^2 H$

$= \frac{1}{2}.\pi 6^2 \times 18$

$= 1\ 017,9\ cm^3$

7.1 Volume can: Area $\times H$

$\odot$

$= \pi 10^2 \times 15$

$= 4\ 712,4\ cm^3$

7.2 External area (without lid):

Perimeter $\times H +$ Area

$\odot$ $\odot$

$= 2\ \pi\ 10 \times 15 + \pi\ 10^2$

$= 1\ 256,6\ cm^2$

7.3 Area plastic lid:

Area $\times H +$ Area

$\odot$ $\odot$

$= 2\ \pi\ 10 \times 1 + \pi\ 10^2$

$= 314,2\ cm^2$

8. Total External Area:

Perimeter $\times H + 2$ Area

$\cap$ over $\square$ $\square$ + ◠

$= (1,7 + 0,8 + 1,7 + \frac{1}{2}.2\ \pi\ 0,4)5 +$

$2(1,7 \times 0,8 + \frac{1}{2}\ \pi\ 0,4^2)$

$= 30,5\ cm^2$

9. Area: $3\times$ ▭ $3 \times l \times b$

$= 3 \times 60 \times 15$

$= 2\ 700 cm^2$

Area: $3\times$ ▱ $3 \times l \times b$

$= 3 \times 60 \times 20$

$= 3\ 600 cm^2$

Total Area: $= 6\ 300\ cm^2$

10. Volume of box: $l \times b \times H$

$= 16 \times 12 \times 4$

$= 768\ cm^3$

Volume: 1 can: $\pi r^2 H$

$= \pi\ 2^2 \times 4$

$= 50,3\ cm^3$

Volume: 12 cans: $= 603,2\ cm^3$

Remaining volume: $= 164,8\ cm^3$

13 *Geometry: General Revision (Std 6)*

1. Angles:

1. Acute angle:

$O^\circ < \hat{B} < 90^\circ$

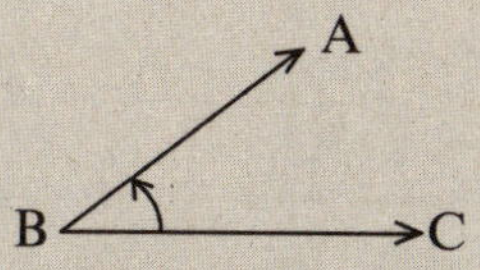

2. Obtuse angle:

$90^\circ < \hat{P} < 180^\circ$

3. Reflex angle:

$180^\circ < \hat{Q} < 360^\circ$

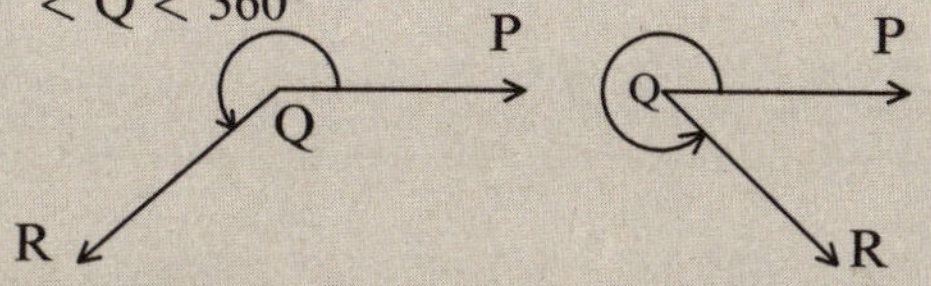

4. Revolution:

$\hat{B} = 360^\circ$

5. Straight angle:

$A\hat{B}C = 180^\circ$

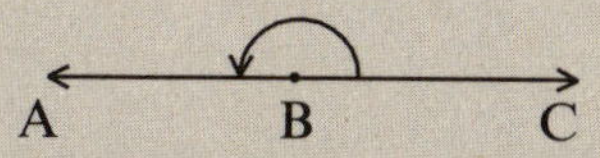

6. Right angle:

$\hat{Y} = 90^\circ$

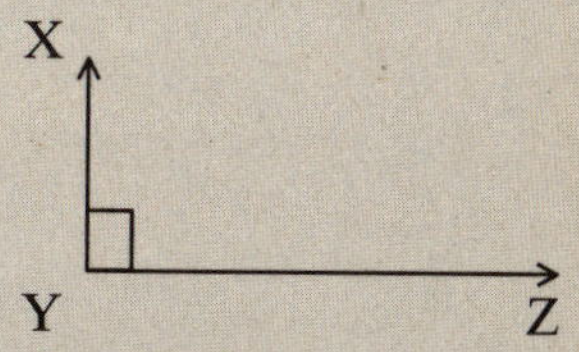

7. Complementary angles:

$\hat{A} + \hat{B} = 90^\circ$; then $\hat{A}$ and $\hat{B}$ are

Complementary angles
Complement of 20° is 70°
Complement of 42° is 48°
Complement of x is $90^\circ - x$
Complement of $90 - 2y$ is $2y$

Complementary adjacent angles

$\hat{A}_1 + \hat{A}_2 = 90^\circ$
(N.B. not necessarily 45° each)

$\hat{A}_1$ and $\hat{A}_2$ are complementary adjacent angles.

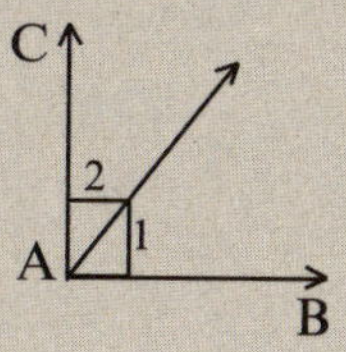

8. Supplementary angles:

$\hat{P} + \hat{Q} = \mathbf{180^\circ}$; then $\hat{P}$ and $\hat{Q}$ **Supplementary** angles.
Supplement of 30° is 150°
Supplement of 110° is 70°
Supplement of 134° is 46°
Supplement of 38° is 142°
Supplement of x is $180^\circ - x$
Supplement of $90 - x$ is $90 + x$

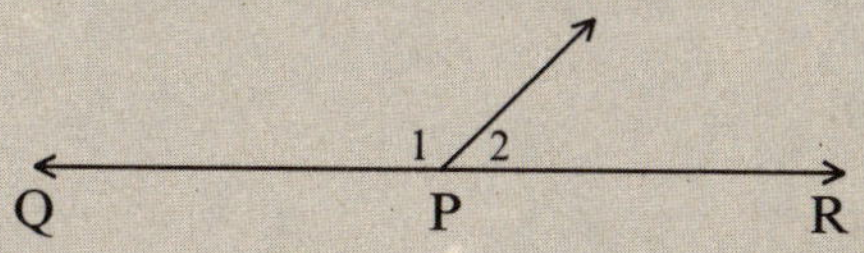

$\hat{P}_1 + \hat{P}_2 = 180^\circ$ (QPR is a straight line)

$\hat{P}_1$ and $\hat{P}_2$ are Supplementary adjacent angles.

9. Vertically opposite angles:

$\hat{O}_1 = \hat{O}_3$ (Vertically opposite angles)

$O_2 = \hat{O}_4$ (Vertically opposite angles)

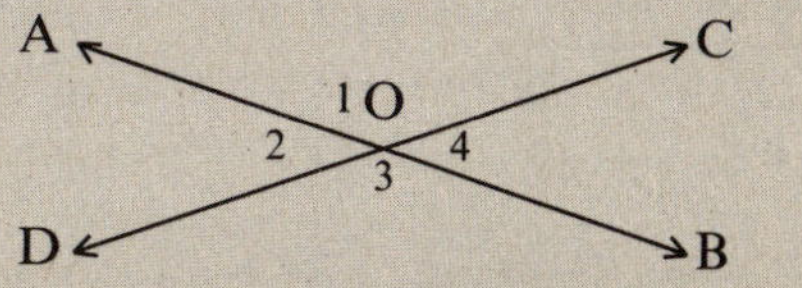

2. Parallel lines:

1. Corresponding angles:

$\hat{1} = \hat{2}$ (Corresponding $\angle^s$ AB||CD)

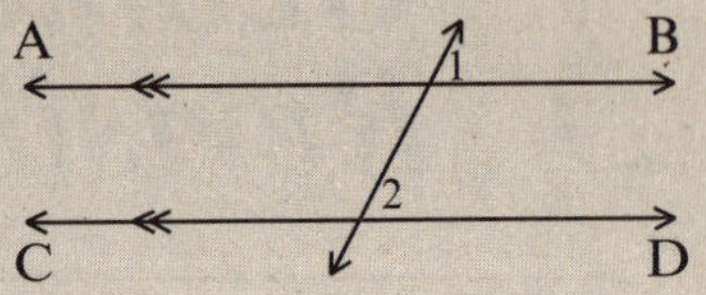

2. Alternate angles:

$\hat{1} = \hat{2}$ (Alternate $\angle^s$; PQ||RS)

$\hat{3} = \hat{4}$ (Alternate $\angle^s$; PQ||RS)

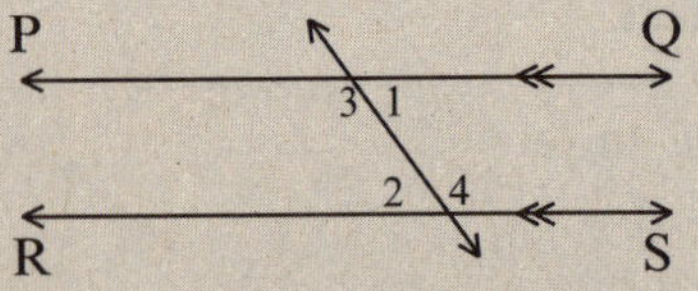

3. Co-interion angles:

$\hat{1} + \hat{3} = 180°$ (co-interior $\angle^s$; AB||DE)
$\hat{2} + \hat{4} = 180°$ (co-interior $\angle^s$; AB||DE)

Co-interior angles are supplementary.

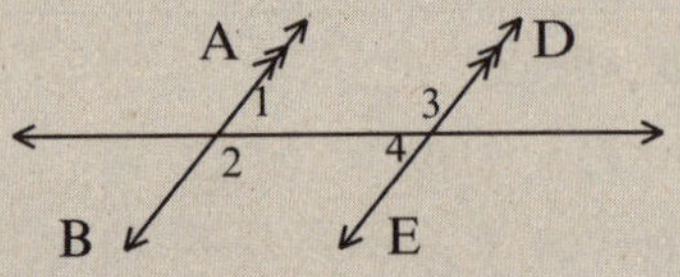

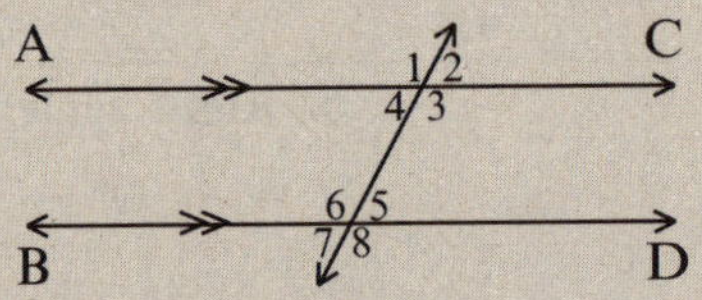

FUN

Corresponding angles
(equal because AC||BD)

$\hat{1}$ and $\hat{6}$

$\hat{4}$ and $\hat{7}$

$\hat{2}$ and $\hat{5}$

$\hat{3}$ and $\hat{8}$

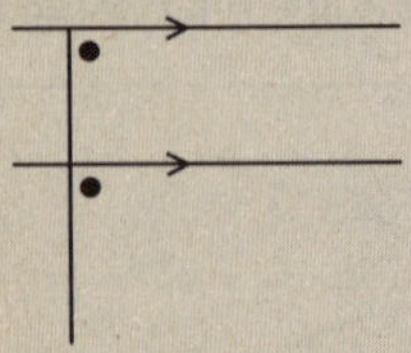

Corresponding angles

Co-interior angles
(supplementary because AC||DB)

$\hat{3}$ and $\hat{5}$

$\hat{4}$ and $\hat{6}$

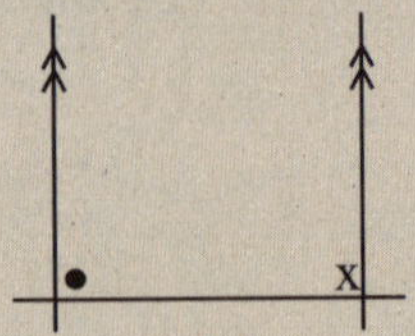

Co-interior angles

Alternate angles
(equal because AC||DB)

$\hat{4}$ and $\hat{5}$

$\hat{3}$ and $\hat{6}$

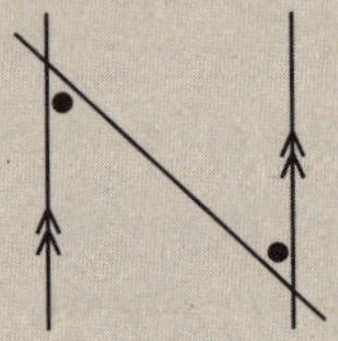

Alternate angles

3. Triangles:

1. Scalene triangle:

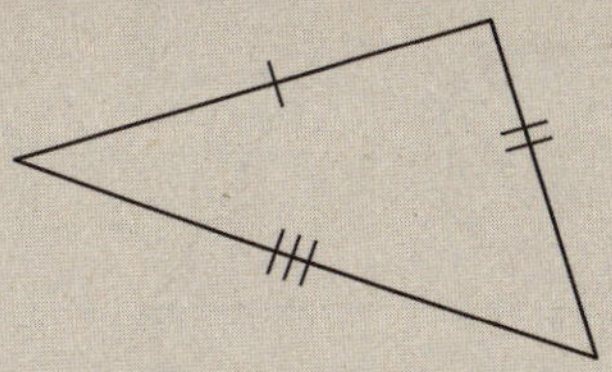

All sides different.

Largest angle opposite longest side.

2. Isosceles triangle:

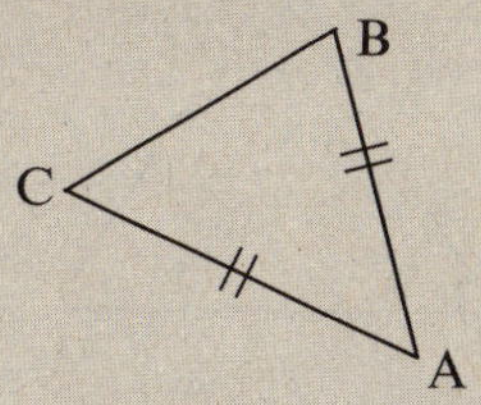

2 sides are equal eg. AB = AC.

Base angles are equal; thus $\hat{B} = \hat{C}$.

3. Equilateral triangle:

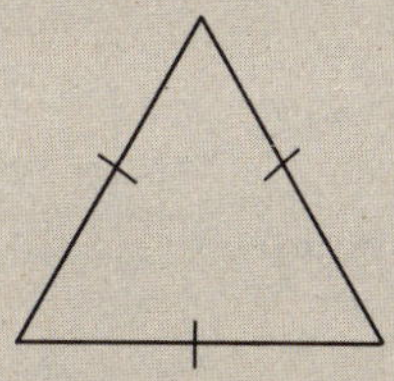

All sides equal.

All angles 60°

4. Acute-angled triangle:

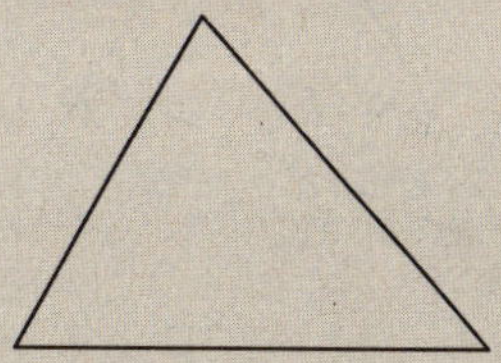

All the angles are acute, i.e. less than 90°

5. Obtuse-angled triangle:

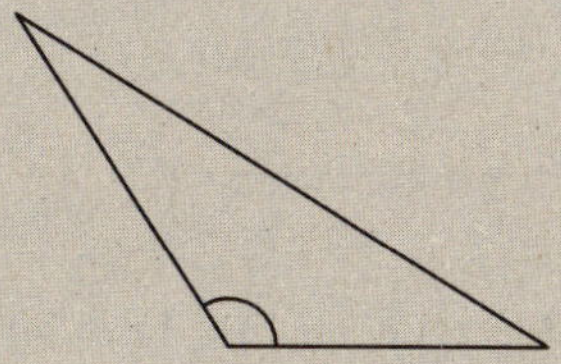

One angle is obtuse.

6. Right-angled triangle:

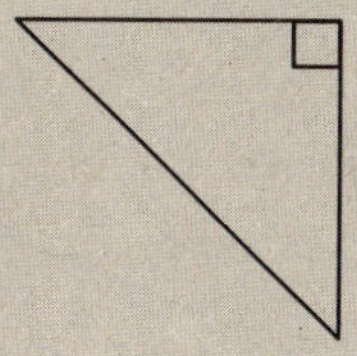

One angle is 90°

4. Properties of triangles:

1. Sum of the interior angles of a triangle is 180°.

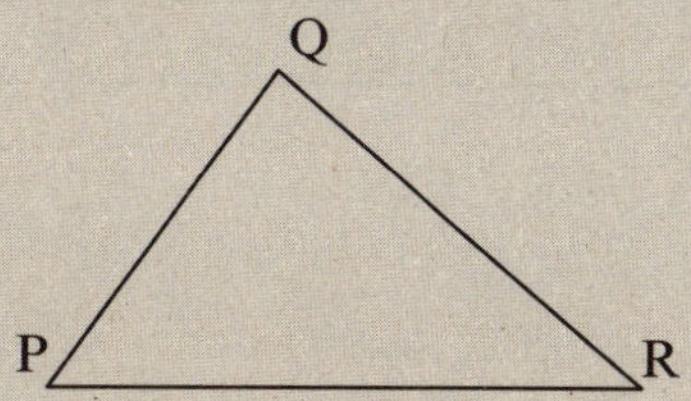

$$\hat{P} + \hat{Q} + \hat{R} = 180°$$

2. Exterior angle of a triangle is equal to the sum of the 2 opposite interior angles.

(*a*)

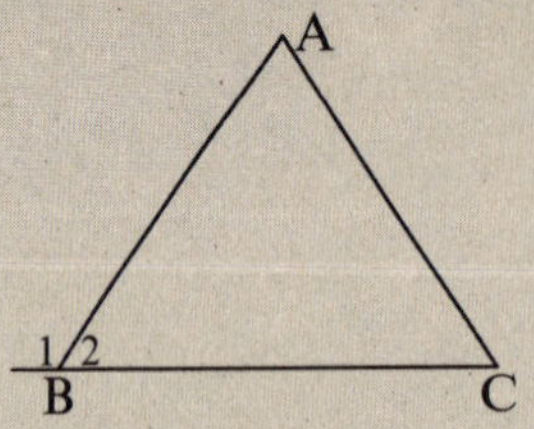

$\hat{B}_1 = \hat{A} + \hat{C}$

(*b*)

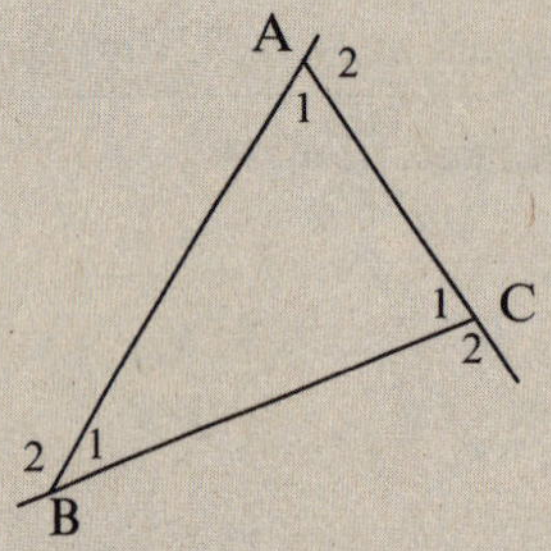

$\hat{A}_2 = \hat{B}_1 + \hat{C}_1$

$\hat{C}_2 = \hat{A}_1 + \hat{B}_1$

$\hat{B}_2 = \hat{A}1 + \hat{C}_1$

(c)

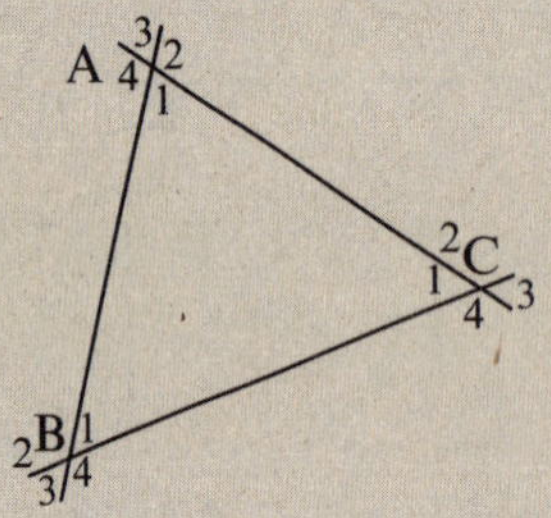

$\hat{A}_1 + \hat{B}_1 + \hat{C}_1 = 180°$ (3∠^{s} of Δ ABC = 180°)

$\hat{C}_1 = \hat{C}_3$ (Vertically opp.∠^{s})

$\hat{B}_2 = \hat{A}_1 + \hat{C}_1$ (Exterior ∠ of Δ ABC)

$\hat{C}_4 = \hat{A}_1 + \hat{B}_1$ (Exterior ∠ of Δ ABC)

$\hat{A}_2 = \hat{C}_1 + \hat{B}_1$ (Exterior ∠ of Δ ABC)

3. The base angles of an isosceles triangle are equal.

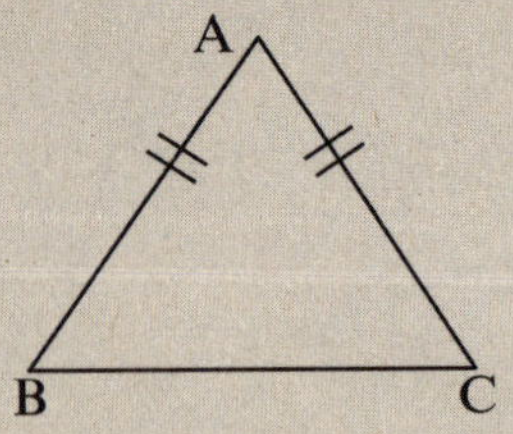

$\hat{B} = \hat{C}$ (AB = AC)

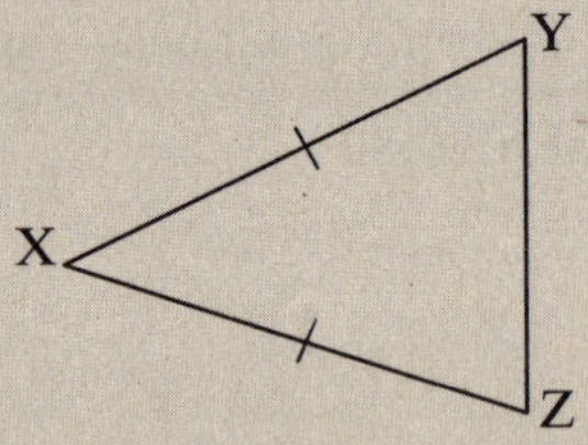

$\hat{Y} = \hat{Z}$ (XY = XZ)

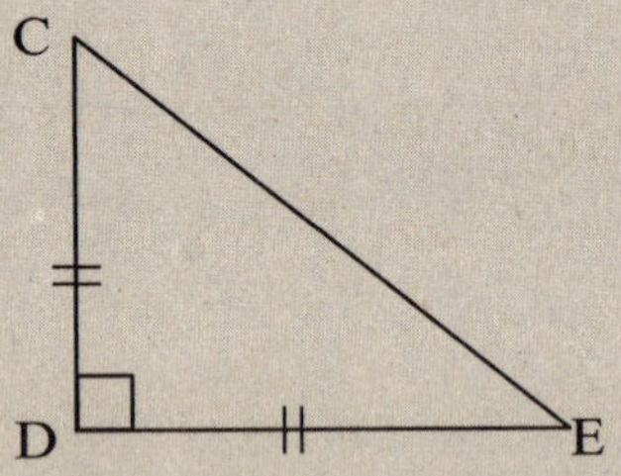

$\hat{C} = \hat{E} = 45°$ (DC = DE; $\hat{D} = 90°$)

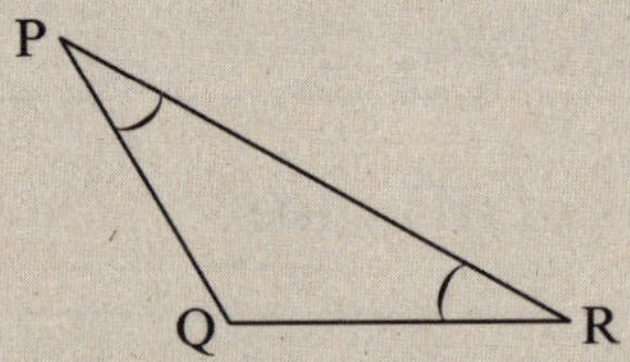

PQ = QR ($\hat{P} = \hat{R}$)

5. Theorem of Pythagoras

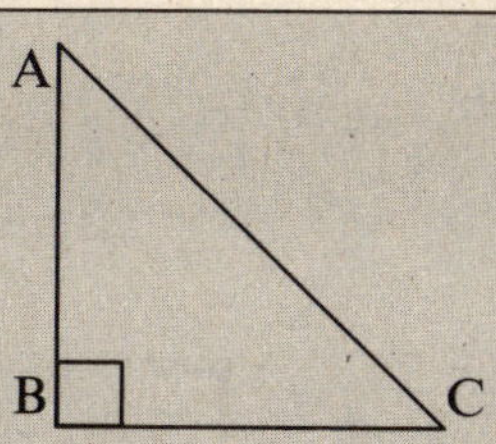

In a **right-angled triangle**, *the* **square** *on the* **hypotenuse** *is equal to* **the sum** *of the* **squares** *on the* **other two sides** *(right-angle sides).*

$AC^2 = AB^2 + BC^2$

Remember:

1. Hypotenuse is opposite 90°-angle.
2. Hypotenuse is always the longest side.
3. Pythagoras uses *sides* only.
4. Pythagoras may only be applied to a 90°-triangle.

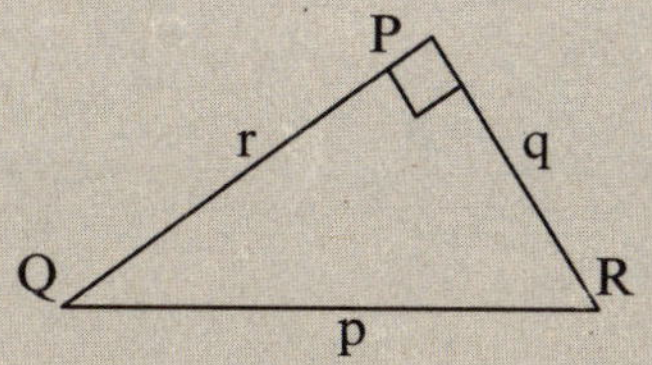

$q^2 = p^2 - r^2$

$r^2 = p^2 - q^2$

$p^2 = q^2 + r^2$

Examples:

1. Calculate x, y and z.

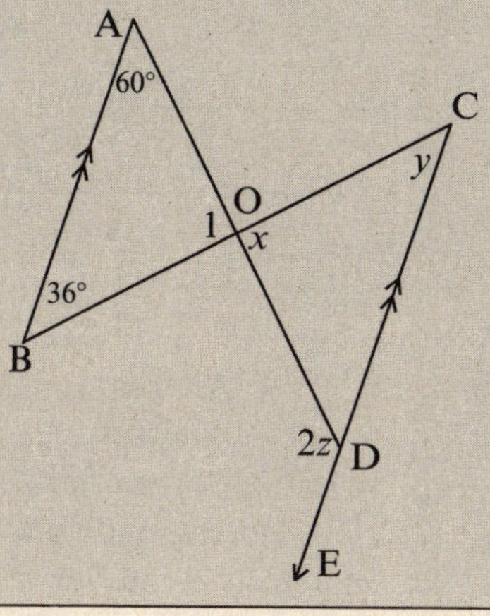

Reasons are very important in the solutions:

$\hat{1} = 180° - (60° + 36°)$ ($\angle^s$ of Δ AOB = 180°)
$= 84°$

$x = \hat{1}$ (Vertically opp. $\angle^s$)
$= 84°$

$y = \hat{B}$ (Alternate. $\angle^s$ AB||CD)
$= 36°$

$2z = x + y$ (Exterior$\angle$ of Δ COD)
$= 120°$

$z = 60°$

2. Calculate x, y and z.

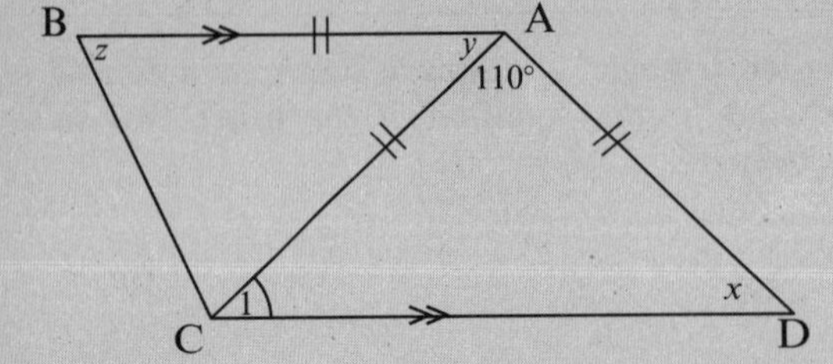

$x = \dfrac{180° - 110°}{2}$ ($3\angle^s$ of Δ ACD = 180°; AC = AD)

$= 35°$

$y = \hat{C}_1$ (Alternate $\angle^s$ AB$\|$CD)

$= 35°$

$z = \dfrac{180° - 35°}{2}$ ($3\angle^s$ of Δ ABC = 180°; AB = AC)

$= 72,5°$

3. Calculate x.

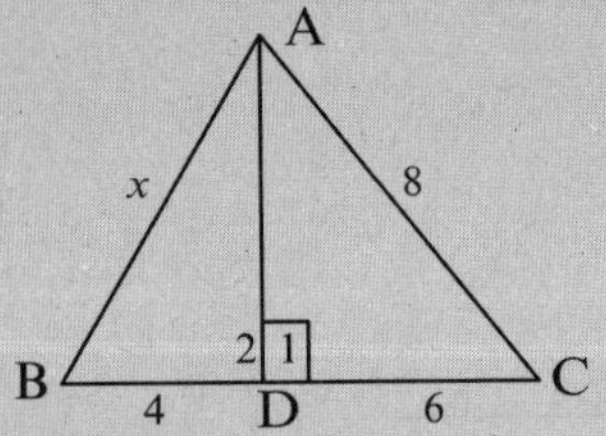

$AD^2 = 8^2 - 6^2$ (Pyth. $\hat{D}_1 = 90°$)

$= 64 - 36$

$= 28$

$x^2 = 4^2 + AD^2$ (Pyth. $\hat{D}_2 = 90°$)

$= 16 + 28$

$= 44$

$\therefore x = \sqrt{44}$

$= 6,63$

Exercise: Revision Std. 6 Geometry

1. Use the diagram and complete:

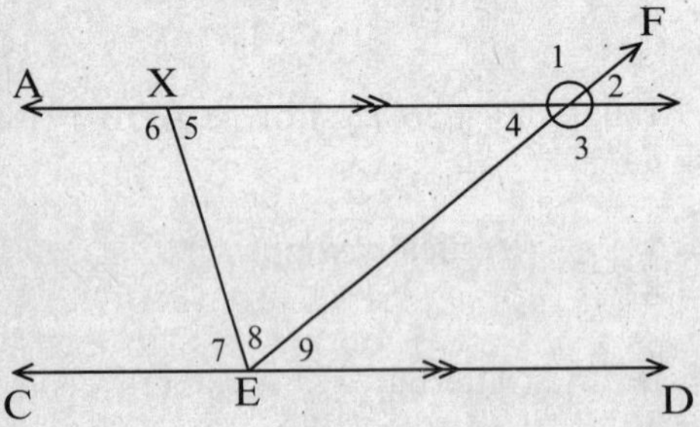

1.1 $\hat{2} = \hat{4}$ (. . . .)

1.2 $\hat{1} = \hat{3}$ (. . . .)

1.3 $\hat{4} = \hat{9}$ (. . . .)

1.4 $\hat{5} = \hat{7}$ (. . . .)

1.5 $\hat{6} + \hat{7} =$ . . . . ° (. . . . .)

1.6 $\hat{2} = \hat{9}$ (. . . .)

1.7 $\hat{7} + \hat{8} + \hat{9} =$ . . . . ° (. . . . .)

1.8 $\hat{4} + \hat{5} + \hat{8} =$ . . . . ° (. . . . .)

2. Use the diagram and complete:

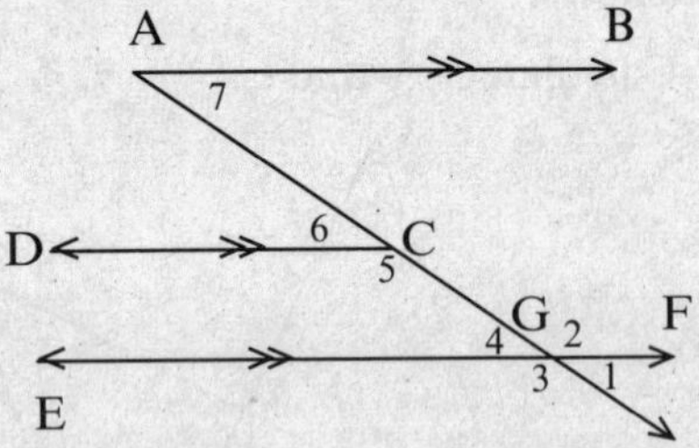

2.1 $\hat{7} =$ . . . . (Alternate $\angle^s$ AB$\|$CD)

2.2 $\hat{7} =$ . . . . (Alternate $\angle^s$; . . . .)

2.3 $\hat{4} +$ . . . . = 180° (Co-interior $\angle^s$; . . . .)

2.4 $\hat{3} =$ . . . . (Corresponding $\angle^s$; . . . .)

2.5 $\hat{7} +$ . . . . = 180° (Co-interior $\angle^s$; . . . .)

3. Calculate x; y and z (step by step, with reasons):

(*a*)

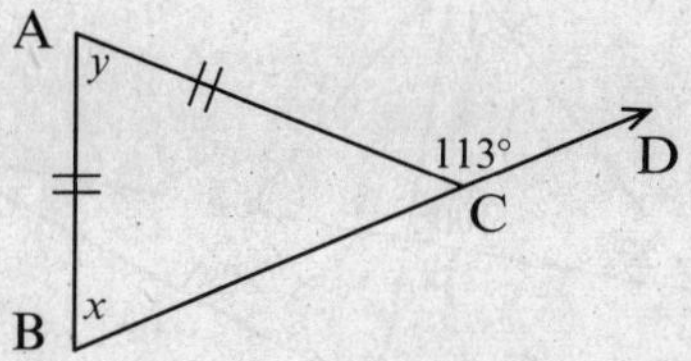

(*b*)

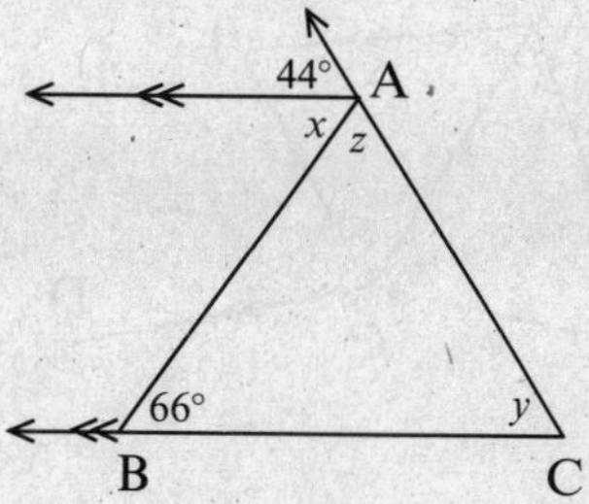

(*c*)

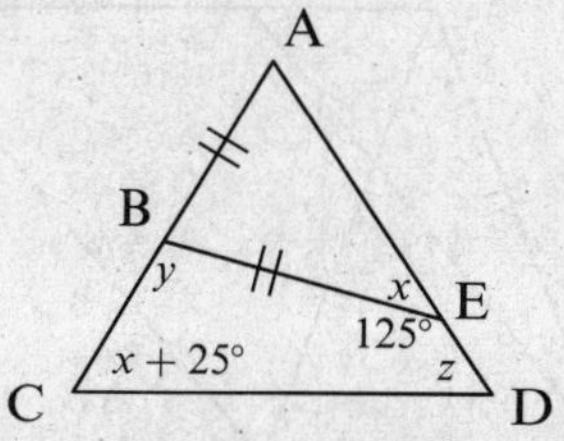

(*d*)

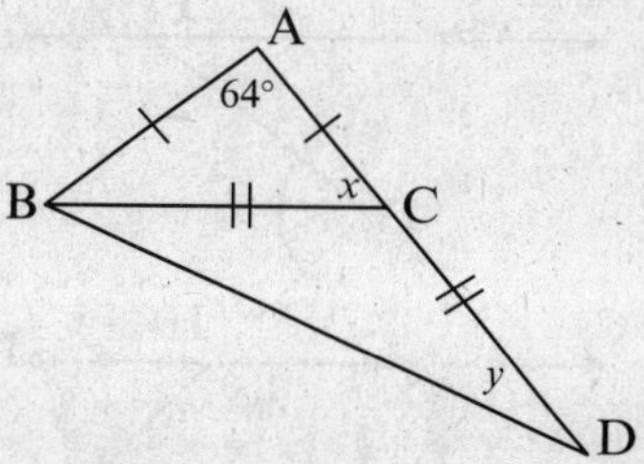

(*e*)

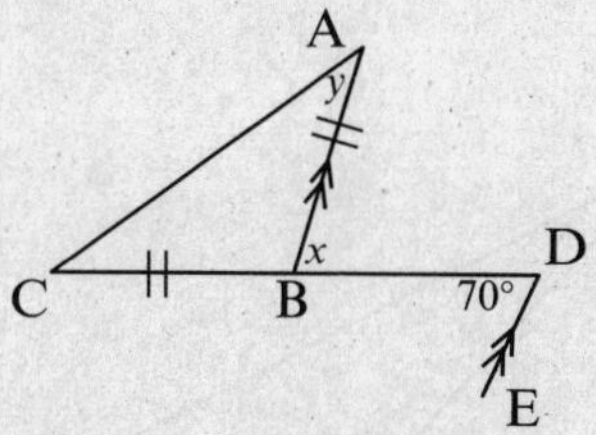

(*f*)

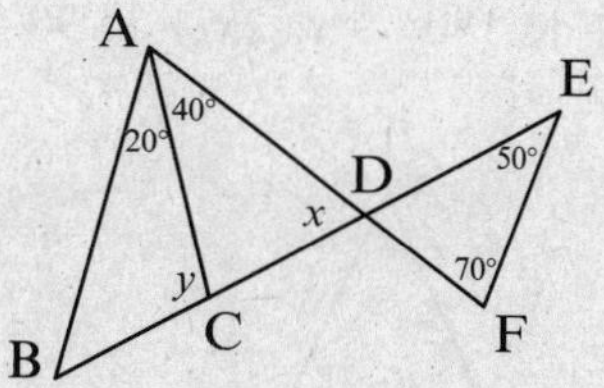

(*g*)

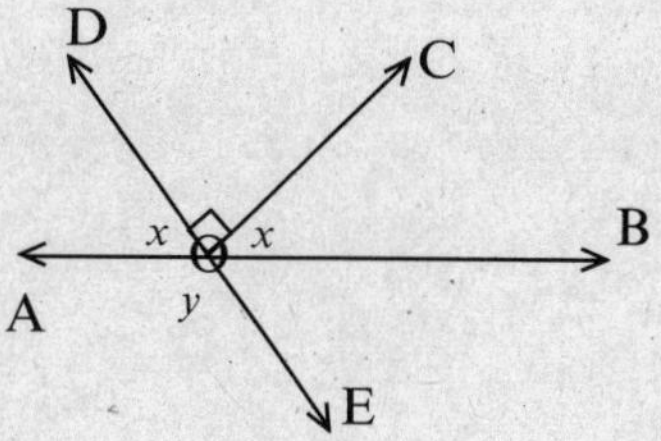

(*h*)

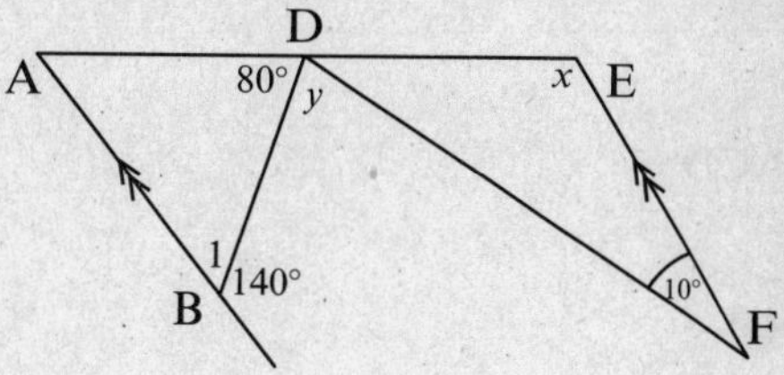

(*i*)

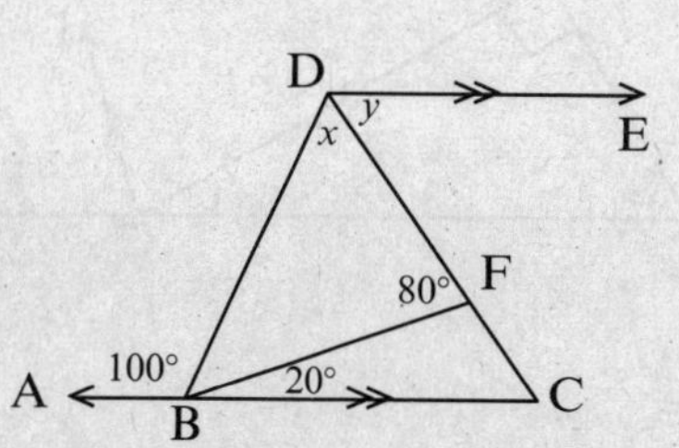

(*j*)

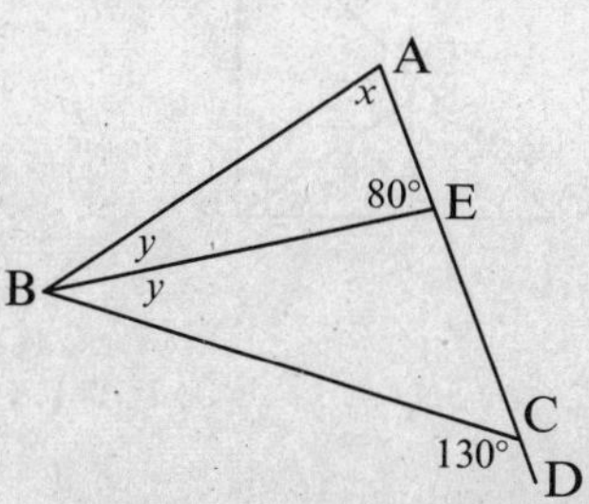

(*k*)

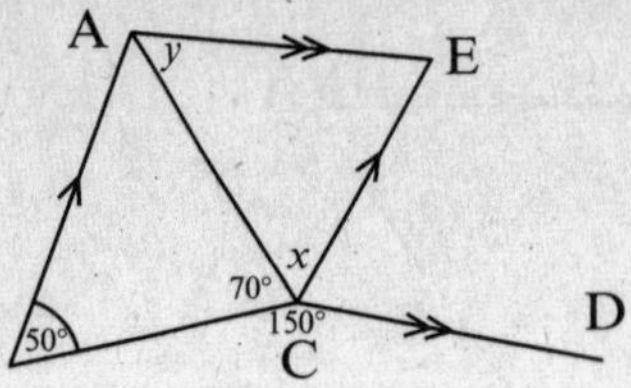

(*l*)

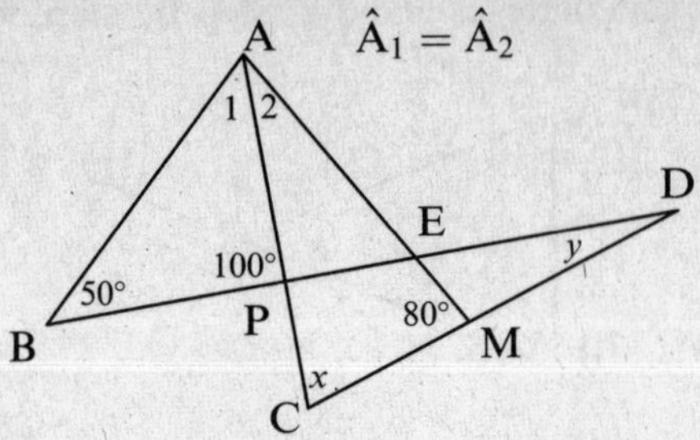

(*m*)

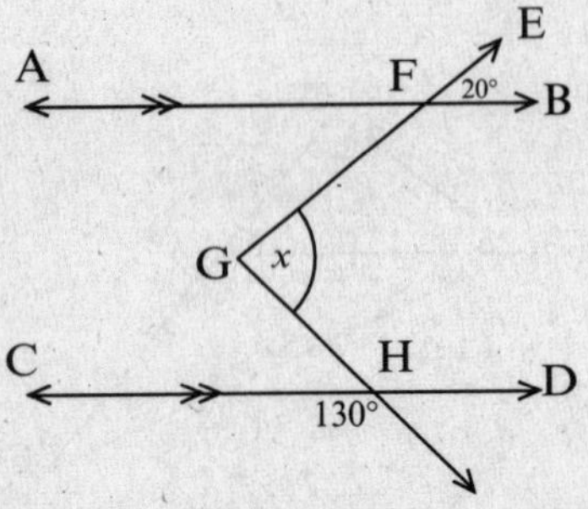

(*n*)

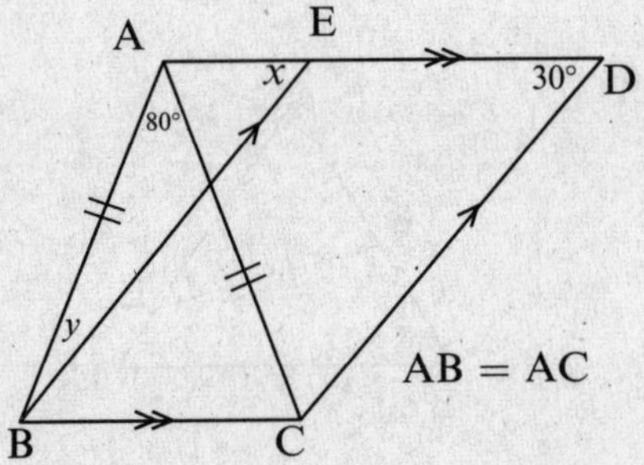

---

4. Express each of the following angles in terms of x and/or y. (with reasons)
(eg. $x + y$; $180° - x$; $2x + y$; $90 + x$; etc.)

4.1

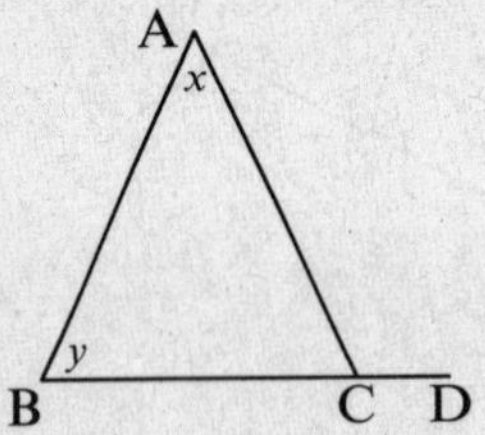

$A\hat{C}D =$ ....(........)

4.2

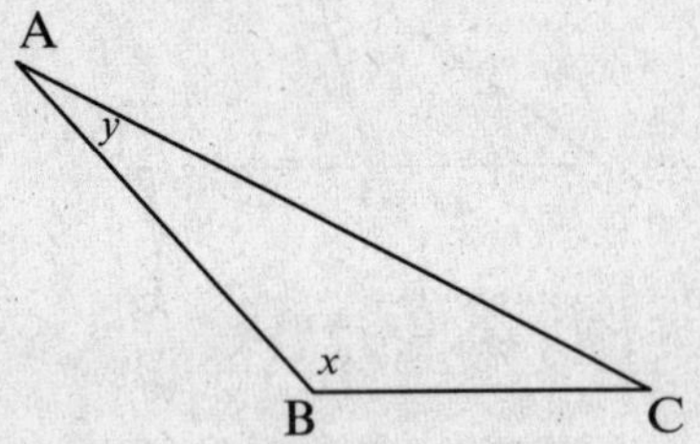

$\hat{C} =$ ....(........)

4.3

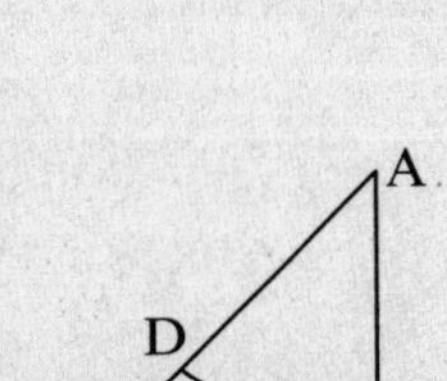

$\hat{C} =$ ....(........)

$D\hat{B}A =$ ....(........)

4.4

$E\hat{C}D =$ ....(........)

$\hat{C}_1 =$ ....(........)

4.5

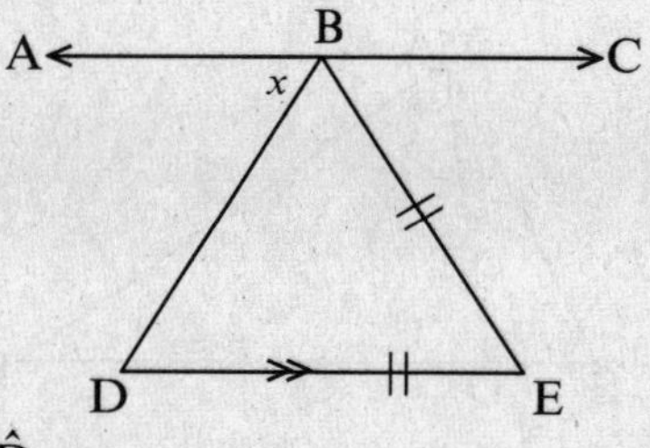

$\hat{D}$ = ...(.........)

$\hat{E}$ = ...(.........)

4.6

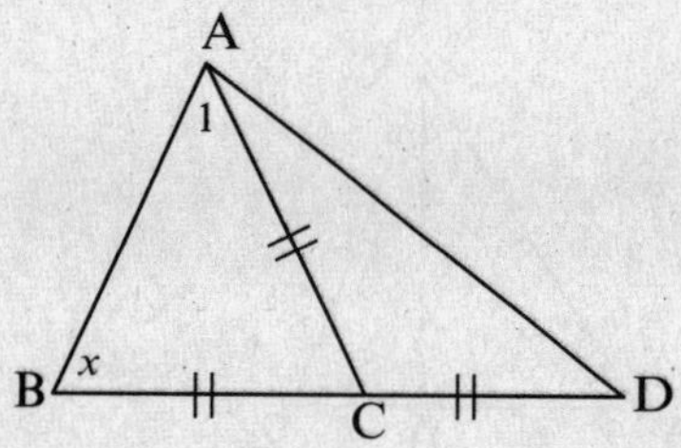

$\hat{A}_1$ = ...(.........)

$A\hat{C}D$ = ...(.........)

$\hat{D}$ = ...(.........)

4.7

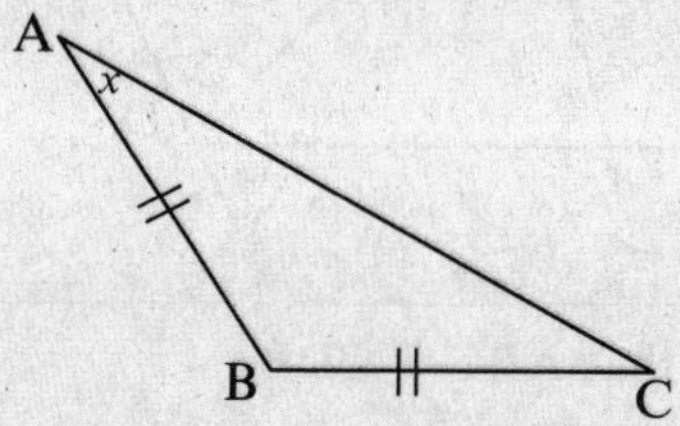

$\hat{B}$ = ...(.........)

4.8

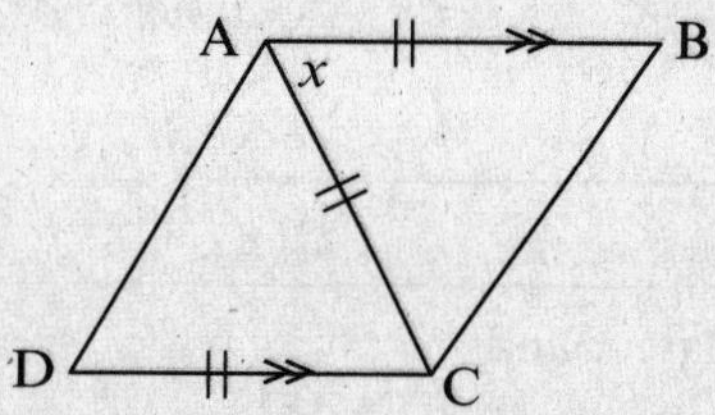

$A\hat{C}D$ = ...(.........)

$\hat{D}$ = ...(.........)

4.9

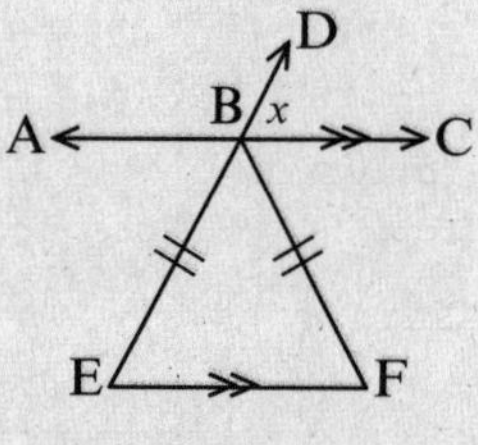

$\hat{E}$ = ...(.........)

$E\hat{B}F$ = ...(.........)

4.10

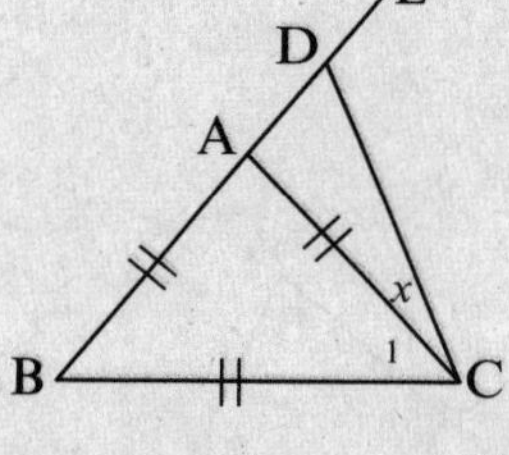

$E\hat{D}C$ = ...(.........)

---

5. Calculate the size of the angles marked by variables (unknown) (completely, with reasons).

5.1

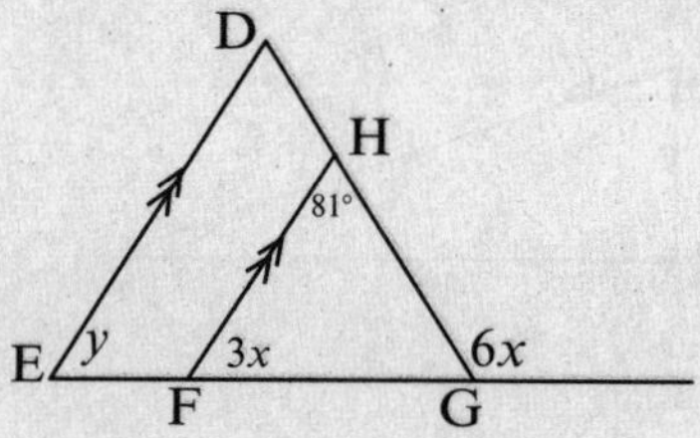

5.2

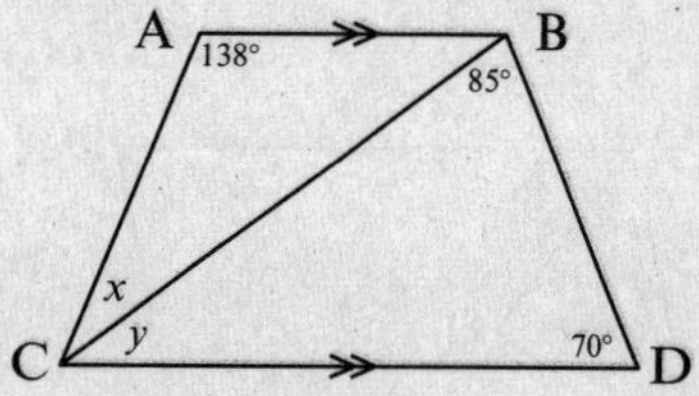

5.3

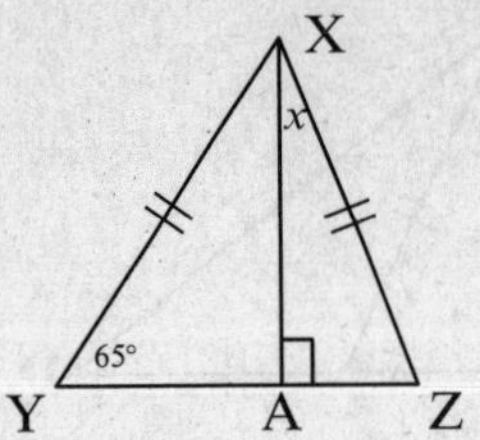

5.4

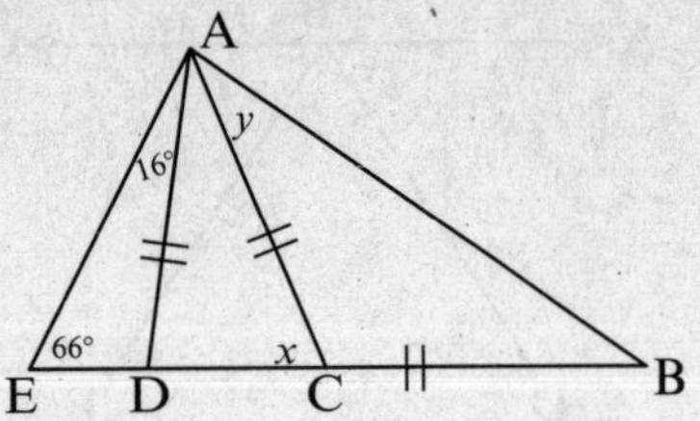

5.5

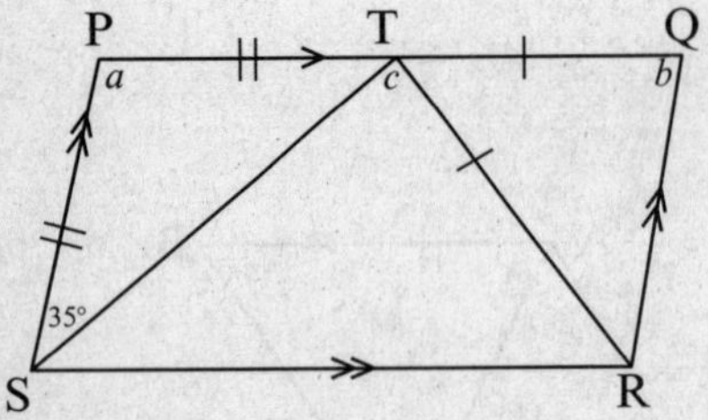

5.6

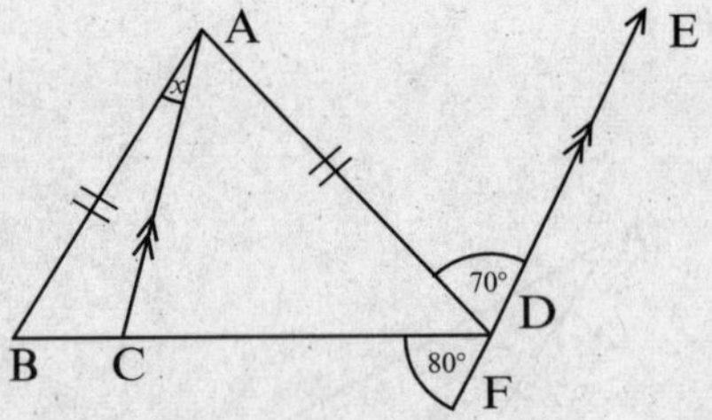

6. Calculate x and y:

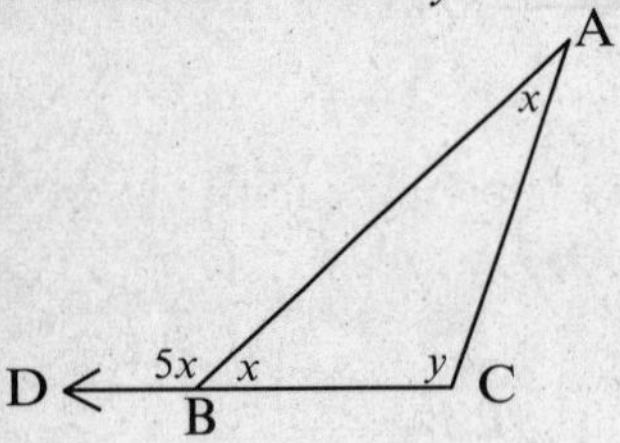

7. Calculate x and y:

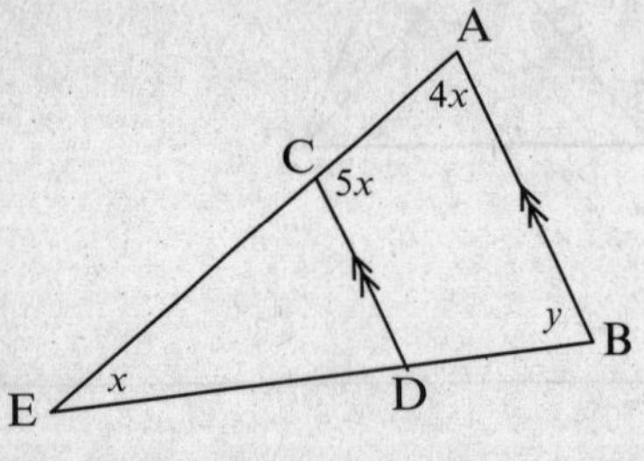

8. Calculate $A\hat{O}B$

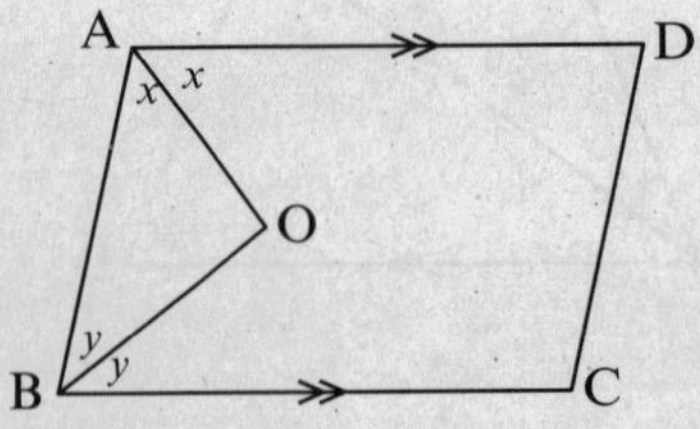

9. Prove that $A\hat{B}C = C\hat{D}B$

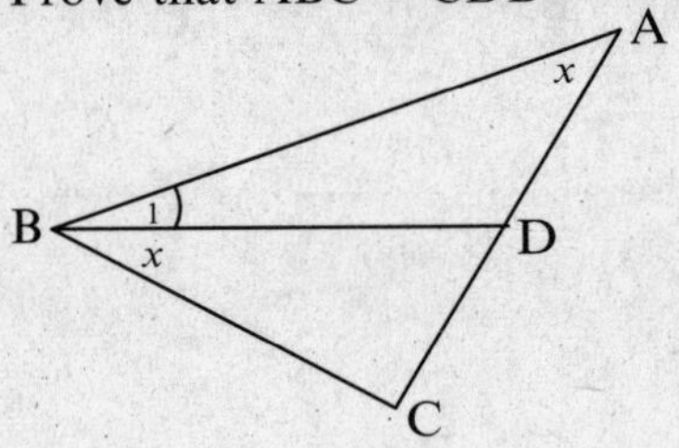

10. Calculate x; a; y

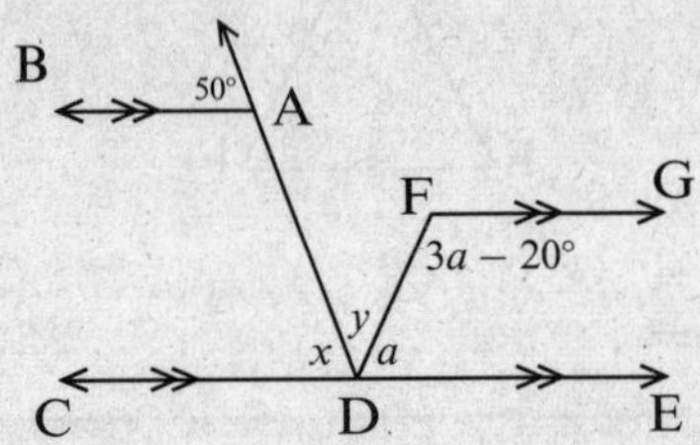

11. Calculate x and y:

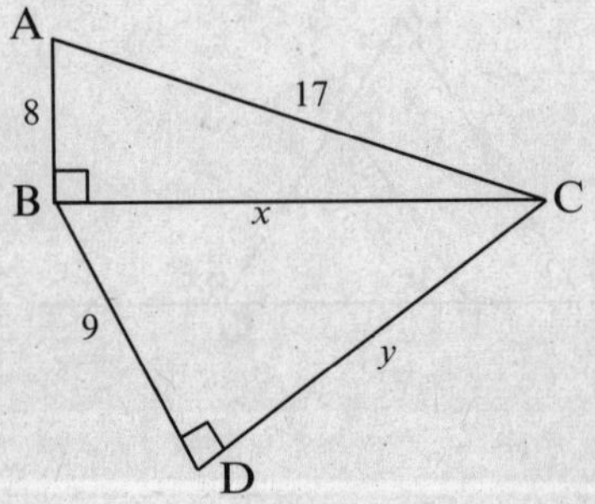

12. Calculate:
(*a*) BE
(*b*) BD
(*c*) BC

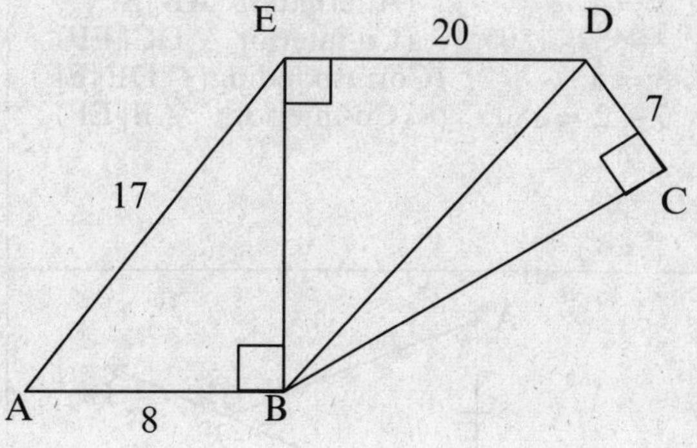

14. Calculate:
(*a*) AC
(*b*) BC
(*c*) XC

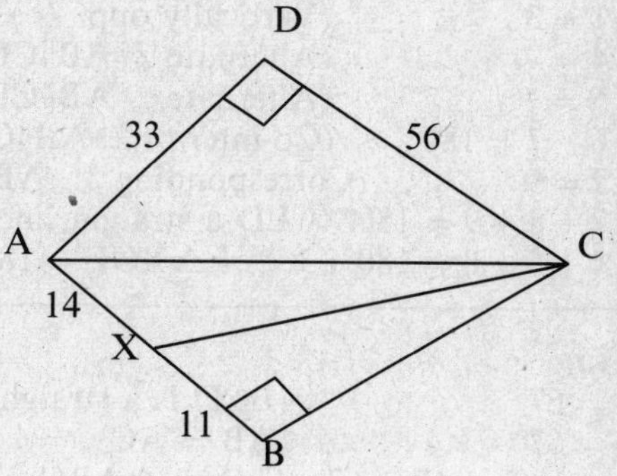

13. Calculate AB:
(AD = 7)

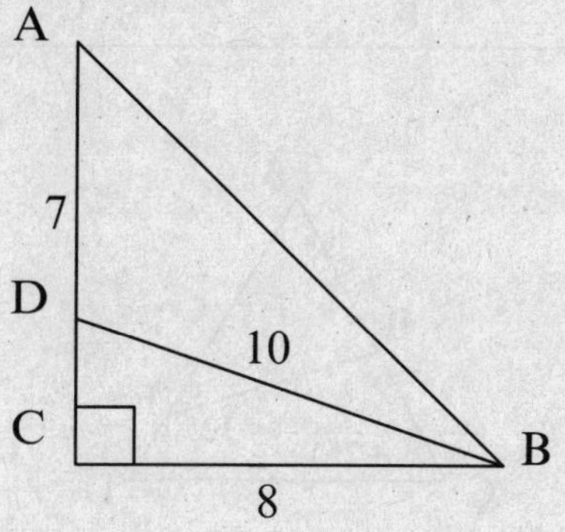

15. Calculate PR:
(PQ = QR)

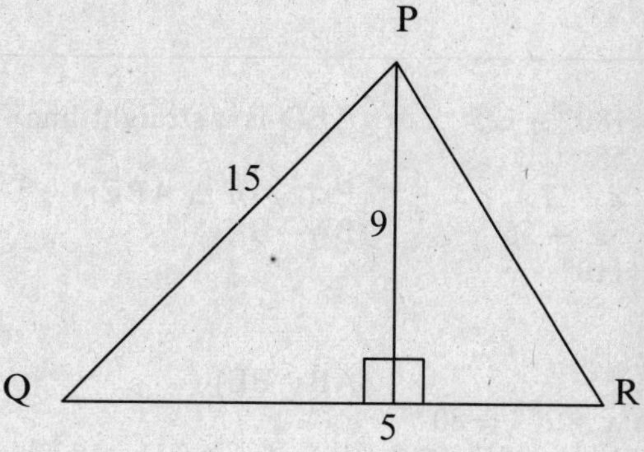

SOLUTIONS

Solutions: Revision Std. 6 Geometry

1. $\hat{2} = \hat{4}$ (Vertically opp. $\angle^s$)
$\hat{1} = \hat{3}$ (Vertically opp. $\angle^s$)
$\hat{4} = \hat{9}$ (Alternate $\angle^s$ AB∥CD)
$\hat{5} = \hat{7}$ (Alternate $\angle^s$ AB∥CD)
$\hat{6} + \hat{7} = 180°$ (Co-interior $\angle^s$ AB∥CD)
$\hat{2} = \hat{9}$ (Corresponding $\angle^s$, AB∥CD)
$\hat{7} + \hat{8} + \hat{9} = 180°$ (CED a straight line)
$\hat{4} + \hat{5} + \hat{8} = 180°$ ($3\angle^s$ of ΔXOE = 180°)

2. $\hat{7} = \hat{6}$ (Alternate $\angle^s$ AB∥CD)
$\hat{7} = \hat{4}$ (Alternate $\angle^s$ AB∥EF)
$\hat{4} + \hat{5} + 180°$ (Co-interior $\angle^s$ DC∥EF)
$\hat{3} = \hat{5}$ (Corresponding($\angle^s$ DE∥EF)
$\hat{7} + \hat{2} = 180°$ (Co-interior $\angle^s$ AB∥EF)

3. (*a*)
$C_1 = 67°$ (BCD is a straight line)
$x = 67°$ (AB = AC)
$y = 180° - (67° + 67°)$ ($3\angle^s$ of ΔABC = 180°)
$= 46°$

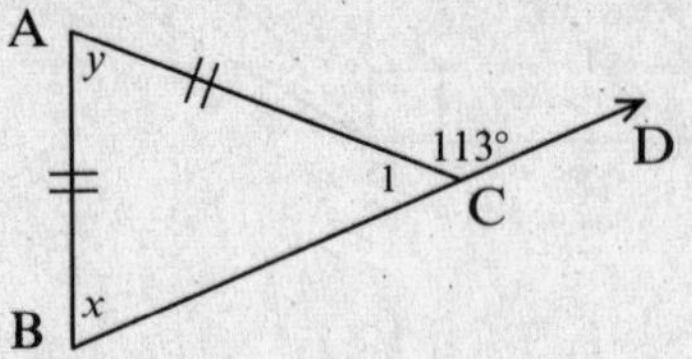

(*b*)
$x = 66°$ (Alternate $\angle^s$ AD∥BC)
$y = 44°$ (Corresponding $\angle^s$ AD∥BC)
$z = 180° - (44° + 66°)$ ($3\angle^s$ of ΔABC = 180°)
$= 70°$

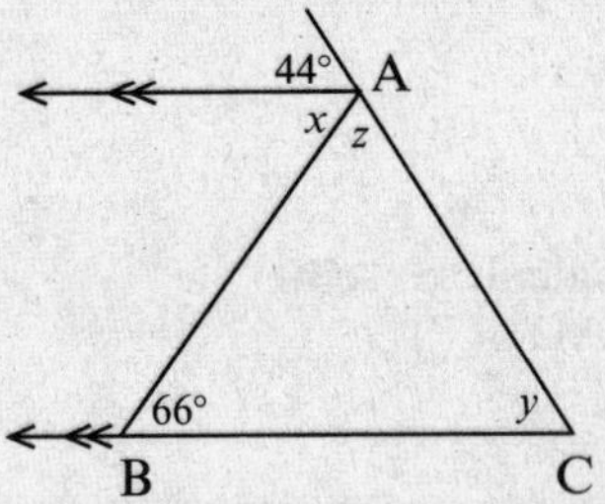

(*c*)
$x = 180° - 125°$ (AED is a straight line)
$= 55°$
$y = x + \hat{A}$ (Ext. $\angle$ of Δ ABE)
$= 55° + 55°$ (BA = BE)
$= 110°$

$\hat{A} = 55°$ (AB = BE)
$\hat{C} = x + 25° = 80°$
$z = 180° - (55° + 80°)$ ($3\angle^s$ of ΔACD = 180°)
$= 45°$

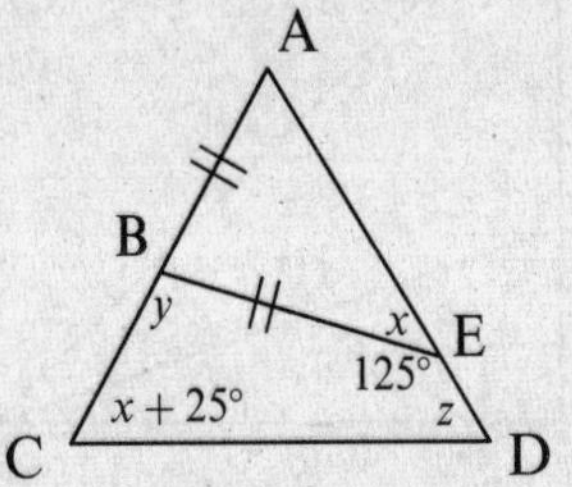

(*d*)
$x = \dfrac{180° - 64°}{2}$ ($3\angle^s$, of Δ ABC = 180°)
(AB = AC)
$= 58°$
$y = \dfrac{58°}{2}$ (Ext $\angle$ of Δ BCD)
(CB = CD)
$= 29°$

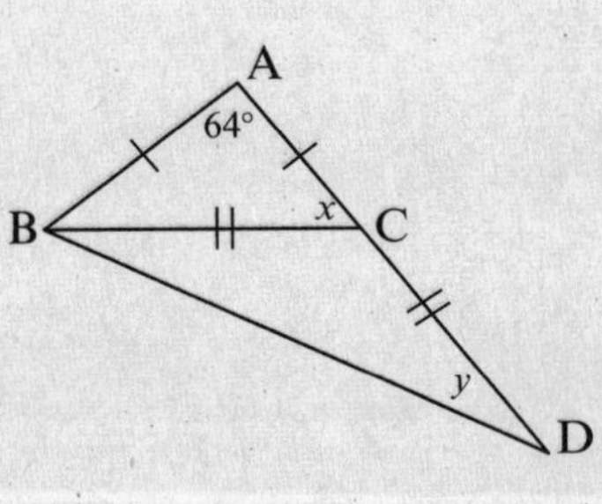

(*e*) $x = 70°$ (Alternate $\angle^s$, AB∥DE)

$y = \dfrac{70°}{2}$ (Ext. ∠ of Δ ABC) (BA = BC)

$= 35°$

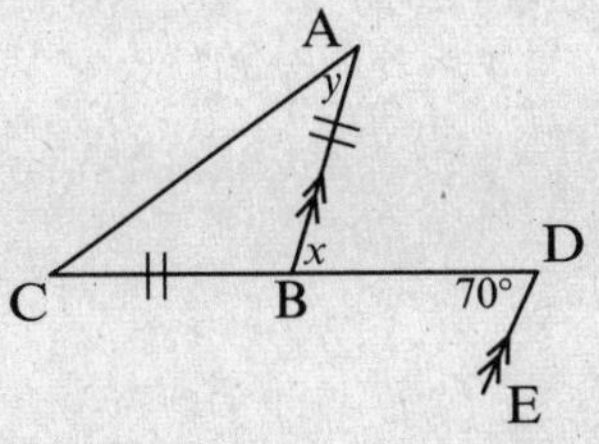

(*f*) $x = \hat{D}_1$ (Vertically opp. $\angle^s$)
$= 180° - (50° + 70°)$ ($3\angle^s$ Δ DEF = 180°)
$= 60°$
$y = 40° + x$ (Ext. ∠ of Δ ACD)
$= 100°$

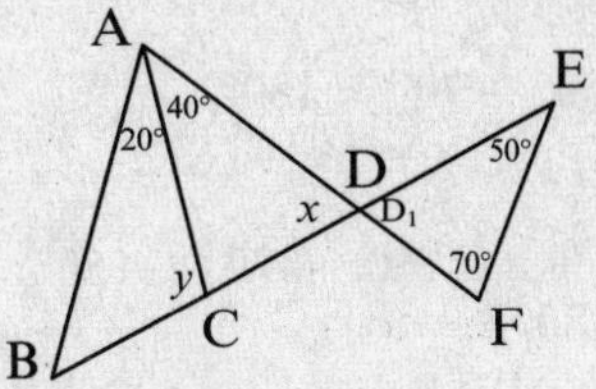

(*g*) $x = \dfrac{180° - 90°}{2}$ (ACB is a straight line)

$= 45°$

$y = 90° + x$ (Vertically opp. $\angle^s$)
$= 135°$

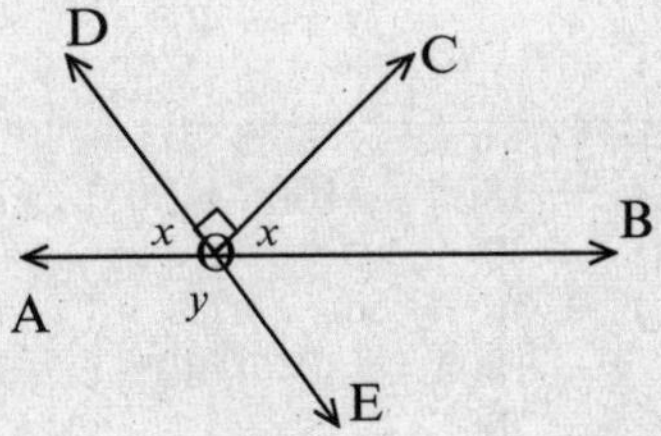

(*h*) $\hat{A} = 140° - 80°$ (Ext. ∠ of Δ ABD)
$= 60°$

$x = 180° - 60°$ (Co-interior $\angle^s$, AB∥EF)
$= 120°$
$y + 80° = x + 10°$ (Ext. ∠ of Δ DEF)
$= 130°$
$\therefore\ y = 50°$

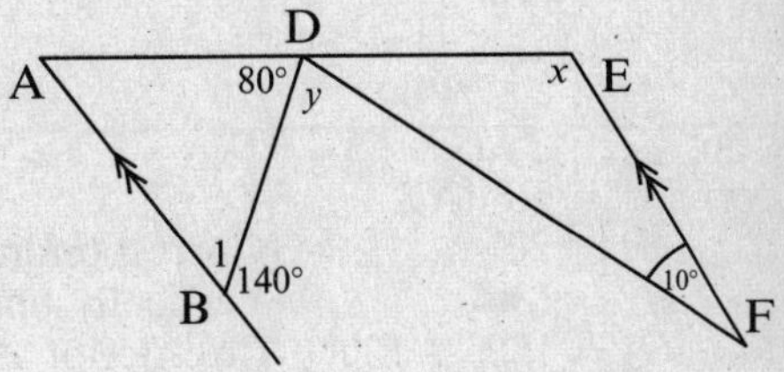

(*i*) $\hat{C} = 80° - 20°$ (Ext. ∠ of Δ BFC)
$= 60°$
$x = 100 - 60°$ (Ext. ∠ of Δ DBC)
$= 40°$
$y = \hat{C}$ (Alternate $\angle^s$, DE∥ABC)
$= 60°$

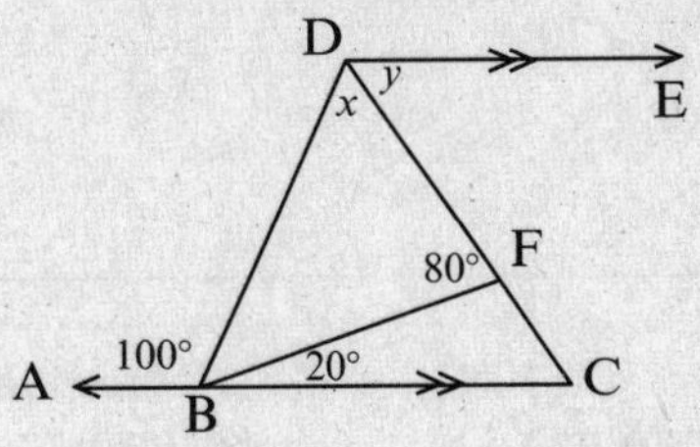

(*j*) $\hat{C}_1 = 50°$ (AEC is a straight line)
$\therefore\ y = 80° - 50°$ (Ext. ∠ of Δ BEC)
$= 30°$
$x = 180° - (80 + 30°)$ ($3\angle^s$, of Δ ABE = 180°)
$= 70°$

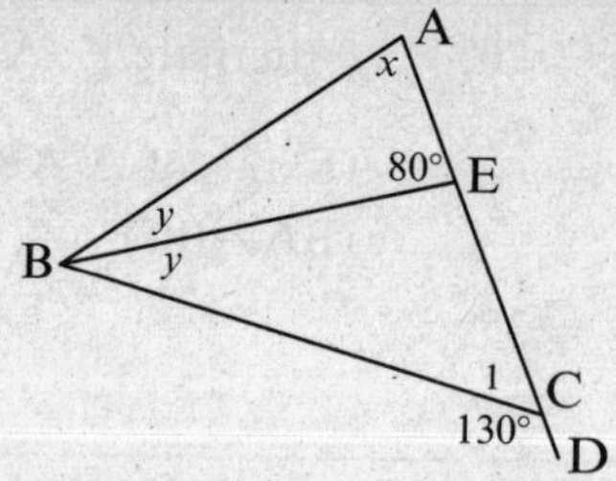

(*k*) $x + 50° + 70° = 180°$ (Co-interior $\angle^s$, AB||CE)
$\therefore\ x = 60°$

$\hat{C}_1 = 360 - (60° + 70° + 150°)$ (Revolution)
$= 80°$

$\hat{C}_1 + x + y = 180°$ (Co-interior $\angle^s$, AE||CD)
$\therefore\ 80° + 60° + y = 180°$
$\therefore\ y = 40°$

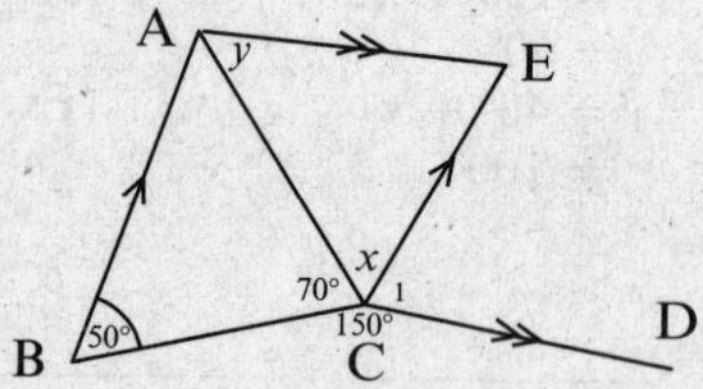

(*l*) $\hat{A}_1 = 180° - (100° + 50°)$ ($3\angle^s$ of Δ ABP = 180)
$= 30°$
$\therefore\ \hat{A}_2 = 30°$
$x = 180° - (30° + 80°)$ ($3\angle^s$ of Δ ACM = 180)
$= 70°$
$\hat{P}_1 = 100°$ (Vertically opp. $\angle^s$)
$\therefore\ y = 180° - (70° + 100°)$ ($3\angle^s$ of Δ PCD = 180)
$= 10°$

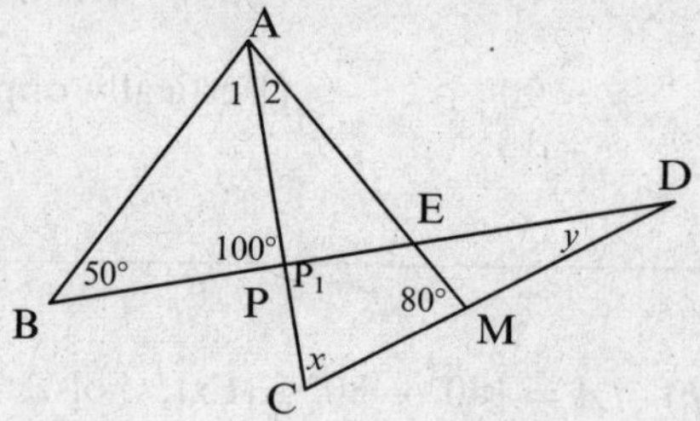

(*m*) Draw EP||AB
$\hat{G}_1 = 20°$ (Corresponding $\angle^s$, AB||EP)
$\hat{H}_1 = 130°$ (Vertically opp. $\angle^s$)
$\hat{G}_2 = 180° - 130°$ (Co-interior $\angle^s$, EP||CD)
$= 50°$

$\therefore\ x = 20° + 50°$
$= 70°$

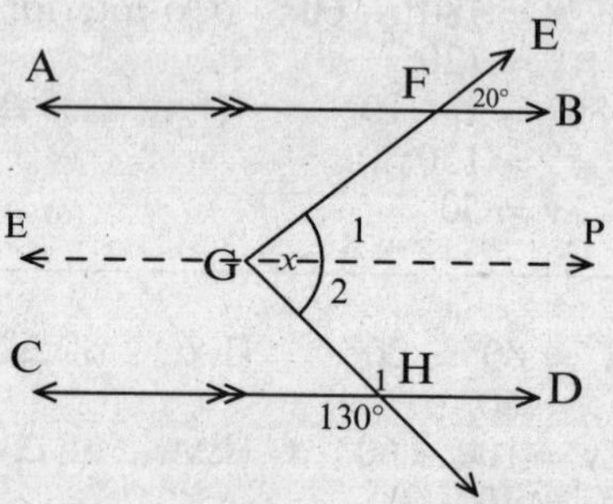

(*n*) $x = 30°$ (Corresponding $\angle^s$, CD||BE)
$\hat{B}_1 = x$ (Alternate $\angle^s$, AE||BC)
$= 30°$

$y + \hat{B}_1 = \dfrac{180° - 80°}{2}$ ($3\angle^s$ of Δ ABC = 180°)
(AB = AC)
$= 50°$
$\therefore\ y = 20°$

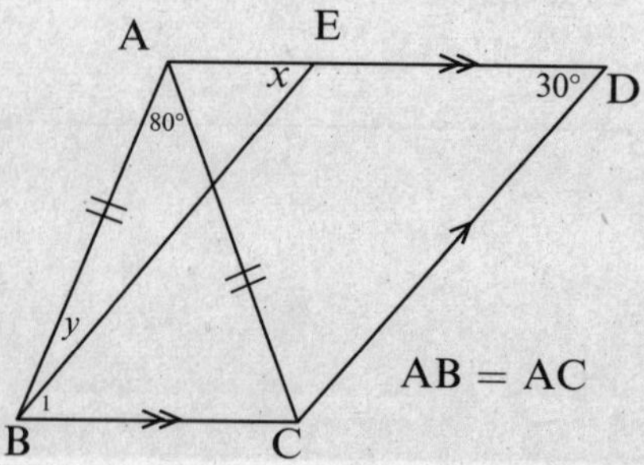

4.1 $A\hat{C}D = x + y$ (Ext. $\angle$ of Δ ABC)

4.2 $\hat{C} = 180° - (x + y)$ ($3\angle^s$ of Δ ABC = 180°)

4.3 $\hat{C} = x$ (DC = DB)
$D\hat{B}A = 90° - x$ ($A\hat{B}A$ = 90°)

4.4 $E\hat{C}D = x$ (Corresponding $\angle^s$, AB∥CE)
$\hat{C}_1 = 90° - x$ ($3\angle^s$ of Δ ABC = 180°)

4.5 $\hat{D} = x$ (Alternate $\angle^s$, AC∥DE)
$\hat{E} = 180° - 2x$ ($3\angle^s$ of Δ BDE = 180°; EB = ED)

4.6 $\hat{A}_1 = x$ (CA = CB)
$A\hat{C}D = 2x$ (Ext. $\angle$ of Δ ABC)
$\hat{D} = \dfrac{180° - 2x}{2}$ ($3\angle^s$ of Δ ACD = 180°; CD = CA)
$= 90° - x$

4.7 $\hat{B} = 180° - 2x$ ($3\angle^s$ of Δ ABC = 180°; BA = BC)

4.8 $A\hat{C}D = x$ (Alternate $\angle^s$, AB∥CD)
$\hat{D} = \dfrac{180° - x}{2}$ ($3\angle^s$ of Δ ACD = 180°; CA = CD)

4.9 $\hat{E} = x$ (Corresponding $\angle^s$, AC∥EF)
$E\hat{B}F = 180° - 2x$ ($3\angle^s$ of Δ BEF = 180°; BE = BF)

4.10 $\hat{B} = \hat{C}_1 = 60°$ (Equilateral Δ)
$\therefore E\hat{D}C = \hat{B} + \hat{C}_1 + x$ (Ext. $\angle$ of Δ DBC)
$= 120° + x$

5.1 $6x = 3x + 81°$ (Ext. $\angle$ of Δ FGH)

$\therefore 6x - 3x = 81°$

$3x = 81°$

$x = 27°$

$y = 3x$ (Corresp. $\angle^s$ DE$\|$HF)

$\therefore y = 81°$

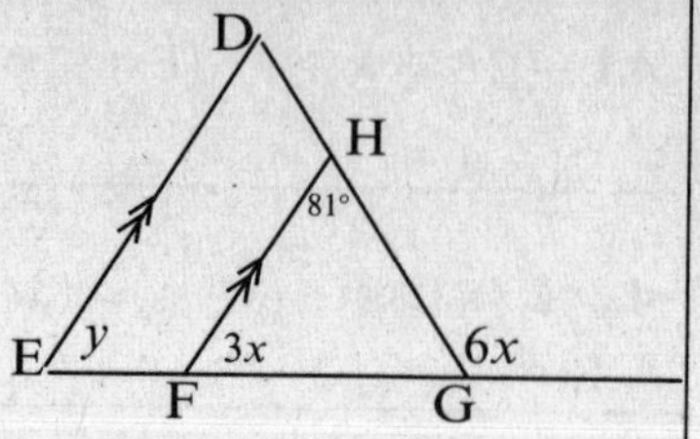

5.2 $y = 180° - (85° + 70°)$ ($3\angle^s$ of Δ BCD)

$= 25°$

$\therefore \hat{B}_1 = 25°$ (Alternate $\angle^s$, AB$\|$CD)

$x = 180° - (138° + 25°)$ ($3\angle^s$ of Δ ABC)

$= 17°$

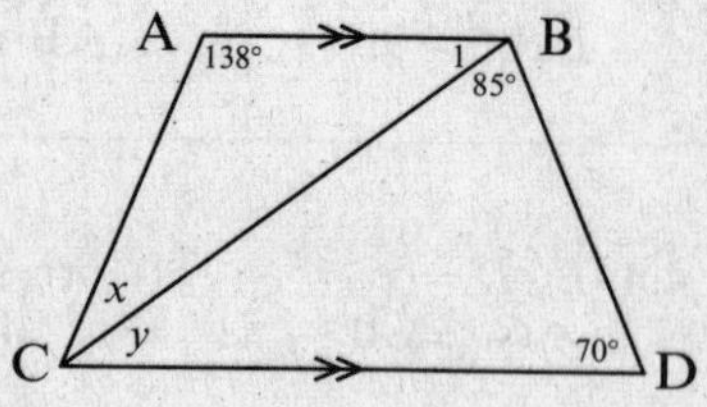

5.3 $\hat{Z} = 65°$ (XY = XZ)

$x = 180° - (90° + 65°)$ ($3\angle^s$ of Δ XAZ = 180°)

$= 25°$

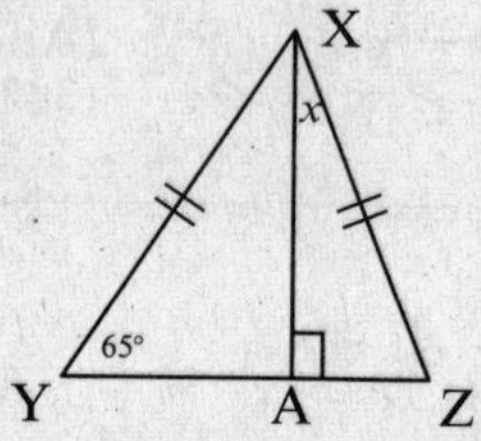

5.4 $\hat{D}_1 = 66 + 16°$ (Ext. $\angle$ of Δ AED)

$= 82°$

$x = 82°$ (AD = AC)

$y = \dfrac{82°}{2}$ (Ext. $\angle$ of Δ ABC; CB = CA)

$= 41°$

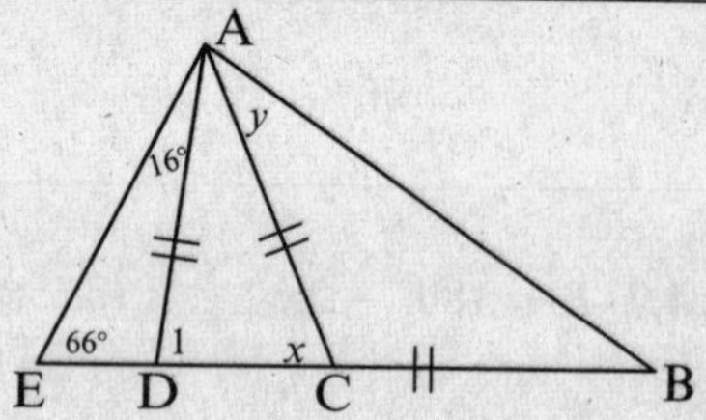

5.5 $\hat{T}_1 = 35°$ (PT = PS)

$a = 180° - (35° + 35°)$ ($3\angle^s$ of Δ PTS)

$= 110°$

$b = 180° - 110°$ (PS$\|$QR, Co-interior $\angle^s$)

$= 70°$

$\hat{T}_1 + c = 70° + 70°$ (Ext $\angle$ of Δ TQR, TQ = TR)

$= 140°$

$\therefore c = 105°$

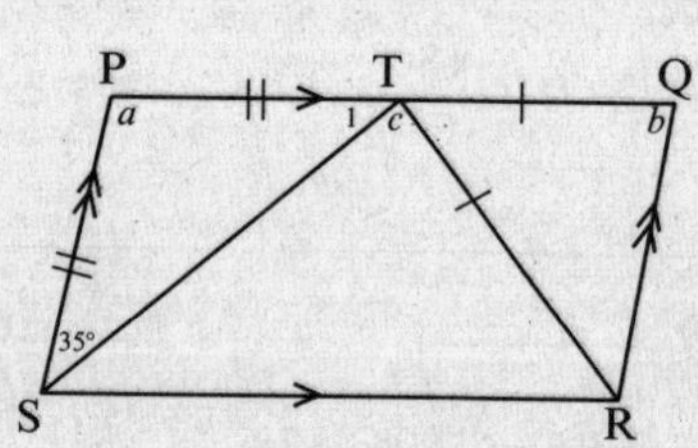

5.6 $\hat{D}_1 = 30°$ (EDF is a straight line.)

$\hat{B} = 30°$ (AB = AC)

$\hat{C}_1 = 80°$ (Alternate $\angle^s$, AC$\|$EF)

$x + 30° = 80°$ (Ext $\angle$ of ABC)

$\therefore x = 50°$

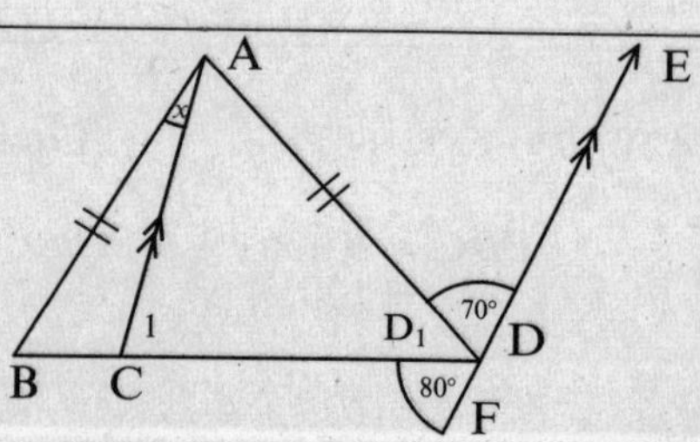

6. $5x + x = 180°$ (D$\hat{B}$C: straight line)
$\therefore\ 6x = 180°$
$x = 30°$

$5x = x + y$ (Ext. ∠ of Δ ABC)
$\therefore\ 150° = 30° + y$
$y = 120°$

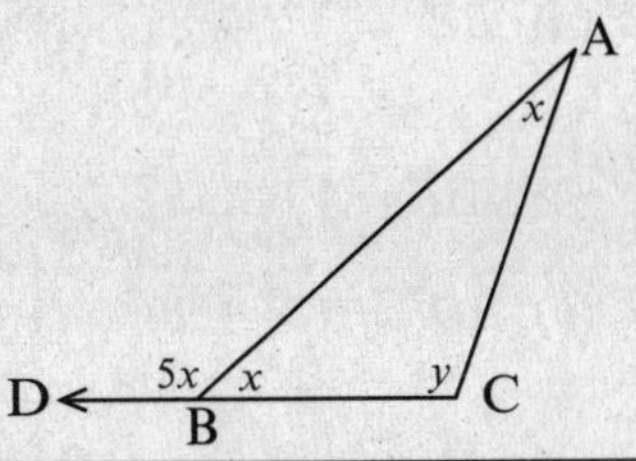

7. $5x + 4x = 180°$ (Co-interior ∠ˢ, AB∥CD)
$9x = 180°$
$x = 20°$

$4x + x + y = 180°$ (3∠ˢ of Δ ABE)
$80° + 20° + y = 180°$
$y = 80°$

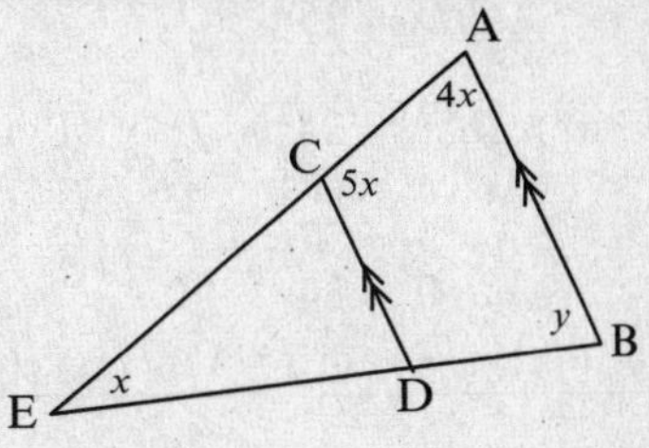

8. $2x + 2y = 180°$ (Co-interior ∠ˢ, AD∥BC)
$\therefore\ x + y = 90°$

$\therefore\ A\hat{O}B = 180° - (x + y)$ (3∠ˢ of Δ AOB)
$= 180° - 90°$o
$= 90°$

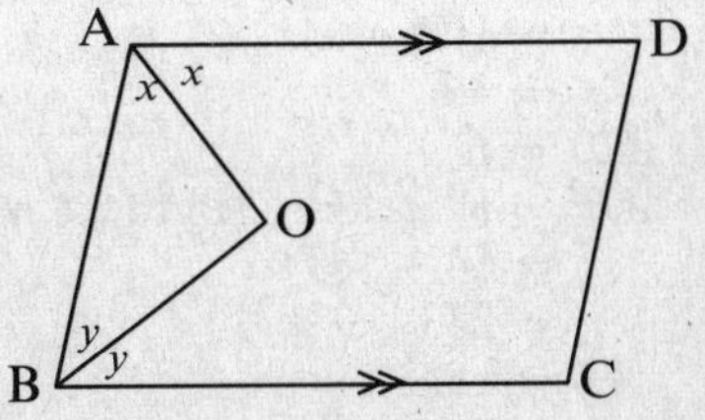

9. $A\hat{B}C = \hat{B}_1 + x$

$C\hat{D}B = \hat{B}_1 + x$ (Ext. ∠ of Δ ABD)
$\therefore\ A\hat{B}C = C\hat{D}B$

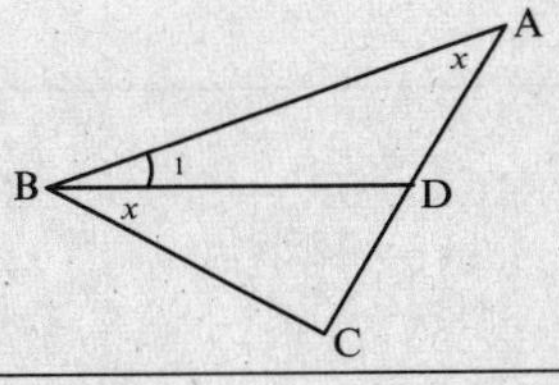

10. $x = 50°$ (Corresponding ∠ˢ, AB∥CD)
$3a - 20 + a = 180°$ (Co-interior ∠ˢ, FG∥DE)
$\therefore\ 4a = 180° + 20°$
$4a = 200°$
$a = 50°$

$x + y + a = 180°$ (CDE, straight line)
$50° + y + 50° = 180°$
$y = 180° - 100°$
$= 80°$

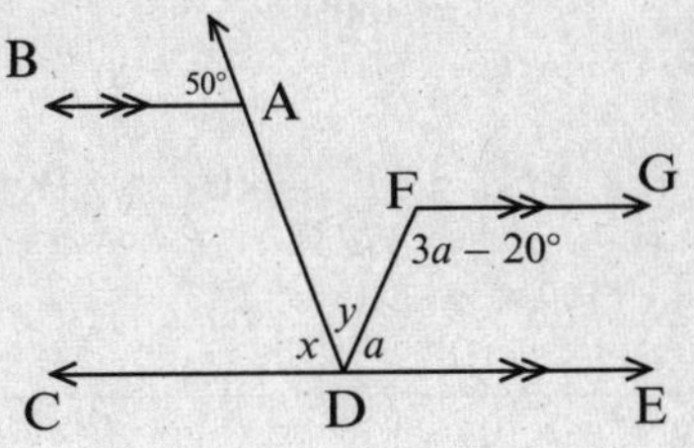

11. $x^2 = 17^2 - 8^2$ (Pyth.; $\hat{B} = 90°$)
$= 289 - 64$
$= 225$
$x = 15$

$y^2 = 15^2 - 9^2$ (Pyth.; $\hat{D} = 90°$)
$= 225 - 81$
$= 144$
$y = 12$

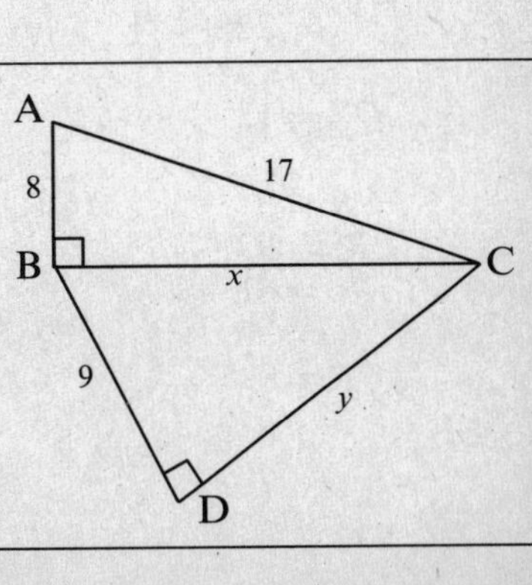

12. (*a*) $BE^2 = 17^2 - 8^2$ (Pyth.; $\hat{B} = 90°$)
$= 289 - 64$
$= 225$
$BE = 15$

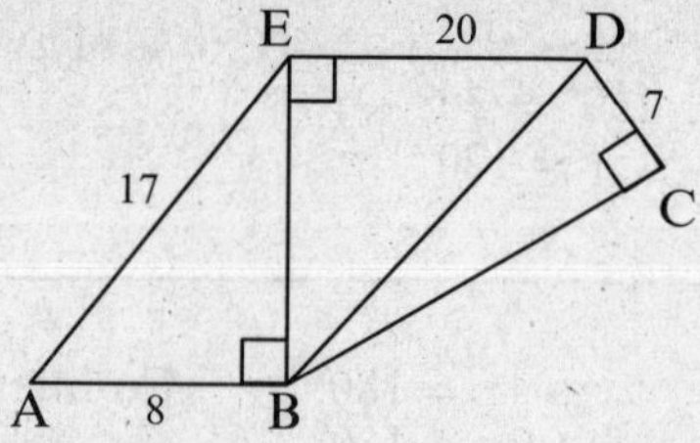

(*b*) $BD^2 = 15^2 + 20^2$ (Pyth.; $\hat{E} = 90°$)
$= 225 + 400$
$= 625$
$\therefore BD = 25$

(*c*) $BC^2 = 25^2 - 7^2$ (Pyth.; $\hat{C} = 90°$)
$= 625 - 49$
$= 576$
$BC = 24$

13. $CD^2 = 10^2 - 8^2$ (Pyth.; $\hat{C} = 90°$)
$= 100 - 64$
$= 36$
$\therefore CD = 6$
$AB^2 = 8^2 + 13^2$ (Pyth.; $\hat{C} = 90°$)
$= 64 + 169$
$= 233$
$AB = 15,3$

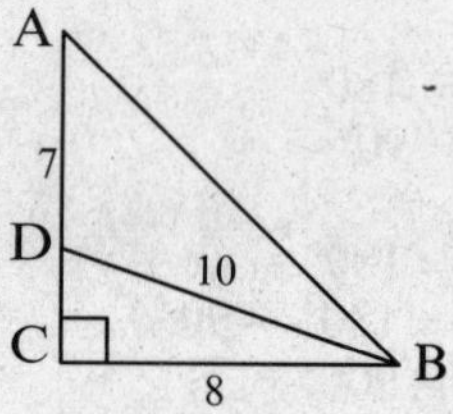

14. (*a*) $AC^2 = 33^2 + 56^2$ (Pyth.; $\hat{D} = 90°$)
$= 4225$
$AC = 65$

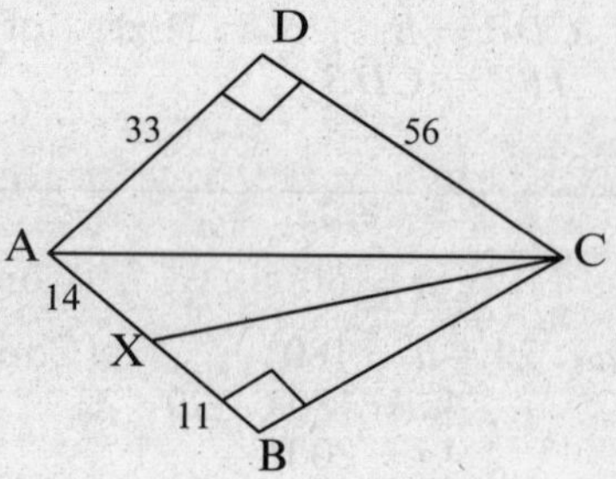

(*b*) $BC^2 = 65^2 - 25^2$ (Pyth.; $\hat{B} = 90°$)
$= 3600$
$BC = 60$

(*c*) $XC^2 = 11^2 + 60^2$ (Pyth.; $\hat{B} = 90°$)
$= 3721$
$XC = 61$

15. $QS^2 = 15^2 - 9^2$ (Pyth.; $\hat{S} = 90°$)
$= 225 - 81$
$= 144$
$QS = 12$

$\therefore SR = 3$ (QR = QP = 15)
$\therefore PR^2 = 9^2 + 3^2$
$= 81 + 9$
$= 90$
$PR = 9,5$

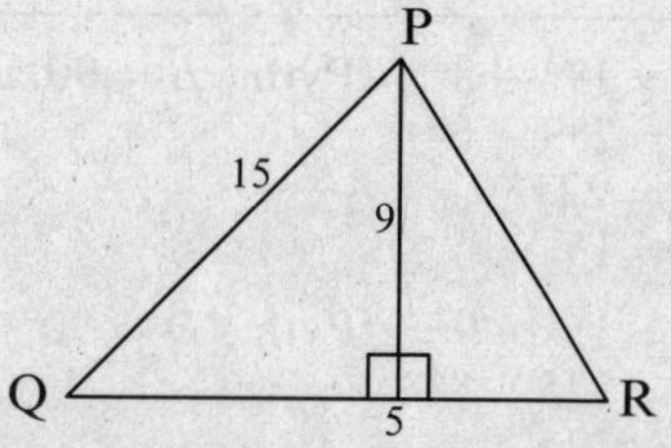

14 *Congruency and Similarity of triangles*

1. Similarity (|||):
Similar triangles fit **inside** one another.
That means that the 2 triangles have:

(1) 3 Pairs of equal angles (equi-angular)
(2) Their corresponding **sides** are **in proportion**

namely $\frac{8}{4} = \frac{6}{3} = \frac{4}{2}$

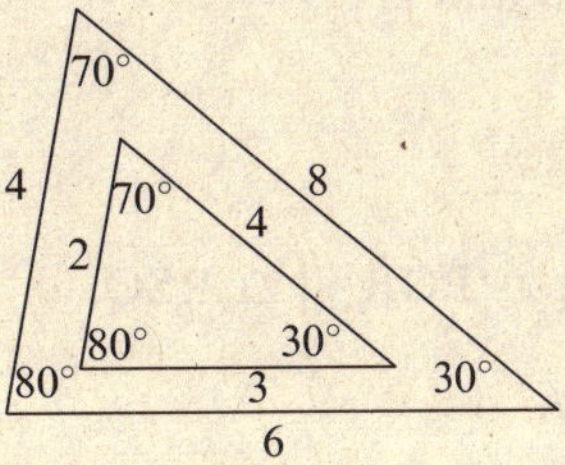

To prove 2 triangles similar:

Only prove 2 pairs of angles equal (with reasons).
(The 3rd pair of angles will automatically be equal because 3 angles of a △ = 180°)

Example of similarity – proof:

(*a*) Proof that: △ ABC ||| △ AED

(*b*) Calculate: ED en CD

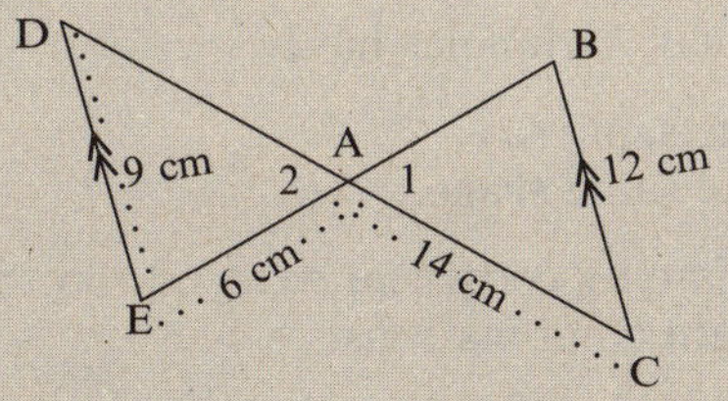

Solution:

(*a*) In △s ABC and AED:

1. $\hat{A}_1 = \hat{A}_2$ (Vertically opp. ∠s)
2. $\hat{B} = \hat{E}$ (Alternate ∠s; DE ∥ BC)

} similarity proof

$\therefore \triangle ABC \parallel\!| \triangle AED \quad (\angle, \angle, \angle)$

(*b*) $\therefore \dfrac{AB}{AE} = \dfrac{BC}{ED} = \dfrac{AC}{AD}$ ...... Because of similarity the corresponding sides are in proportion

$\dfrac{AB}{6} = \dfrac{12}{9} = \dfrac{14}{AD}$ ...... Fill in all the given lenghts.

$\dfrac{AB}{6} = \dfrac{12}{9} \Rightarrow 9AB = 72$

$AB = 8\ cm$

$\therefore EB = 14\ cm$

$\dfrac{12}{9} = \dfrac{14}{AD} \Rightarrow 12AD = 9 \times 14$

$= 126$

$\therefore AD = 10{,}5\ cm$

$\therefore CD = 14 + 10{,}5$

$= 24{,}5\ cm$

Exercise 1: Similarity

1. (*a*) Prove that △ BCD ||| △ ACE.

   (*b*) Calculate *a*, *b* and *c*.

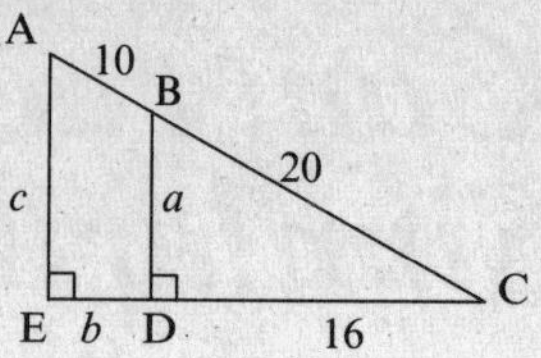

2. (*a*) Prove that △ PMN ||| △ LMK.

   (*b*) Calculate *d* and *e*.

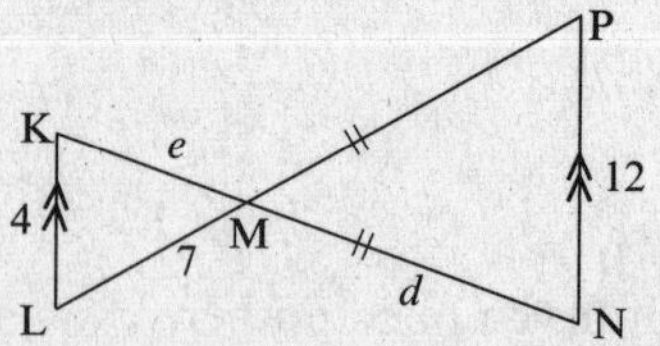

3. (*a*) Prove that △ PQR ||| △ RSQ.

   (*b*) Calculate *f*.

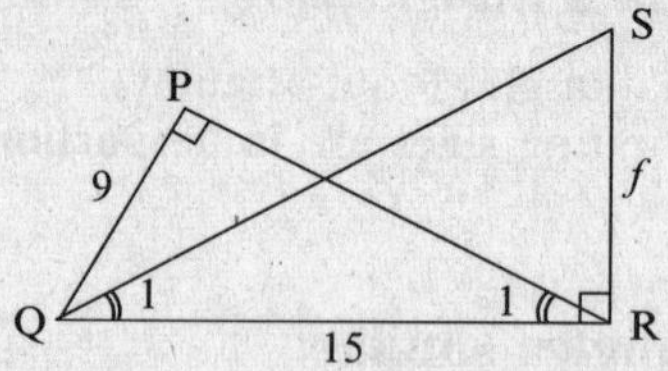

2. Congruency (≡):

2 Congruent triangles fit exactly **on top of one another**.

That means that 2 triangles have:

(1) Exactly the same **size**.
(2) Exactly the same **shape**.

To prove 2 triangles **congruent** we need to prove that the sides and angles of the two triangles are equal, in certain combinations.

Congruency – conditions:

S, S, S. (side, side, side)

S, ∠, S. (side, included angle; side)

∠, ∠, S. (angle; angle; any side)
RHS; (90°; hypotenuse, any other side)

Examples of congruent triangles:

1.

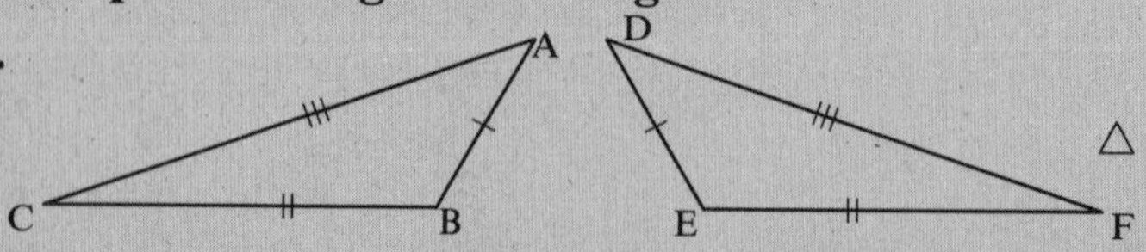

△ ABC ≡ △ DEF (S, S, S)

2.

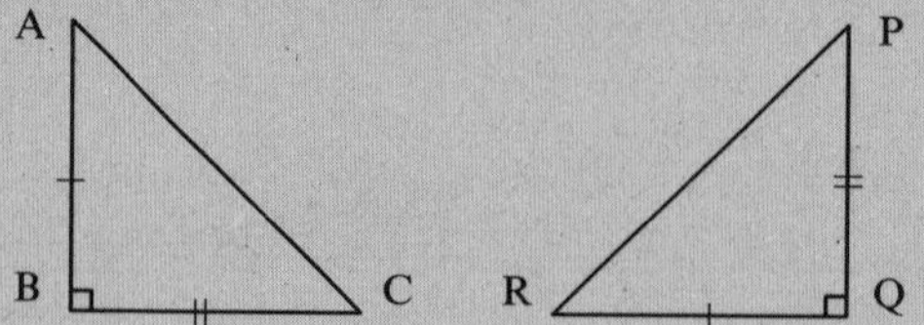

△ ABC = △ RQP (S, ∠, S)
(Note the order of the names of the triangles)

3.

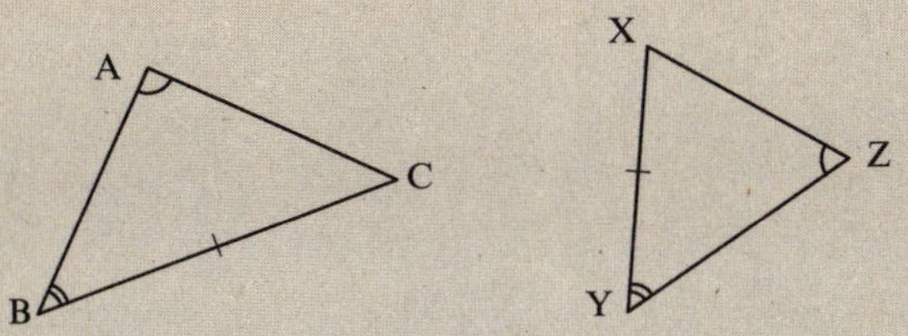

△ ABC ≡ △ ZYX (∠, ∠, S)
(Note that the equal sides are opposite the equal angles.)

4.

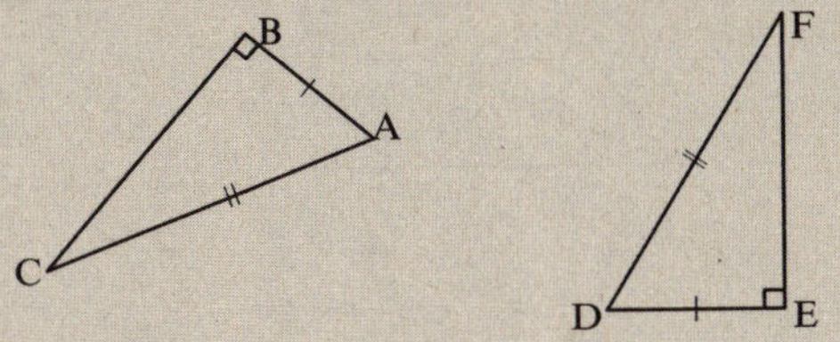

△ ABC ≡ △ DEF (RHS)

The proof for congruency has a very definite pattern:

N.B.: (1) According to one of the **conditions for congruency, 3 pairs** of equal **angles/sides** are used in the proof **(with reasons)**.

(2) Following the proof the **remaining 3 pairs of angles/sides** may be assumed to be **equal**.

(3) The **order** of the names of the triangles are very important.

eg. △ ABC ≡ △ PQR implies –

$\hat{A} = \hat{P}$ $\quad$ $AB = PQ$

$\hat{B} = \hat{Q}$ $\quad$ and $\quad$ $BC = QR$

$\hat{C} = \hat{R}$ $\quad$ $AC = PR$

Look carefully at the following examples.

Examples of proofs and deductions for congruency:

1. In △ˢ ABC and ADC:

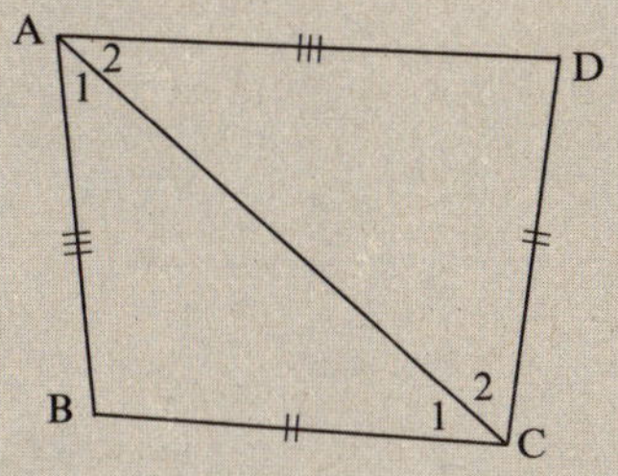

1. $AB = AD$ (given)
2. $CB = CD$ (given)
3. AC is common

∴ △ ABC ≡ △ ADC (S, S, S)

Therefore: $\hat{A}_1 = \hat{A}_2$
$\hat{B} = \hat{D}$
$\hat{C}_1 = \hat{C}_2$

2. In △ˢ ABC and CDE:

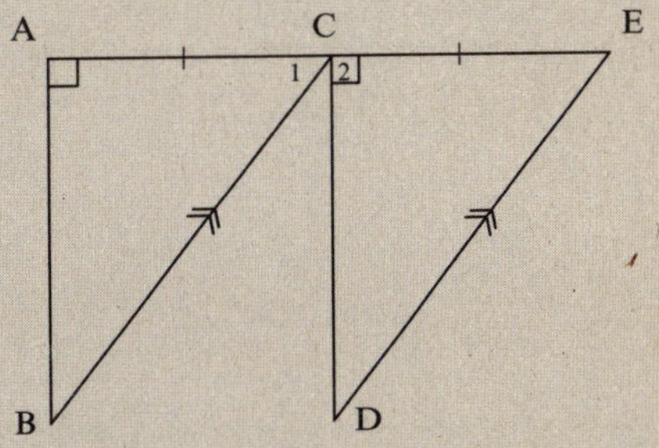

1. $\hat{A} = \hat{C}_2$ (= 90°, given)
2. $\hat{E} = \hat{C}_1$ (Corresp. ∠ˢ, BC ∥ DE)
3. $AC = CE$ (given)

∴ △ ABC ≡ △ CDE (∠, ∠, S)

(NB ↑)

Therefore: $AB = CD$
$\hat{B} = \hat{D}$
$BC = DE$

3. Use the diagram and prove that **RQ ⊥ PS**.

Proof: In $\triangle^s$ RPT and RST:

1. $RP = RS$ (given)
2. $PT = TS$ (given)
3. RT is common

$\therefore \triangle RPT \equiv \triangle RST$ (S, S, S)

$\therefore \hat{R}_1 = \hat{R}_2$

Now: In $\triangle^s$ RPQ and RSQ:

1. $RP = RS$ (given)
2. RQ is common
3. $\hat{R}_1 = \hat{R}_2$ (proved above)

$\therefore \triangle RPQ \equiv \triangle RSQ$ (S, ∠, S)

$\therefore \hat{Q}_1 = \hat{Q}_2$

$\therefore RQ \perp PS$ (PQS is a straight line)

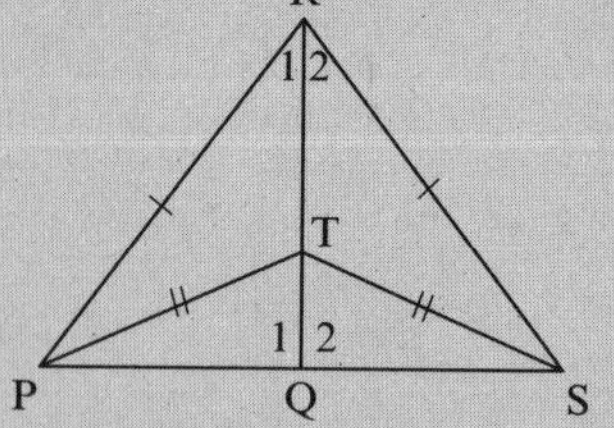

Exercise 2: Congruency

Decide whether the following pairs of triangles are congruent or not, write the triangles in the right order, and give the appropriate condition for congruency.

Eg.

$\triangle ABC \equiv \triangle YZX$ (S, ∠, S)

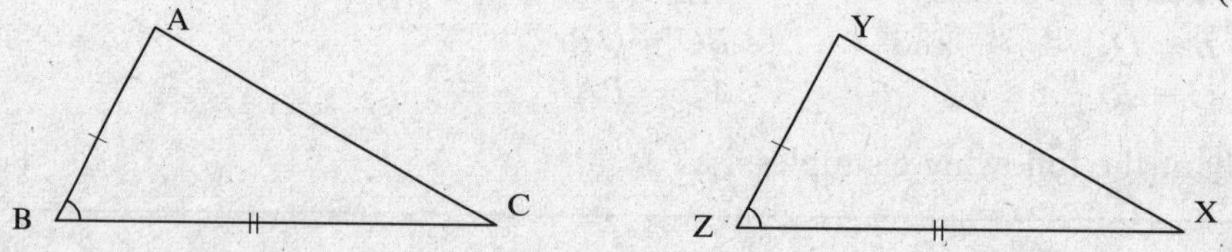

1.

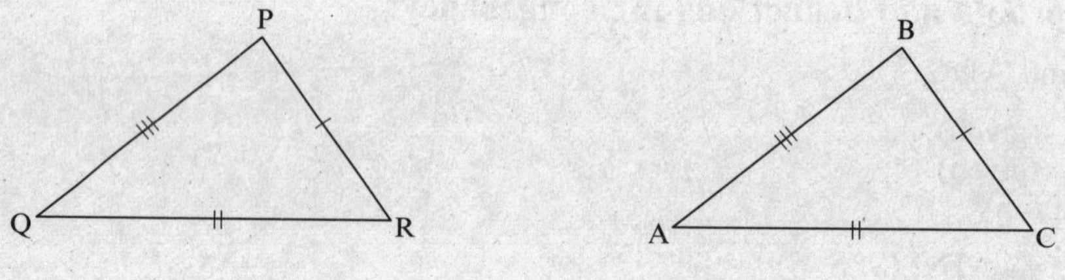

2.

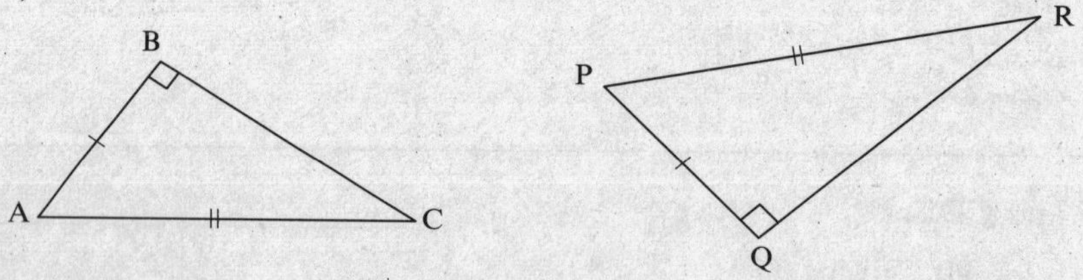

3.

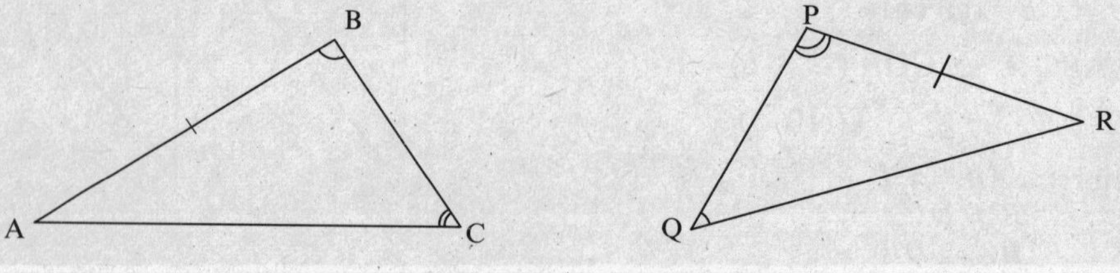

4.

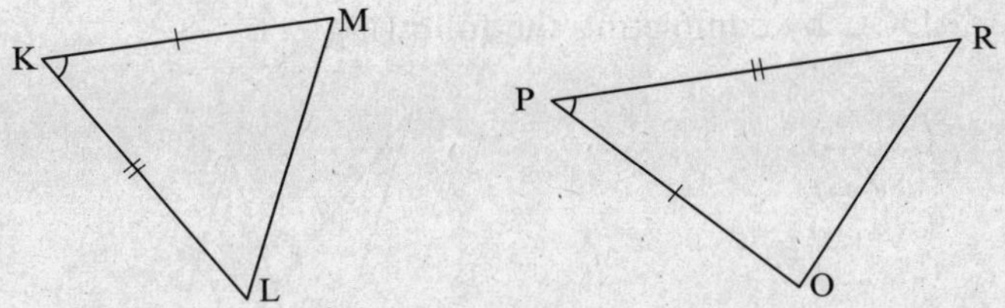

5.

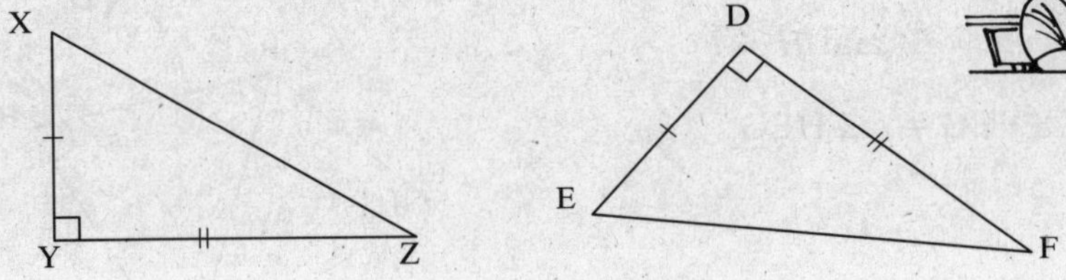

6.

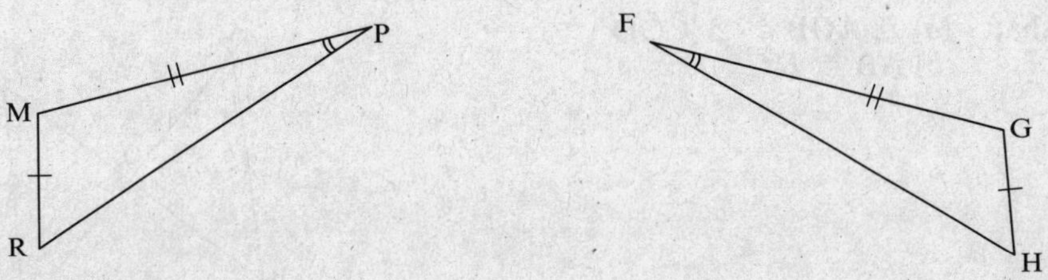

7.

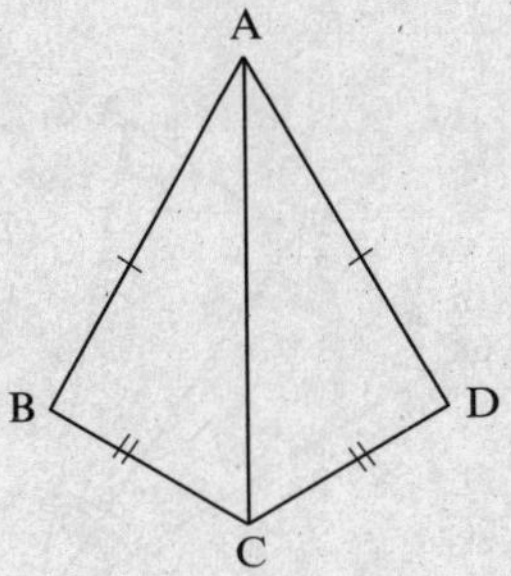

8.

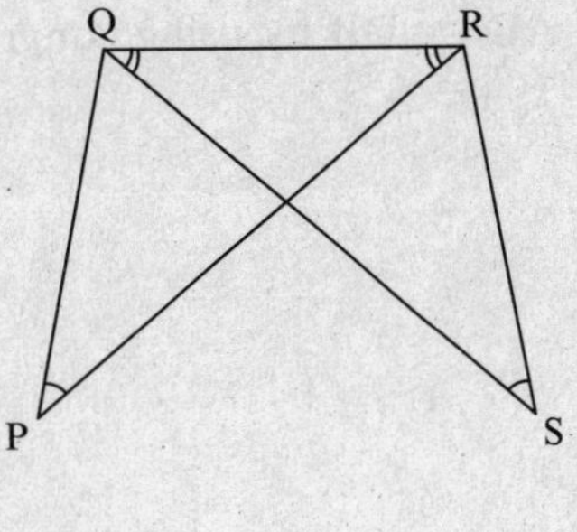

9.

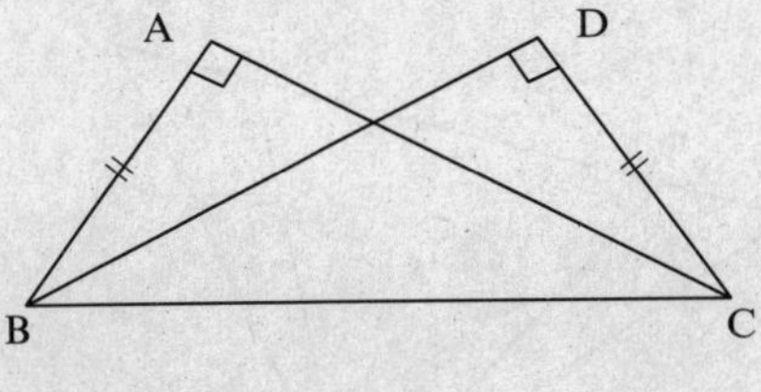

10.

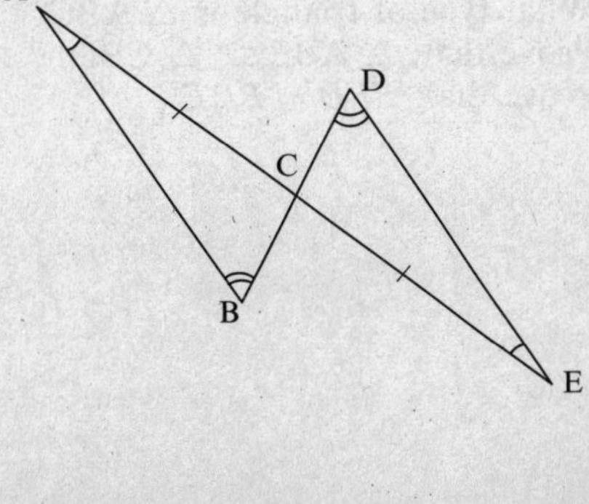

Exercise 3:

1. Prove that $\triangle$ AOB $\equiv$ $\triangle$ DOC by completing the following:

In $\triangle^s$ AOB and DOC:

1. $AO = \ldots\ldots$ $(\ldots\ldots)$
2. $BO = \ldots\ldots$ $(\ldots\ldots)$
3. $A\hat{O}B = \ldots\ldots$ $(\ldots\ldots)$

$\therefore$ $\triangle$ AOB $\equiv$ $\triangle$ …… $(\ldots\ldots)$

$\therefore$ $\hat{A} = \ldots\ldots$

$\hat{B} = \ldots\ldots$

$AB = \ldots\ldots$

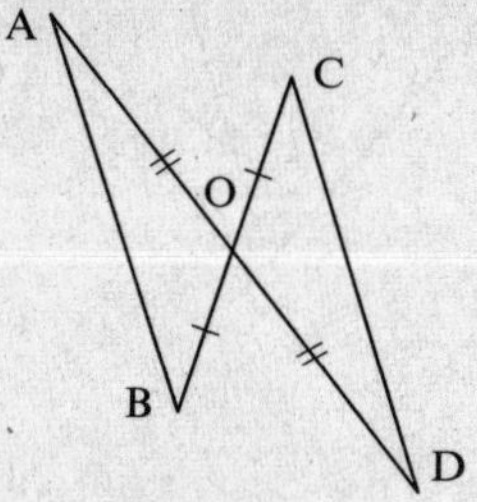

2. In the diagram $\hat{E}_1 = \hat{E}_2$ and $\hat{H} = \hat{F}$

Prove that: $\triangle$ FEG $\equiv$ $\triangle$ HEG

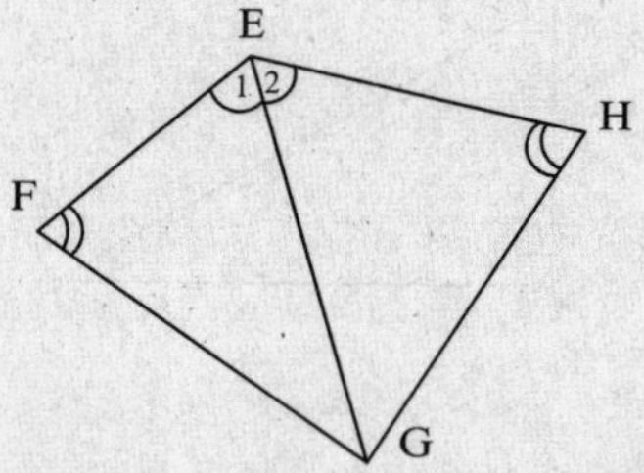

3. Prove that: (*a*) $\triangle$ AOB $\equiv$ $\triangle$ COD
(*b*) AB = DC

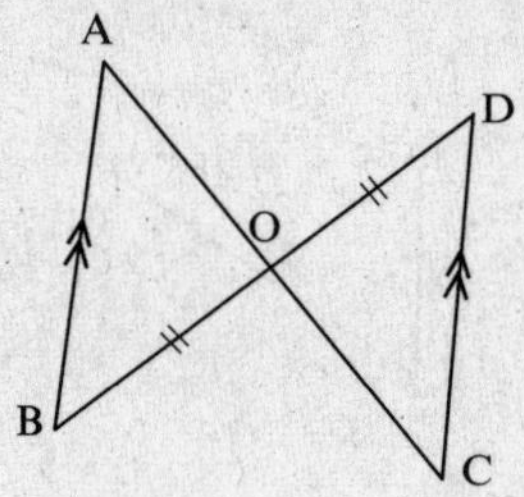

4. Prove that: AD = DB by using congruency.
($\hat{C}_1 = \hat{C}_2$)

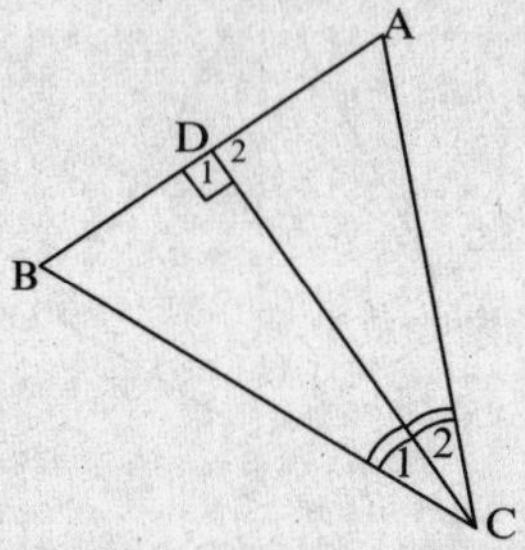

5. (*a*) What type of triangle is $\triangle$ ABC?
(*b*) Prove that: $\triangle$ ABE $\equiv$ $\triangle$ CBD
(*c*) Prove that: $B\hat{E}D = B\hat{D}E$

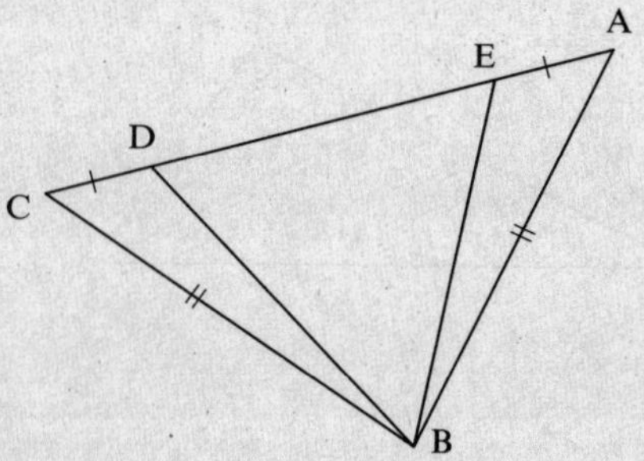

6. In the diagram △ DBE is an **equilateral triangle**.

$\hat{1} = \hat{2}$

Prove that: (*a*) △ ABE ≡ △ CDB
(*b*) AD = EC

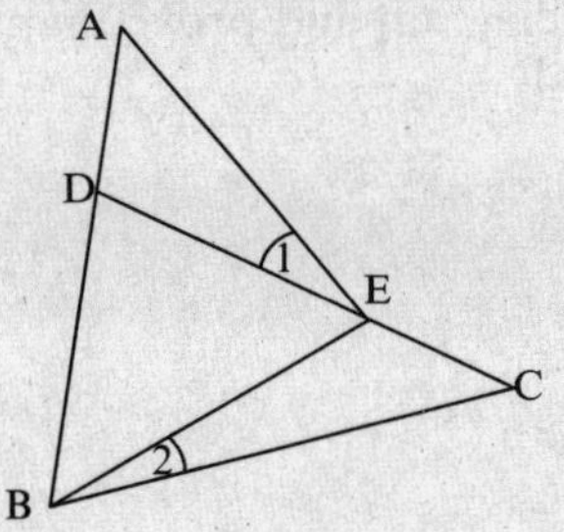

7. Prove that: (*a*) △ ABD ≡ △ BAC
(*b*) △ AOD ≡ △ BOC

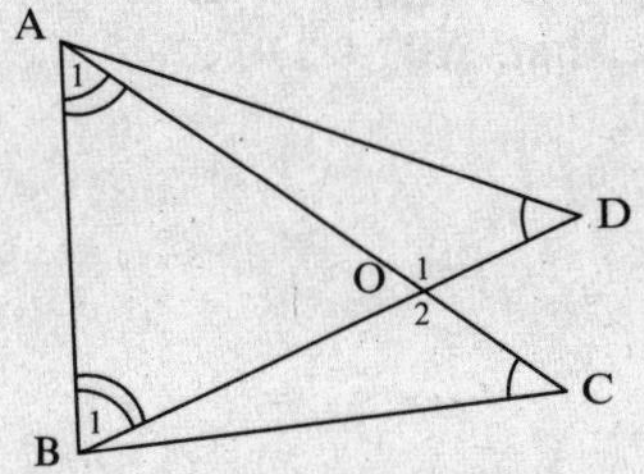

8. In the diagram:

AB = AC and DC = AE.

Prove that: (*a*) △ ACD ≡ △ BAE
(*b*) $B\hat{A}D = B\hat{E}C$

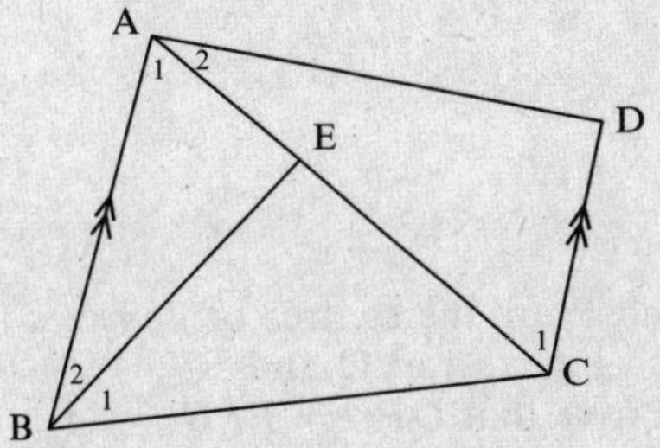

9. If AB = DC and $A\hat{B}C = D\hat{C}B$,

prove that: AC = DB.

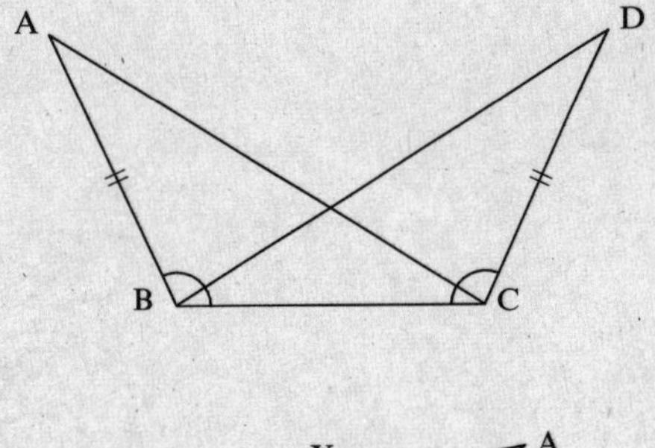

10. If AB = XW, $V\hat{X}W = 90° = A\hat{B}V$,

prove that: △ ABV ≡ △ WXV

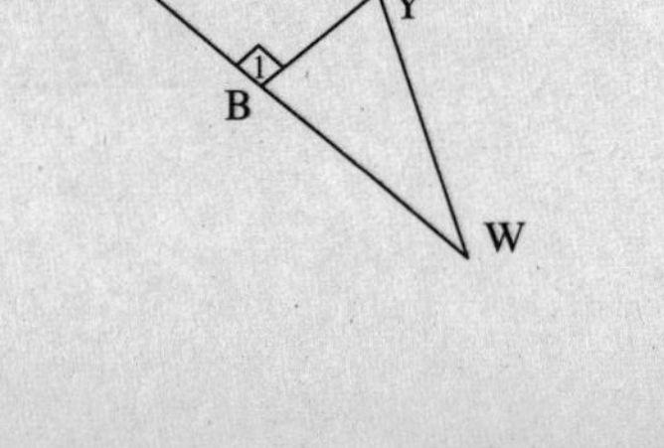

11. Use the diagram and prove that:
AB = AD

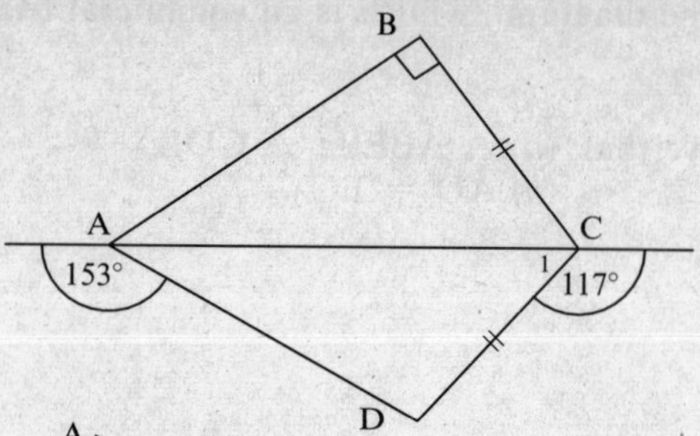

12. In the diagram:
AO = OC and OD = OB
Prove that: AB = DC

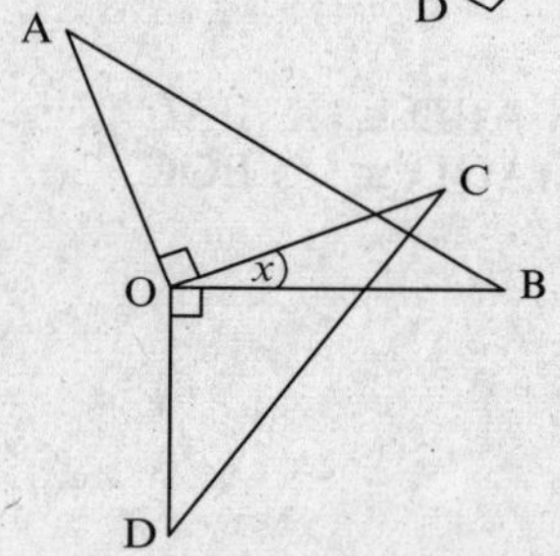

13. Prove the theorem: The base angles of an isosceles triangle are equal.

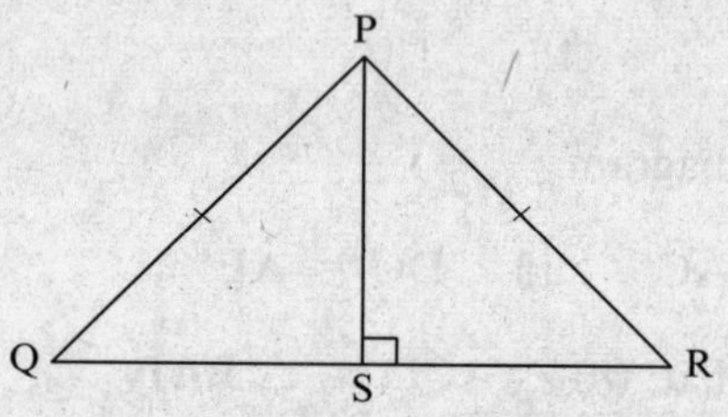

14. O and P are the centres of 2 circles which intersect at Q and R:
(*a*) Prove that $Q\hat{P}O = R\hat{P}O$.
(*b*) Prove that OP bisects QR at T.

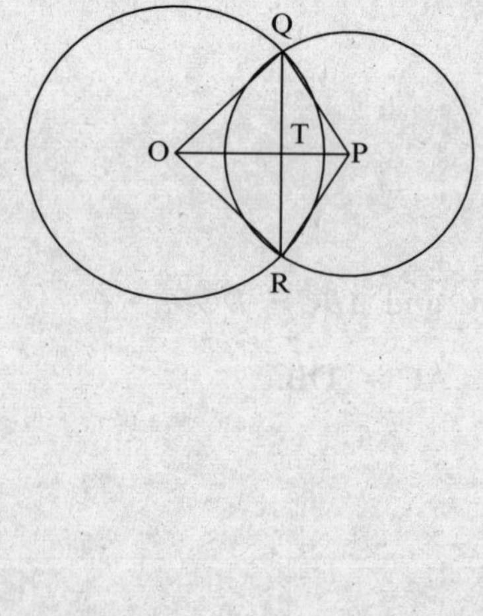

SOLUTIONS

Exercise 1: Similarity Solutions:

1. (*a*) In $\triangle^s$ BCD and ACE:

1. $\hat{C}$ is common
2. $\hat{E} = \hat{D}$ (= 90°; given)

$\therefore$ $\triangle$ BCD ||| $\triangle$ ACE ($\angle$, $\angle$)

(*b*) $a^2 = 20^2 - 16^2$ (Pyth.)

$= 400 - 256$

$= 144$

$\therefore$ **a = 12**

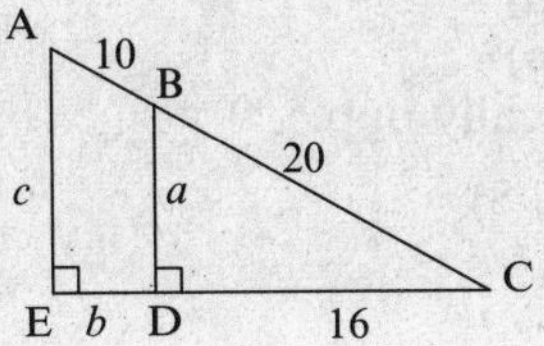

From similarity:

$$\frac{BC}{AC} = \frac{CD}{CE} = \frac{BD}{AE}$$

$$\frac{20}{30} = \frac{16}{CE} = \frac{12}{c} \; *$$

$\therefore$ $20CE = 16 \times 30$

$CE = 24$

$\therefore$ $\boldsymbol{b = 8}$

From similarity: *

$$\frac{20}{30} = \frac{12}{c}$$

$20c = 12 \times 30$

$\boldsymbol{c = 18}$

---

2. (*a*) In $\triangle^s$ PMN and LMK:

1. $\hat{K} = \hat{N}$ (Alternate $\angle^s$, KL || PN)
2. $\hat{L} = \hat{P}$ (Alternate $\angle^s$, KL || PN)

$\therefore$ $\triangle$ PMN ||| $\triangle$ LMK ($\angle$, $\angle$)

(*b*) **From similarity:**

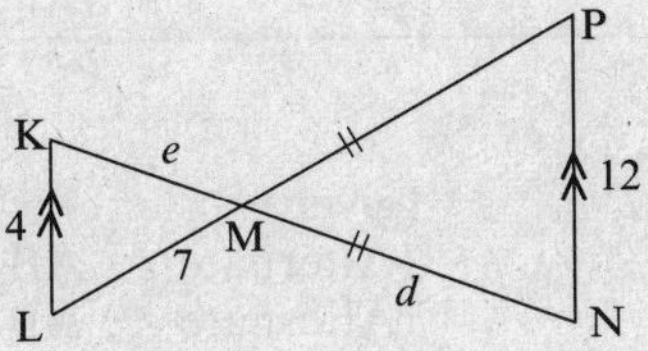

$$\frac{PM}{LM} = \frac{MN}{MK} = \frac{PN}{LK}$$

$$\frac{d}{7} = \frac{d}{e} = \frac{12}{4} \left(= \frac{3}{1}\right)$$

$\therefore$ $\boldsymbol{d = 21}$ and $\boldsymbol{e = 7}$

---

3. (*a*) In $\triangle^s$ PQR and RSQ

1. $\hat{P} = S\hat{R}Q$ (= 90°; given)
2. $\hat{Q}_1 = \hat{R}_1$ (given)

$\therefore$ $\triangle$ PQR ||| $\triangle$ RSQ ($\angle$; $\angle$)

(*b*) $PR^2 = 15^2 - 9^2$ (Pyth; $\hat{P} = 90°$)

$= 225 - 81$

$= 144$

$PR = 12$

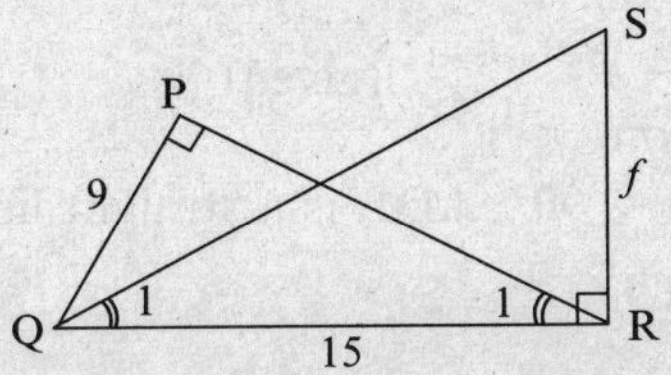

From similarity:

$$\frac{PQ}{RS} = \frac{QR}{SQ} = \frac{PR}{RQ}$$

$$\frac{9}{f} = \frac{12}{15}$$

$12f = 9 \times 15$

$\boldsymbol{f = 11{,}25}$

Exercise 2. Solutions

1. △ PQR ≡ △ BAC (S, S, S)
2. △ ABC ≡ △ PQR (90°; RHS; Side)
3. Not congruent.
4. △ KLM ≡ △ PRO (S, ∠, S)
5. △ XYZ ≡ △ EDF (S, ∠, S)

6. Not congruent.
7. △ ABC ≡ △ ADC (S, S, S)
8. △ PQR ≡ △ SRQ (∠, ∠, S)
9. △ ABC ≡ △ DCB (RHS)
10. △ ABC ≡ △ EDC (∠, ∠, S)

Exercise 3. Solutions

1. In △ˢ AOB and DOC:
1. $AO = OD$ (given)
2. $BO = OC$ (given)
3. $A\hat{O}B = C\hat{O}D$ (vertically opp. ∠ˢ)

△ AOB ≡ △ DOC (S, ∠, S)

∴ $\hat{A} = \hat{D}$
$\hat{B} = \hat{C}$
$AB = CD$

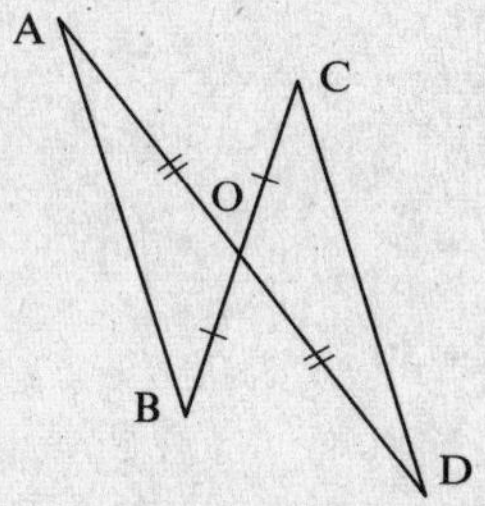

2. In △ˢ FEG and HEG:
1. $\hat{E}_1 = \hat{E}_2$ (given)
2. $\hat{F} = \hat{H}$ (given)
3. EG is common

∴ △ FEG ≡ △ HEG (∠, ∠, S)

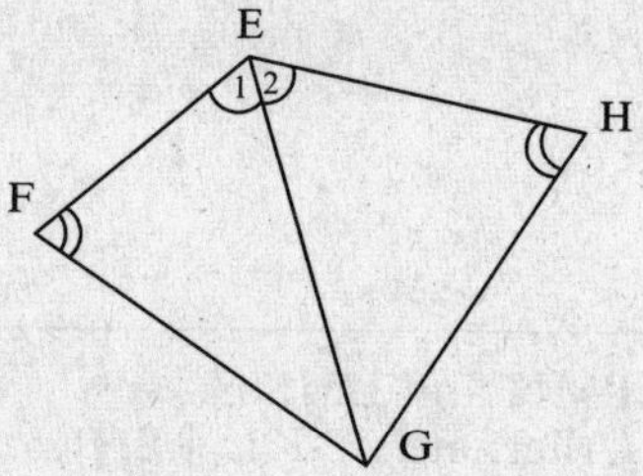

3. (*a*) In △ˢ AOB and COD
1. $BO = OD$ (given)
2. $\hat{A} = \hat{C}$ (Alternate ∠ˢ, AB ∥ CD)
3. $\hat{B} = \hat{D}$ (Alternate ∠ˢ, AB ∥ CD)

∴ △ AOB ≡ △ COD (∠, ∠, S)

(*b*) From congruency: AB = CD

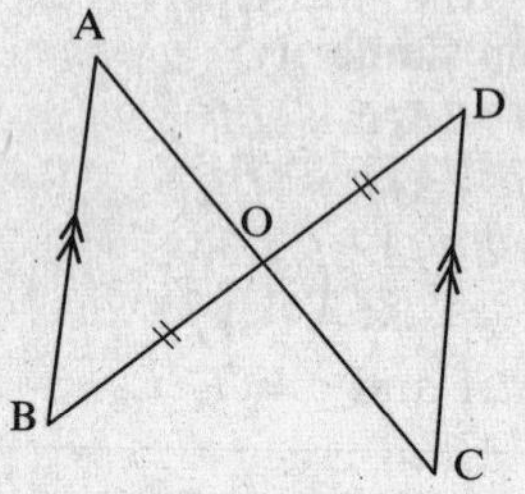

4. In △ˢ ACD and BCD:
1. $\hat{C}_1 = \hat{C}_2$ (given)
2. CD is common
3. $\hat{D}_1 = \hat{D}_2$ (= 90°; ADB is a straight line ∠)

∴ △ ACD ≡ △ BCD (∠, ∠, S)

∴ AD = BD

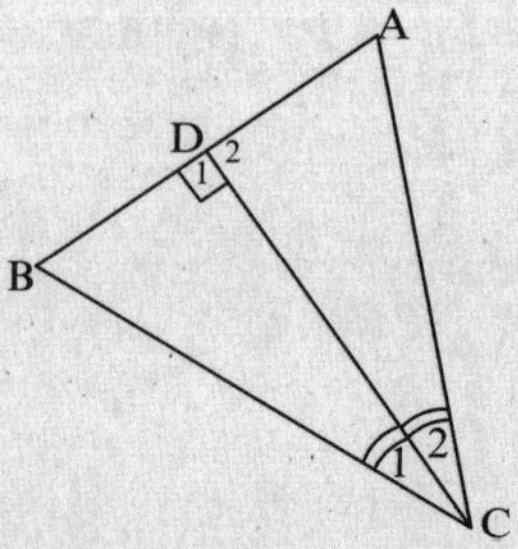

5. (*a*) $\triangle$ ABC: isosceles $\triangle$: $\hat{A} = \hat{C}$

(*b*) In $\triangle^s$ ABE and CBD:

1. $\hat{A} = \hat{C}$ (BA = BC)
2. $CD = AE$ (given)
3. $BC = BA$ (given)

$\therefore$ $\triangle$ ABE $\equiv$ $\triangle$ CBD (S, $\angle$, S)

(*c*) From congruency: BE = BD

$\therefore$ $B\hat{D}E = B\hat{E}D$

(base angles of isosceles $\triangle$ BED)

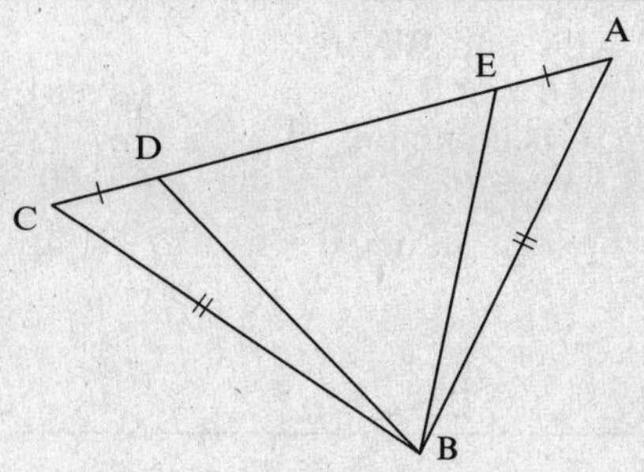

6. (*a*) In $\triangle^s$ ABE and CDB:

1. $BE = DB$ ($\triangle$ DBE; equilateral)
2. $B\hat{D}C = A\hat{B}E$ (= 60°; equilateral $\triangle$)
3. $A\hat{E}B = D\hat{B}C$ ($60° + \hat{1} = 60° + \hat{2}$)

$\therefore$ $\triangle$ ABE $\equiv$ $\triangle$ CDB ($\angle$, $\angle$, S)

(*b*) From congruency:

$AB = DC$

$\therefore$ $AB - DB = DC - DE$ (equilateral $\triangle$)

$\therefore$ $AD = EC$

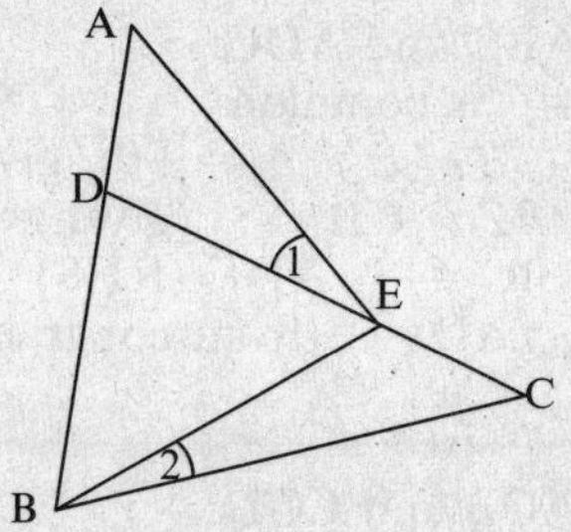

7. (*a*) In $\triangle^s$ ABD and BAC:

1. AB is common
2. $\hat{A}_1 = \hat{B}_1$ (given)
3. $\hat{C} = \hat{D}$ (given)

$\therefore$ $\triangle$ ABD $\equiv$ $\triangle$ BAC ($\angle$, $\angle$, S)

(*b*) From congruency: AD = BC

In $\triangle^s$ AOD and BOC:

1. $AD = BC$ (From congruency (*a*))
2. $\hat{D} = \hat{C}$ (given)
3. $\hat{O}_1 = \hat{O}_2$ (Vertically opp. $\angle^s$)

$\therefore$ $\triangle$ AOD $\equiv$ $\triangle$ BOC ($\angle$, $\angle$, S)

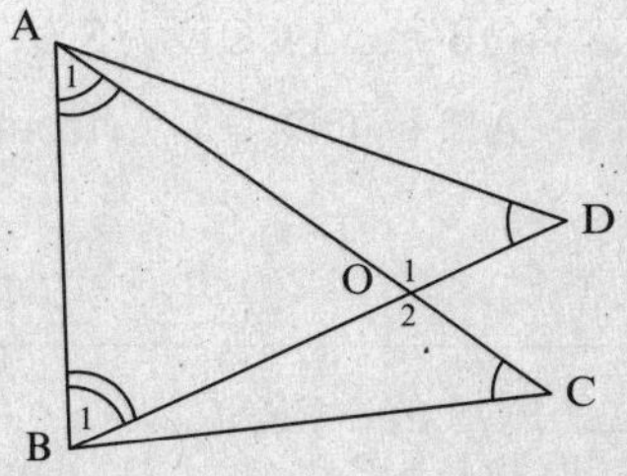

8. (*a*) In $\triangle^s$ ABE and ACD:

1. $AB = AC$ (given)
2. $\hat{A}_1 = \hat{C}_1$ (Alternate $\angle^s$ AB ∥ CD)
3. $AE = CD$ (given)

$\therefore$ $\triangle$ ABE $\equiv$ $\triangle$ CAD (S, $\angle$, S)

(*b*) $\hat{A}_2 = \hat{B}_2$ (from congruency (*a*))

$\therefore$ $B\hat{A}D = \hat{A}_1 + \hat{A}_2$

$B\hat{E}C = \hat{A}_1 + \hat{B}_2$ (Ext. $\angle$ of $\triangle$ *ABE*)

$\therefore$ $B\hat{A}D = B\hat{E}C$

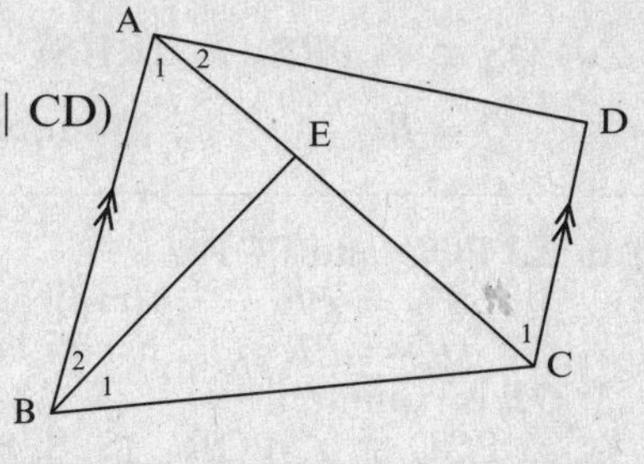

9. In $\triangle^s$ ABC and DCB:

1. $AB = CD$ (given)
2. $A\hat{B}C = D\hat{C}B$ (given)
3. BC is common

$\therefore$ $\triangle$ ABC $\equiv$ $\triangle$ DCB (S, $\angle$, S)

$\therefore$ AC = DB (from congruency)

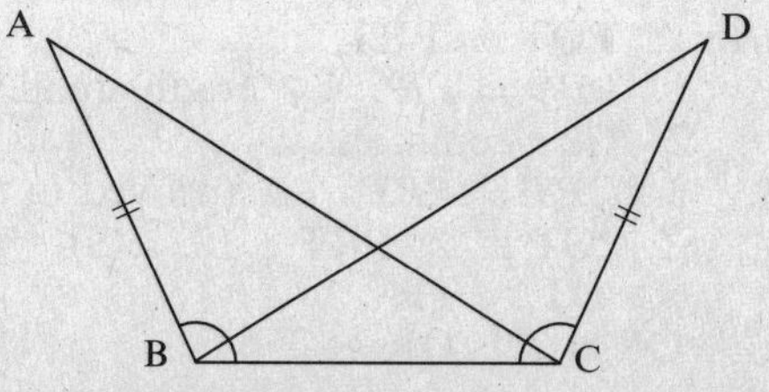

10. In $\triangle^s$ ABV and WXV:

1. $AB = XW$ (given)
2. $\hat{V}$ is common
3. $\hat{X}_1 = \hat{B}_1$ (= 90° given)

$\therefore$ $\triangle$ ABV $\equiv$ $\triangle$ WXV ($\angle$, $\angle$, S)

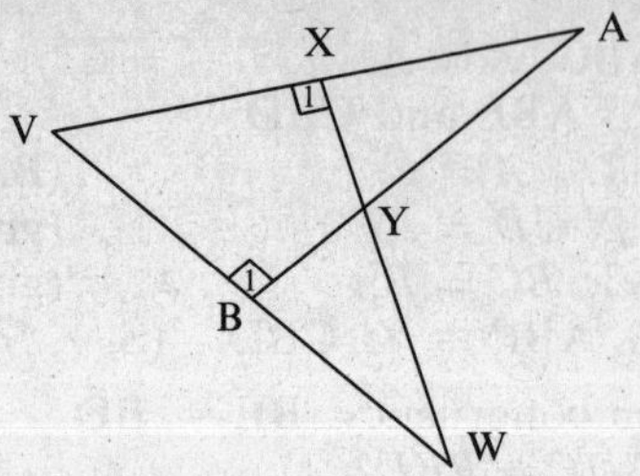

11. $\hat{C}_1 = 63°$ (straight line $\angle$)

$\hat{D} = 153 - 63°$ (Ext. $\angle$ of $\triangle$ ACD)

$= 90°$

In $\triangle^s$ ABC and ADC:

1. AC is common
2. $\hat{B} = \hat{D}$ (= 90°, given and proved)
3. $BC = CD$ (given)

$\therefore$ $\triangle$ ABC $\equiv$ $\triangle$ ADC (RHS)

$\therefore$ AB = AD (from congruency)

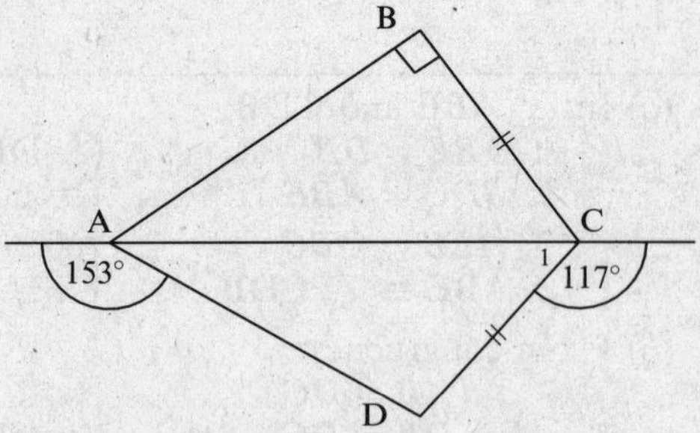

12. In $\triangle^s$ AOB and COD:

1. $AO = OC$ (given)
2. $OB = OD$ (given)
3. $A\hat{O}B = C\hat{O}D$ (both: 90° + x)

$\therefore$ $\triangle$ AOB $\equiv$ $\triangle$ COD (S, $\angle$, S)

$\therefore$ AB = CD (from congruency)

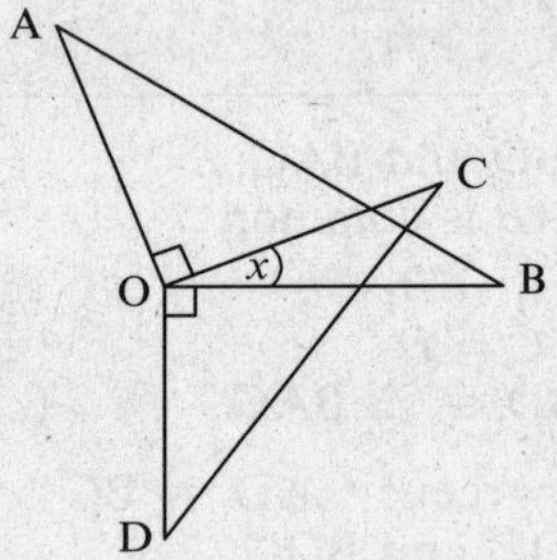

13. In $\triangle^s$ PQS and PRS:

1. $PQ = PR$ (given)
2. PS is common
3. $\hat{S}_1 = \hat{S}_2$ (= 90°, QSR straight line)

$\therefore$ $\triangle$ PQS $\equiv$ $\triangle$ PRS (RHS)

$\therefore$ $\hat{Q} = \hat{R}$ (from congruency)

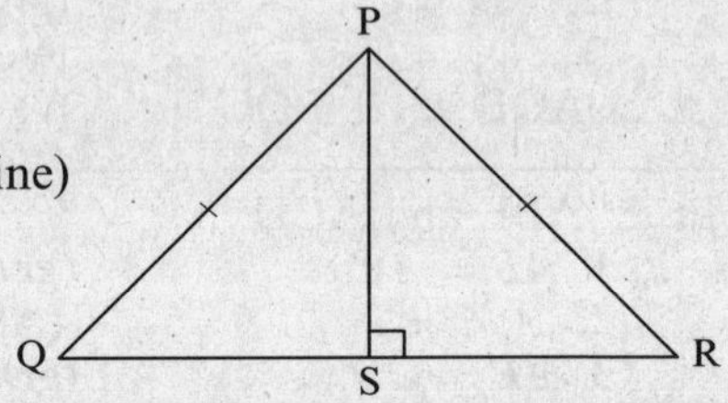

14. (*a*) In $\triangle^s$ POQ and POR:

1. $PQ = PR$ (radii small ⊙)
2. $OQ = PR$ (radii large ⊙)
3. OP is common

$\therefore$ $\triangle$ POQ $\equiv$ $\triangle$ POR (S, S, S)

$\therefore$ $Q\hat{P}O = R\hat{P}O$ (from congruency)

(*b*) In $\triangle^s$ PQT en PRT:

1. $PQ = PR$ (radii small ⊙)
2. PT is common
3. $Q\hat{P}T = R\hat{P}T$ (proved in (*a*))

$\therefore$ $\triangle$ PQT $\equiv$ $\triangle$ PRT (S, $\angle$, S)

$\therefore$ QT = TR

$\therefore$ OP bisects QR at T

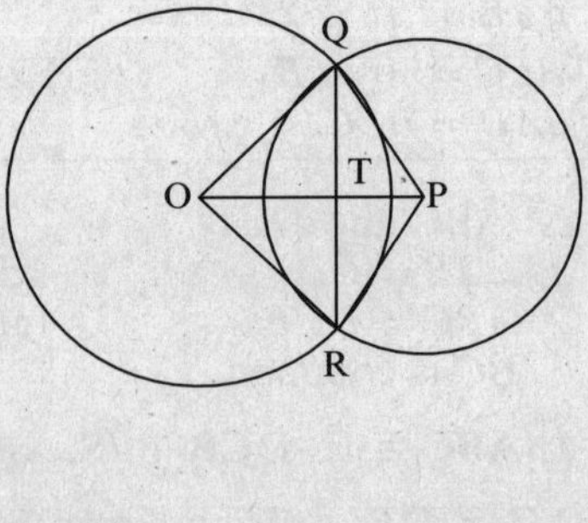

15 *Quadrilaterals*

Special Quadrilaterals:

1. Parellelogram

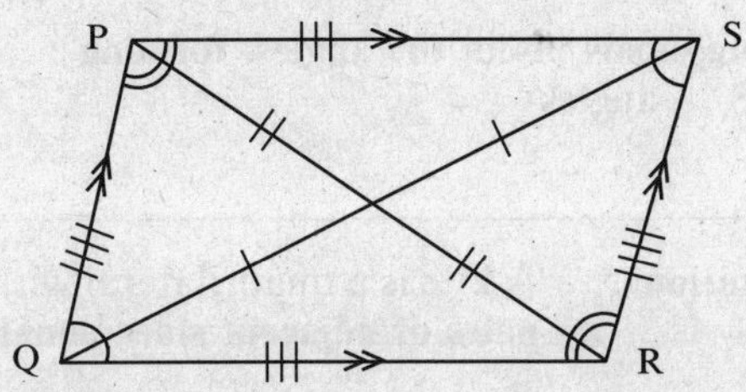

Definition: A quadrilateral with **both pairs of opposite sides parallel.**

Properties:
- 2 pairs opposite sides =
- 2 pairs opposite angles =
- Diagonals bisect one another
- Diagonals bisect area of parallelogram.

2. Rectangle

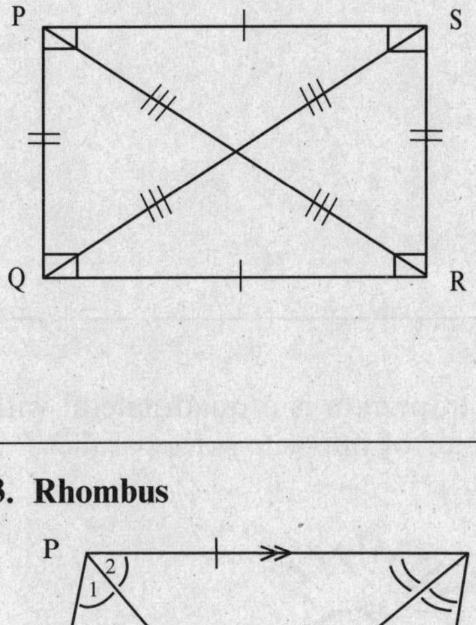

Definition: A rectangle is a parallelogram with **one 90°-angle**

Properties:
- 2 Pairs opposite sides =
- 2 Pairs opposite angles ||
- Diagonals bisect one another

} from parallelogram

- **All angles are 90°.**
- **Diagonals are equal. (PR = QS)**

3. Rhombus

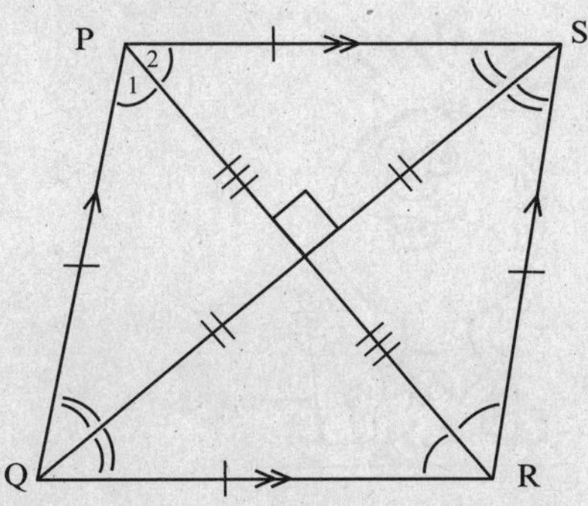

Definition: A rhombus is a paralleogram with 2 adjacent sides equal:

Properties:
- 2 pairs opposite sides: ||
- 2 pairs opposite angles =
- Diagonals bisect one another.

} from parallelogram

- **All the sides =**
- **Diagonals bisect** one another – at 90°.
- **Diagonals bisect the angles $\hat{P}_1 = \hat{P}_2$, etc.**

4. Square

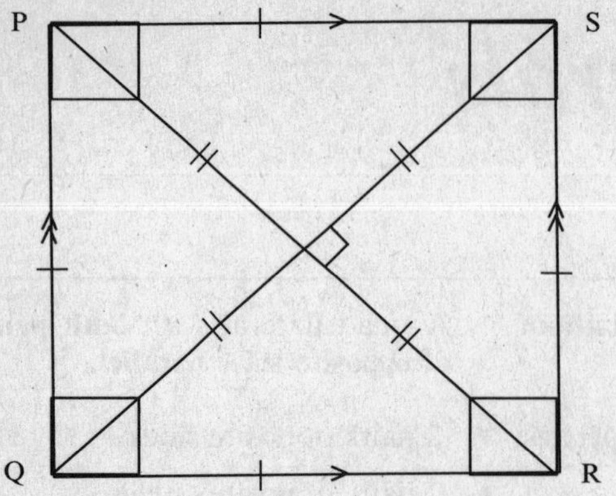

Definition: A square is a rectangle with **2 adjacent sides equal.**

Properties:

- 2 pairs opposite sides ||
- Diagonals bisect one another.
- All angles are 90°.
- Diagonals are equal.

} from rectangle

- **All sides are equal.**
- **Diagonals cut at 90°.**
- **Diagonals bisect the angles, forming 45° – angles**

5. Kite

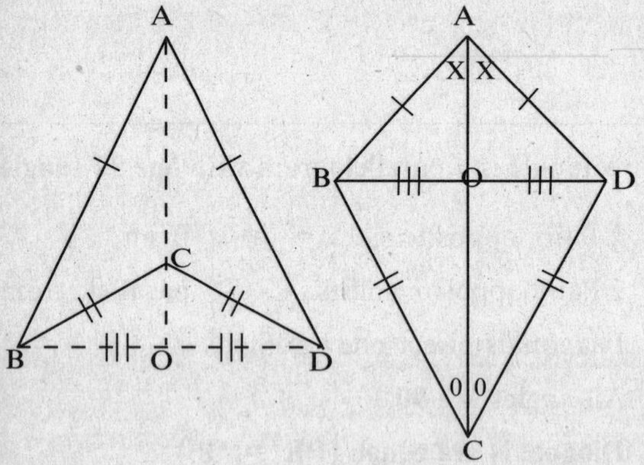

Definition: A kite is a quadrilateral with **two pairs of adjacent sides equal**.

Properties:

- Diagonals cut at 90°.
- One diagonal bisects the other. BO = OD
- Diagonal AC bisects the angles.

6. Trapezium

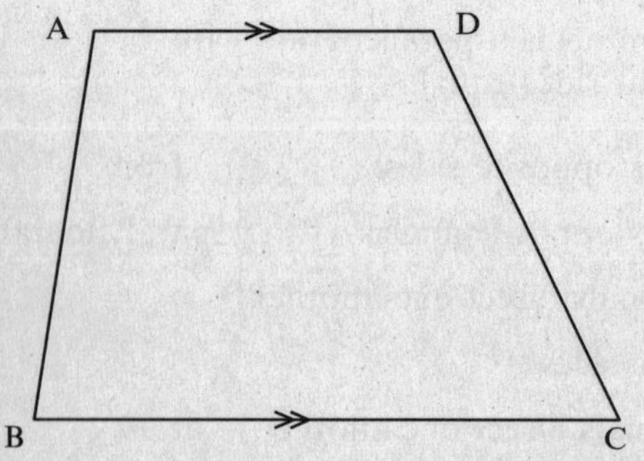

Definition: A trapezium is a quadrilateral with 1 pair of opposite sides parallel.

Area formulas:

1. Parallelogram ($||^m$)

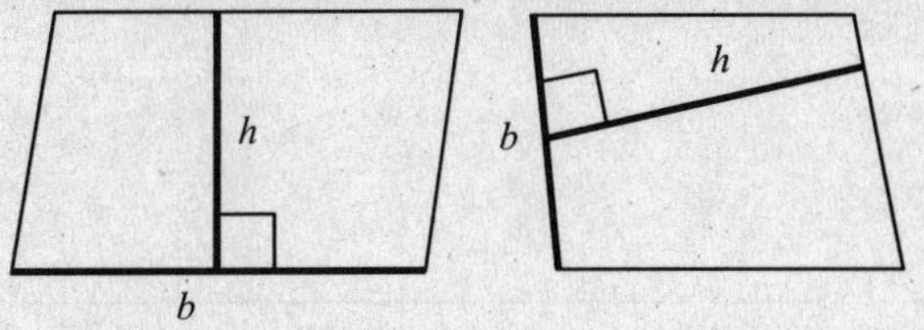

Area $= b \times h$
$=$ base $\times$ perpendicular height

Base: a side of the parallelogram
Perpendicular height: height (altitude) on **this** side (base)

2. Triangle

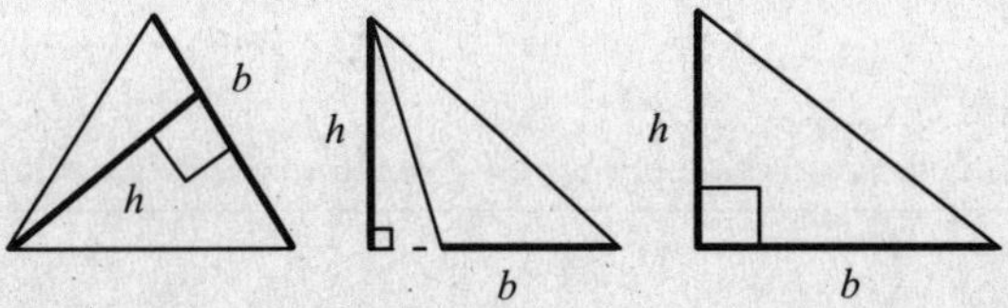

Area $= \frac{1}{2}\, b \times h$
$= \frac{1}{2}$ base $\times$ perpendicular height
(a side) (on this side)

3. Square

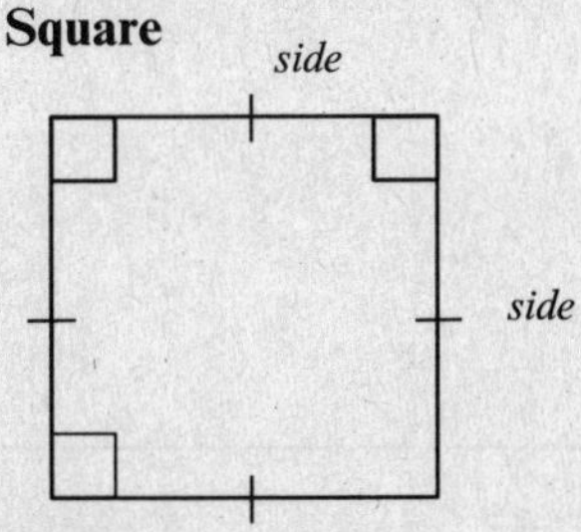

Area $= side \times side$
$= side^2$

4. Rectangle

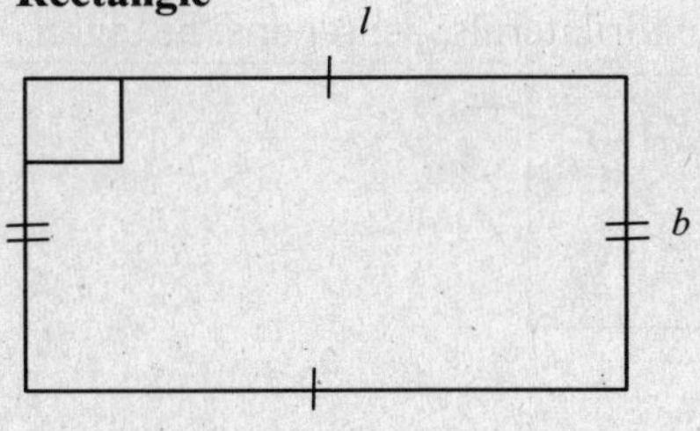

Area $= l \times b$
$=$ length $\times$ breadth

5. Rhombus

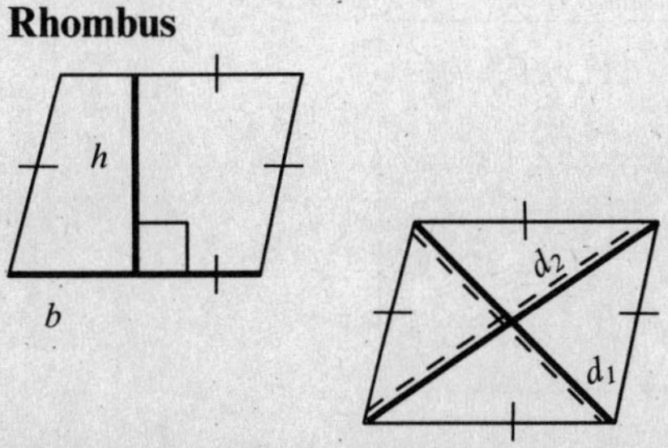

Area $= b \times h$ (like parallelogram)
$=$ base $\times$ perpendicular height

or

Area $= \frac{1}{2}\, d_1.d_2$
$= \frac{1}{2}$ product of the diagonals

6. Kite

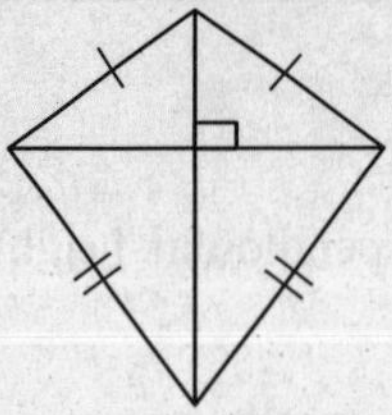

Area $= \frac{1}{2}$ product of the diagonals

7. Trapezium

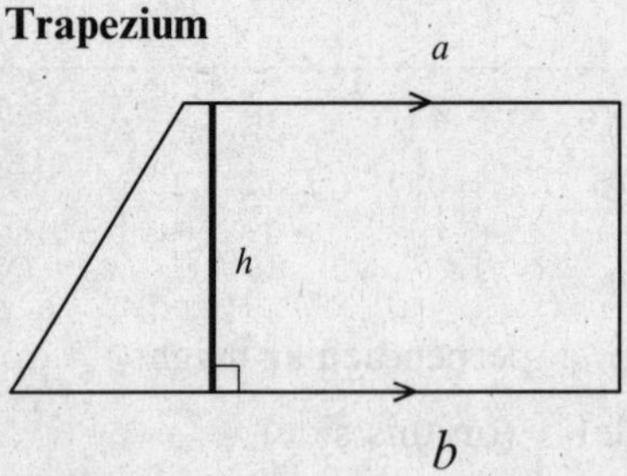

Area $= \frac{h}{2}$. sum of parallel sides

$= \frac{h}{2} \cdot (a + b)$

8. Circle

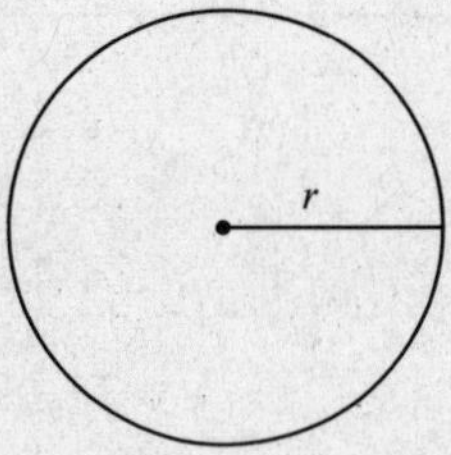

Area $= \pi r^2$

Circumference $= 2\pi r$

Polygons:

Important formulas concerning polygons namely triangles, quadrilaterals, pentagons, hexagons, etc:

Sum of the interior angles of a polygon with n sides $= (n–2)180°$

Sum of the exterior angles of any polygon $= 360°$

A **regular polygon** is a polygon with equal sides and therefore equal angles too.

Every interior angle of a regular polygon with n sides $= \frac{(n-2)180°}{n}$

Every exterior angle of a regular polygon with n sides $= \frac{360°}{n}$

Polygons:

Regular polygon – all the sides and all the angles are equal.

Pentagon (5 sides)

Hexagon (6 sides)

Heptagon (7 sides)

Octagon (8 sides)

Examples: Polygons

1. Determine the sum of the interion angles of an octagon.

Solution:

$$\begin{aligned}\text{Sum of interior } \angle^{s} &= (n-2)180^\circ \\ &= (8-2)180^\circ \\ &= 6 \times 180^\circ \\ &= 1080^\circ\end{aligned}$$

2. Determine the size of each interior angle of a regular polygon with 12 sides .

Solution:

$$\begin{aligned}\text{Each interior angle} &= \frac{(n-2)180^\circ}{n} \\ &= \frac{(12-2)180}{12} \\ &= \frac{10 \times 180}{12} \\ &= 150^\circ\end{aligned}$$

3. Determine the number of sides of a polygon if every interior angle is 144°.

Solution:

$$\begin{aligned}\text{Every exterior angle is } &= 180^\circ - 144^\circ \text{ (straight line)} \\ &= 36^\circ\end{aligned}$$

Sum of exterior angle is 360°

$$\text{Number of sides: } = \frac{360^\circ}{36^\circ} = 10$$

4. ABCDE is a regular pentagon. Find x

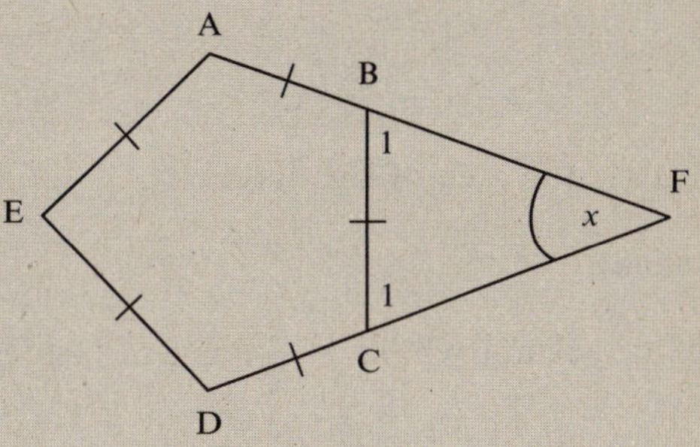

Solution:

$$\text{Every exterior angle of pentagon: } \frac{360^\circ}{5} = 72^\circ$$

$$\hat{B}_1 = \hat{C}_1 = 72^\circ$$

$$\begin{aligned}x &= 180^\circ - (72+72) \text{ (3}\angle^{s} \text{ of } \triangle BCF = 180^\circ \\ &= 36^\circ\end{aligned}$$

Examples: Quadrilaterals

1. Find x.

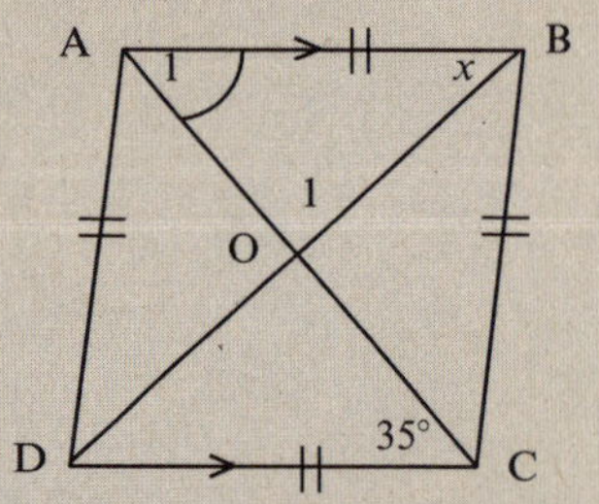

Solution:

ABCD is a rhombus
(AB=AD; AB||DC and AD||BC)

$\therefore \hat{A}_1 = 35°$ (alternate $\angle^s$, AB||DC)
$\hat{O}_1 = 90°$ (diagonals of rhombus intersect at 90°)
$\therefore\ x = 180° - (90+35)$ ($3\angle^s$ of $\triangle AOB = 180°$)
$= 55°$

2. Find x.

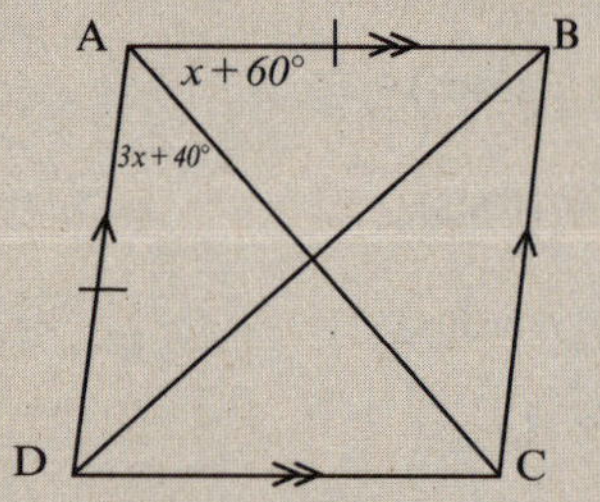

Solution:

ABCD is a rhombus (AB = AD and AB||DC

$x + 60 = 3x + 40$ (diagonals of rhombus bisect the angles)
$\therefore\ -2x = -20$
$x = 10$

Examples: Areas

1.

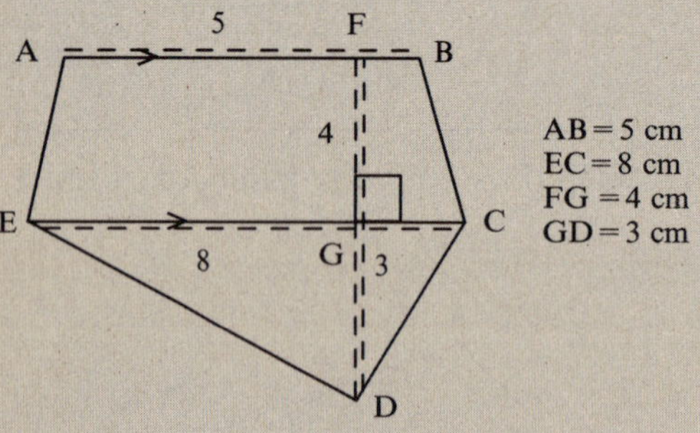

Calculate the area of the diagram.

Solution:

Area trapezium ABCE $= \frac{h}{2}$ (sum of ||-sides)
$= \frac{4}{2}(5+8)$
$= 2\ (13)$
$= 26\ cm^2$

Area $\triangle$DEC $= \frac{1}{2}\, b.h.$
$= \frac{1}{2} \cdot 8.3$
$= 12\ cm^2$

$\therefore$ Area of diagram: 38 cm^2

2.

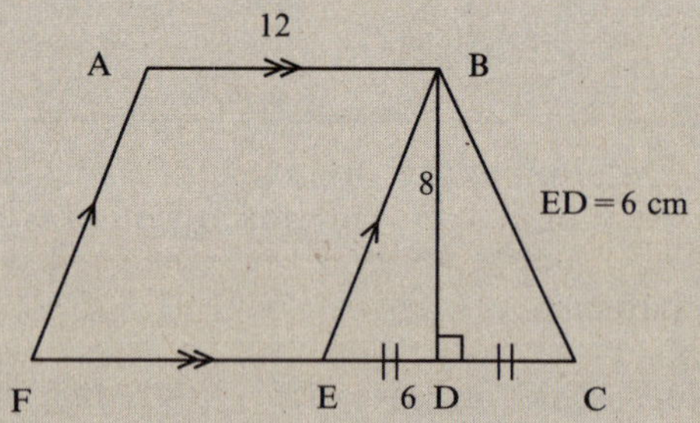

Calculate the area of the diagram.

Solution:

Area $||^m$ ABEF = base × height
$= 12 \times 8$
$= 96\ cm^2$

Area BEC $\triangle = \frac{1}{2}\, b.h$
$= \frac{1}{2} \cdot 12 \cdot 8$
$= 48\ cm^2$

$\therefore$ Area of diagram: $96 + 48 = 144\ cm^2$

Exercise 1: Quadrilaterals

1. Decide which of the following quadrilaterals has the property given below:

parallelogram; rectangle; rhombus; square; kite.

1.1 Diagonals always equal.

1.2 Diagonals always perpendicular.

1.3 Diagonals bisect the area of the quadrilateral.

1.4 Diagonals bisect the angles.

2. Identify each of the following quadrilaterals:

2.1

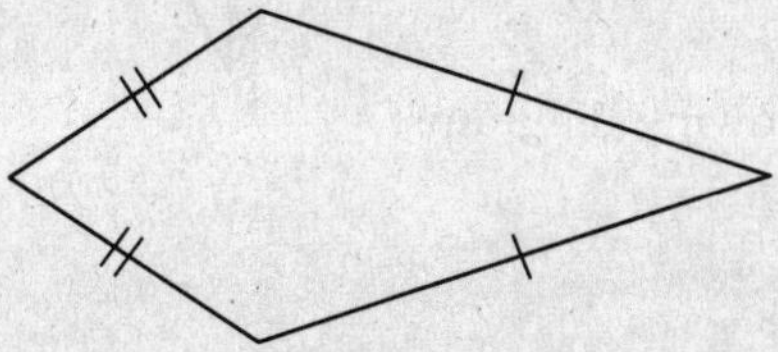

2.2

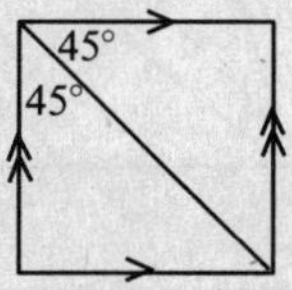

2.3

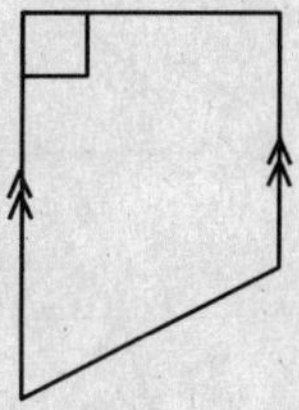

2.4

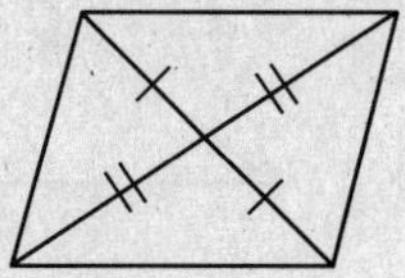

2.5

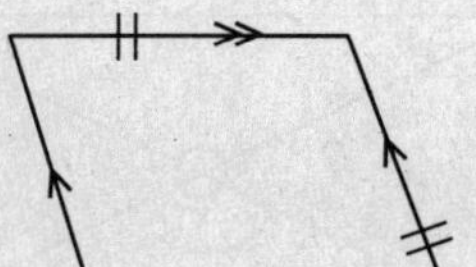

2.6

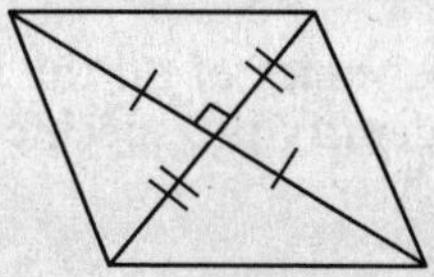

2.7

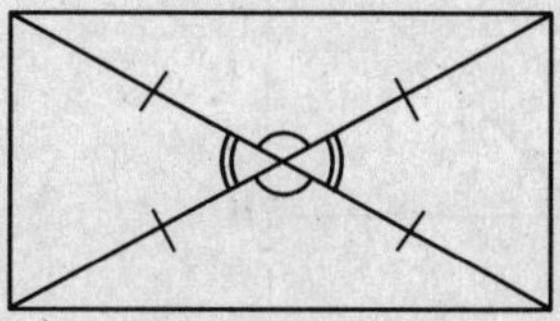

2.8

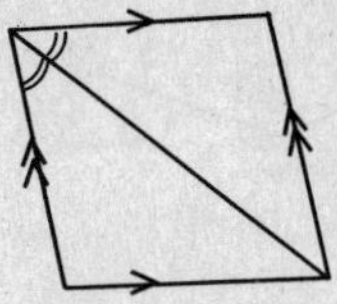

2.9

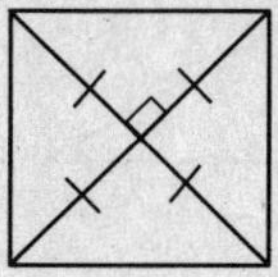

2.10

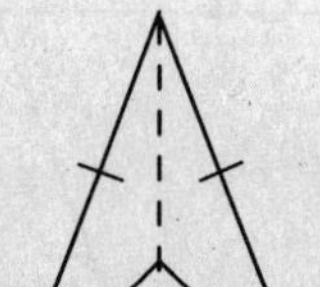

3. Determine, with reasons, the values of **a – g**.

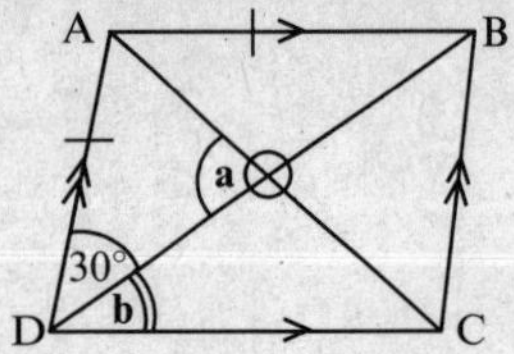

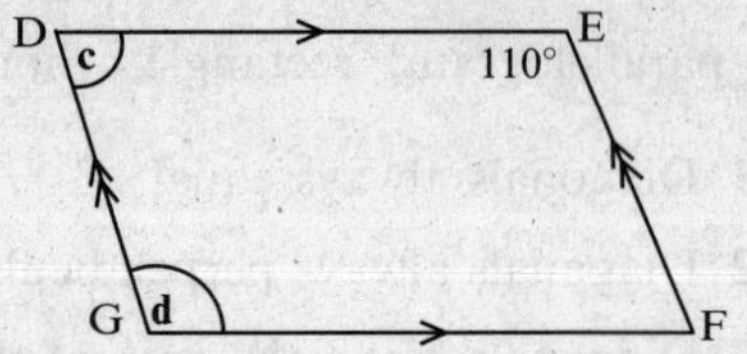

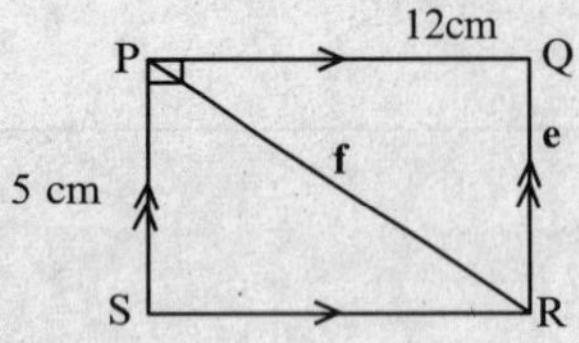

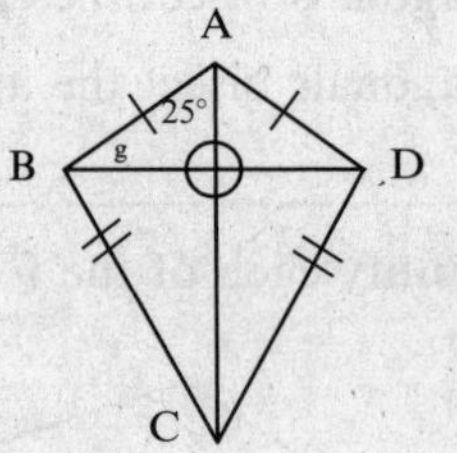

4. Determine the **values** of all the **variables** in the following diagrams:
(**Hint:** First identify the specific quadrilateral)

4.1

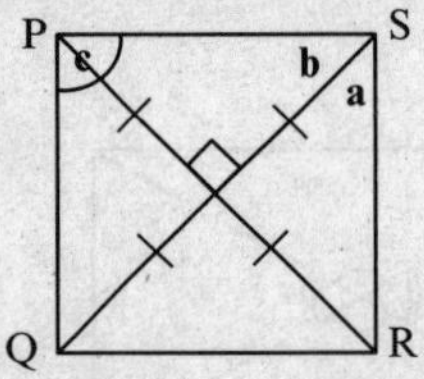

4.5

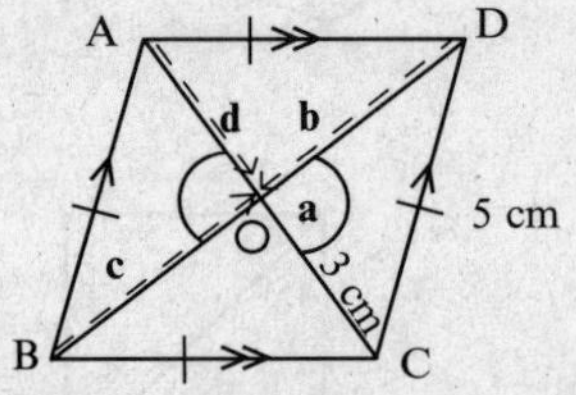

4.2

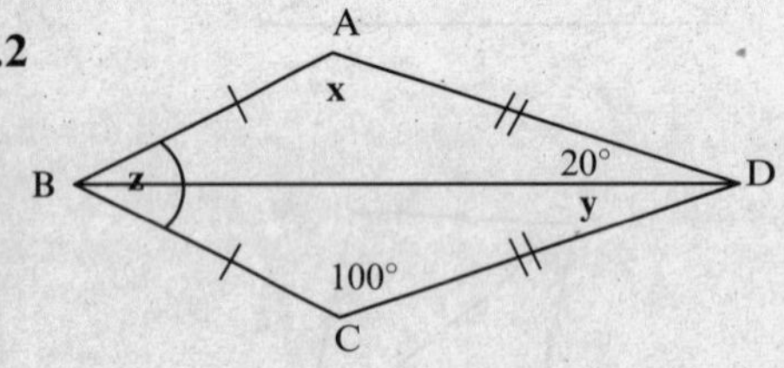

4.6

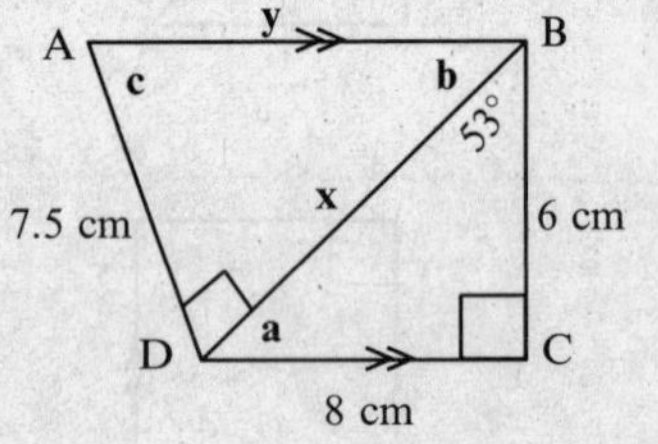

4.3

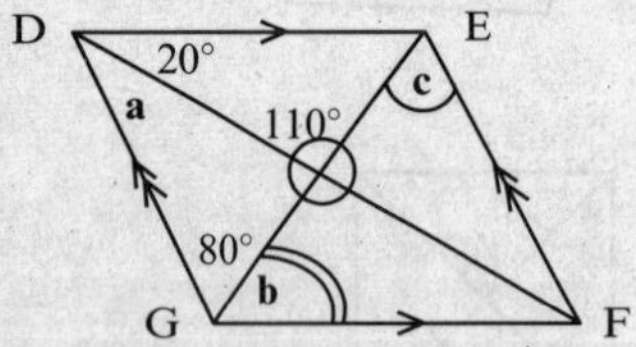

4.7

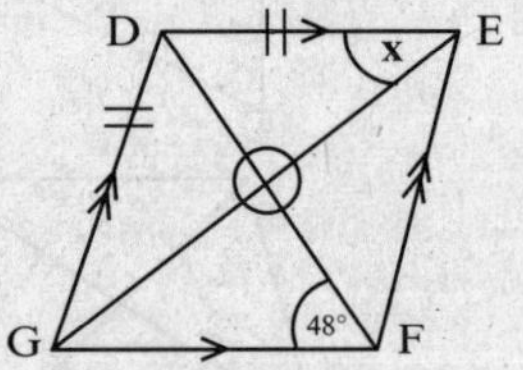

4.4

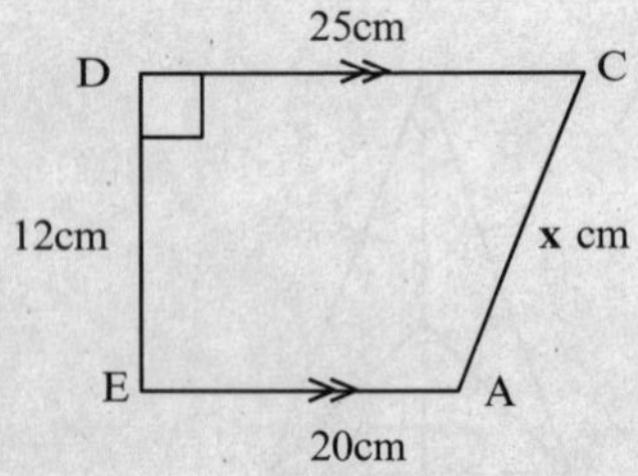

4.8

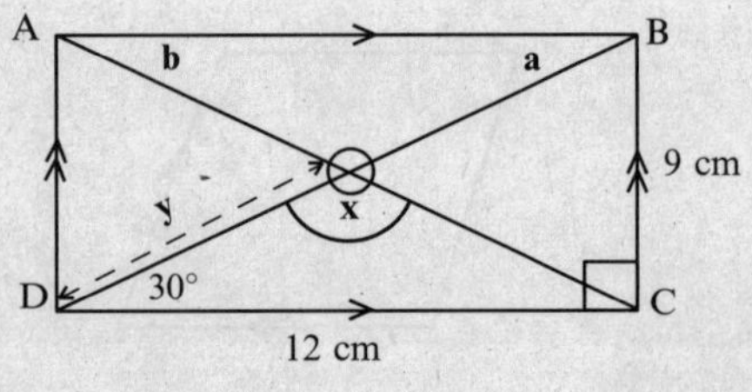

5. Determine the value of x:
(with reasons)

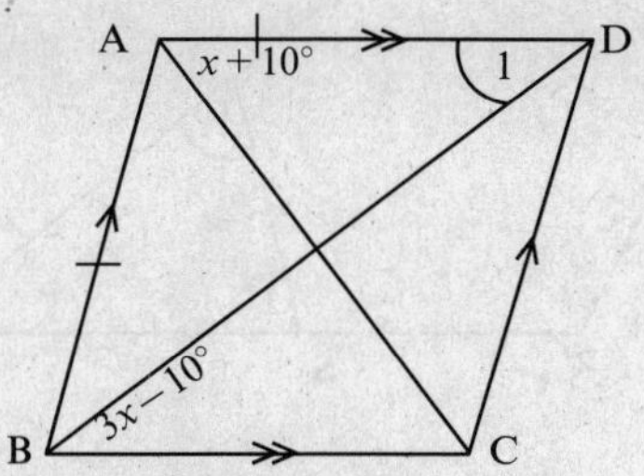

6. ABCD is a parallelogram.
BF = BE and AC‖FE. Calculate x

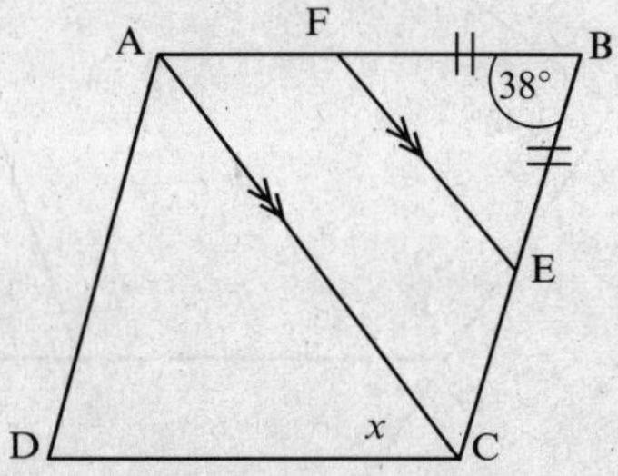

7. ABCD is a parallelogram.
Calculate $\hat{A}$:

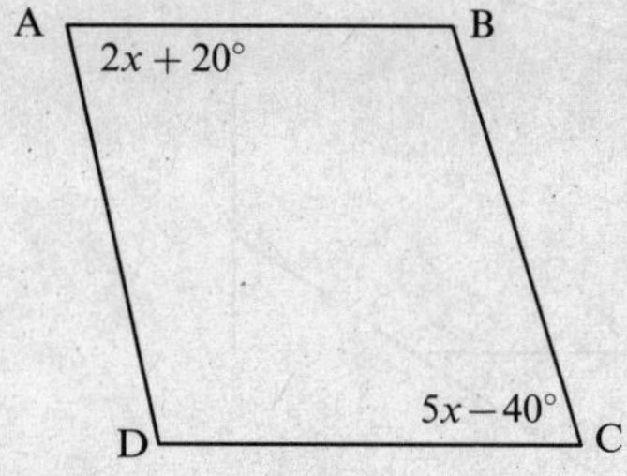

8. DEFG is a rhombus.
Find x.

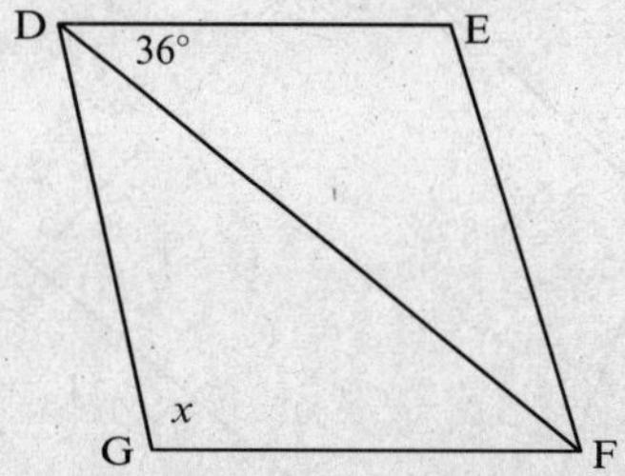

Exercise 2: Polygons

1. Calculate, with reasons, the value of x (and / or y) in each of the following polygons:

1.1

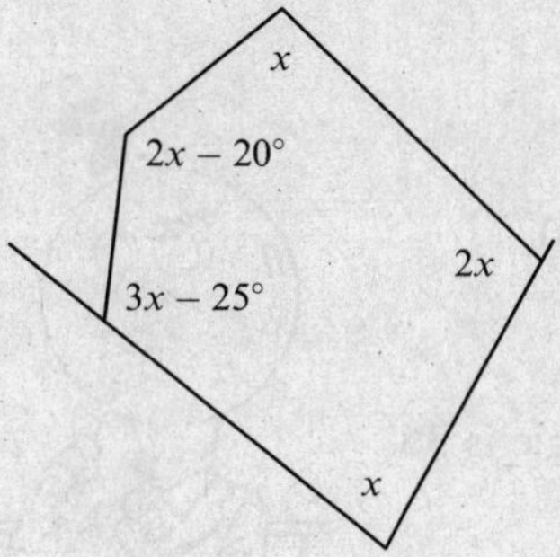

1.2

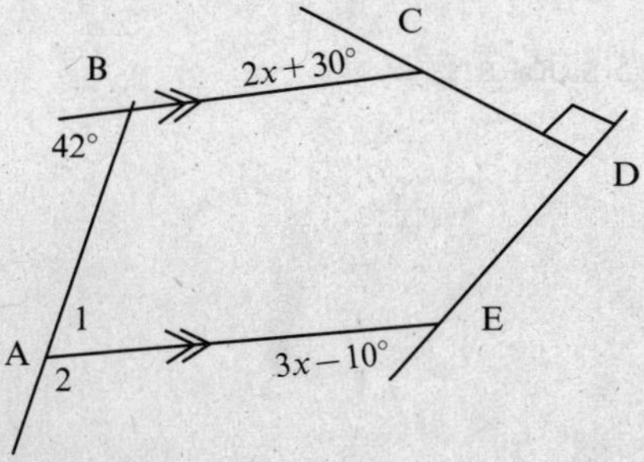

1.3

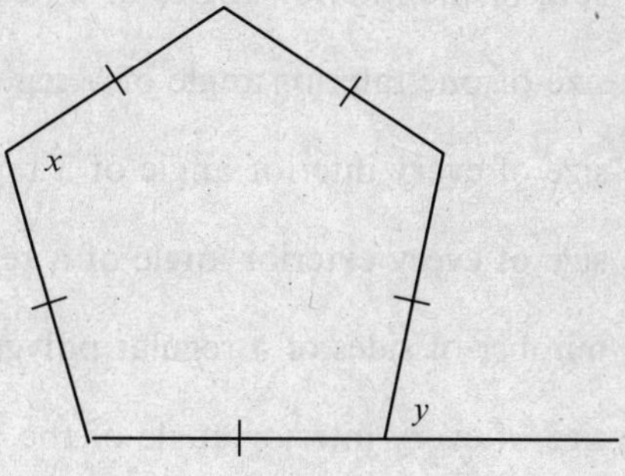

1.4

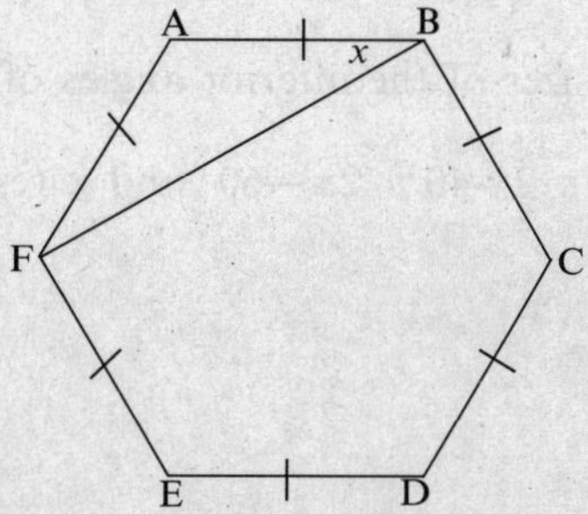

1.5

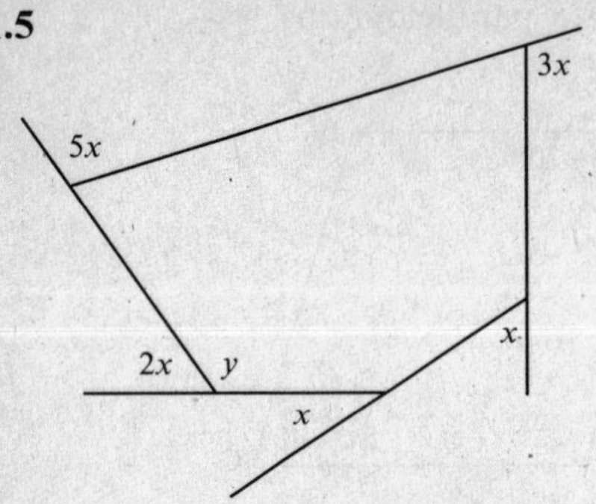

1.6

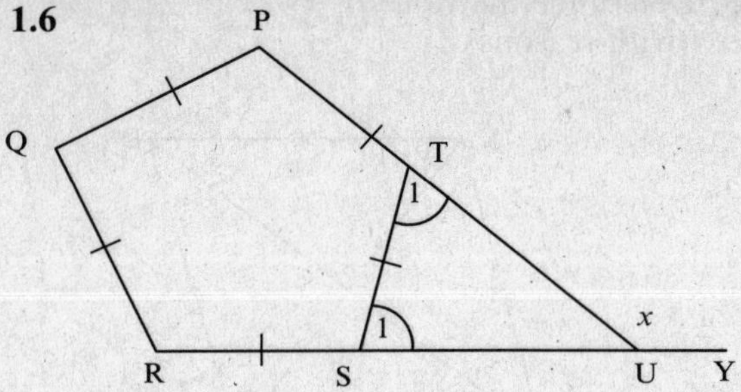

1.7

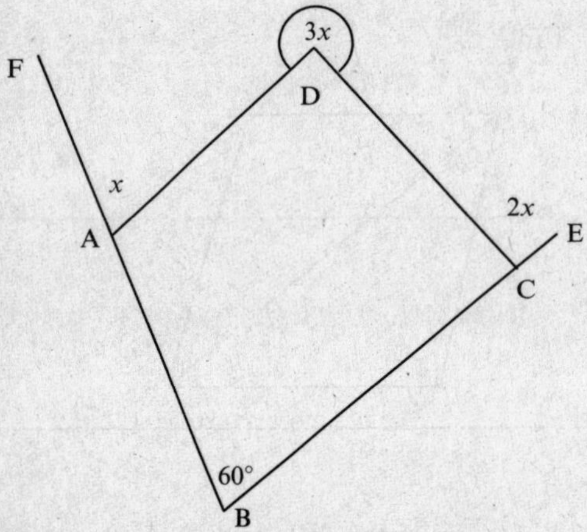

1.8

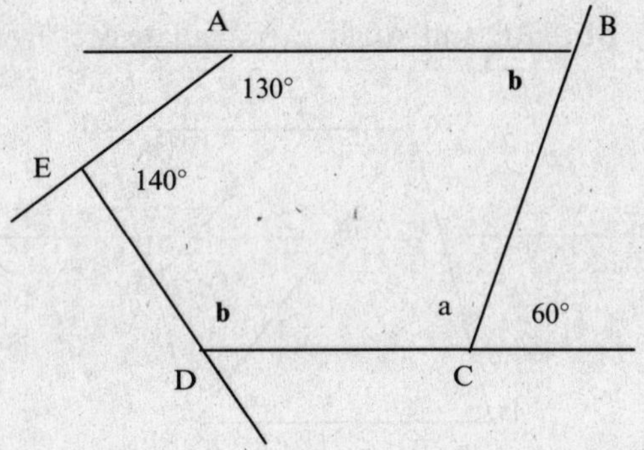

2. Calculate

2.1 the sum of the interior angles of a hexagon.

2.2 the size of one interior angle of a regular pentagon.

2.3 the size of every interior angle of a regular nonagon (9 sides).

2.4 the size of every exterior angle of a regular octagon.

2.5 the number of sides of a regular polygon if every exterior angle is 36°.

2.6 the size of every interior angle of the polygon in 2.5.

2.7 the number of sides of a regular polygon if every interior angle is 135°.

2.8 x, if the sizes of the interior angles of a polygon with 5 sides are:

x; $x+10°$; $2x$-40°; $2x-60°$ and x respectively.

Exercise 3. Areas

1. Complete:

The area of (formulas)

1.1 A parallelogram is . . . .

1.2. A rectangle is . . . .

1.3. A square is . . . .

1.4 A rhombus is . . . .

1.5 A kite is . . . .

1.6 A trapezium is . . . .

2. Write down the area of each of the following diagrams, in terms of *a*; *b*; and/or *c*.

2.1

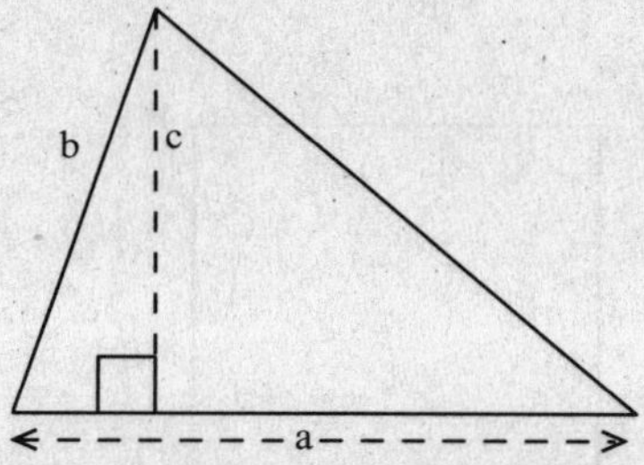

2.2

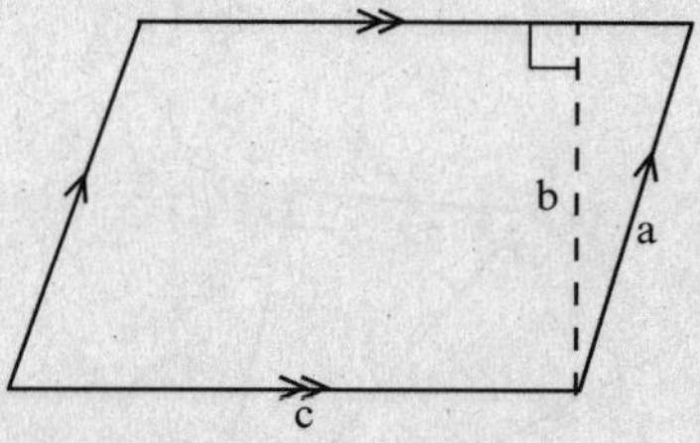

2.3

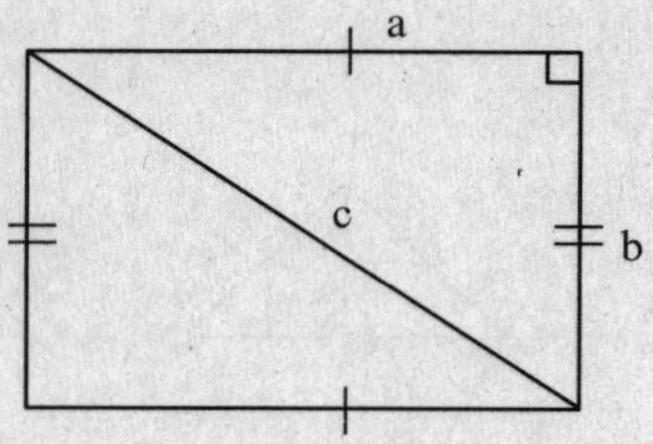

2.4

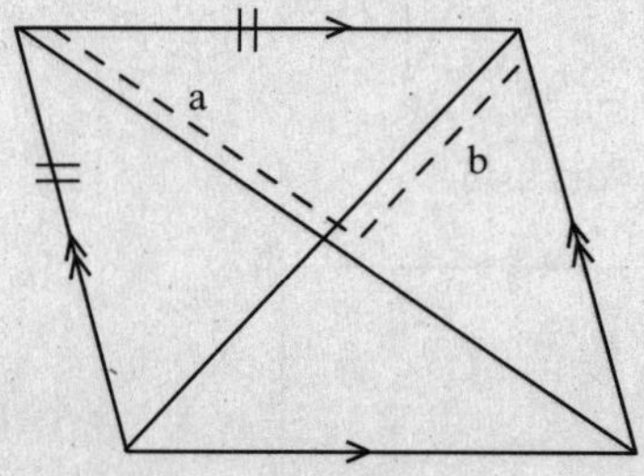

2.5

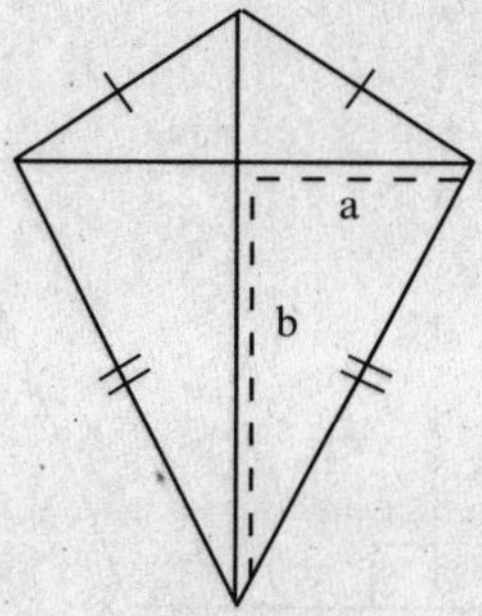

2.6

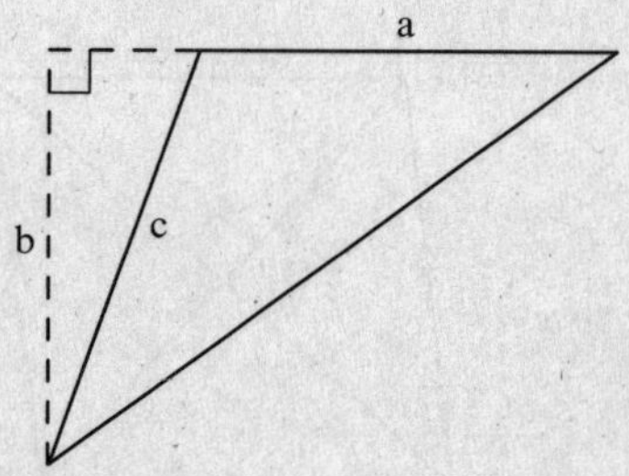

2.7

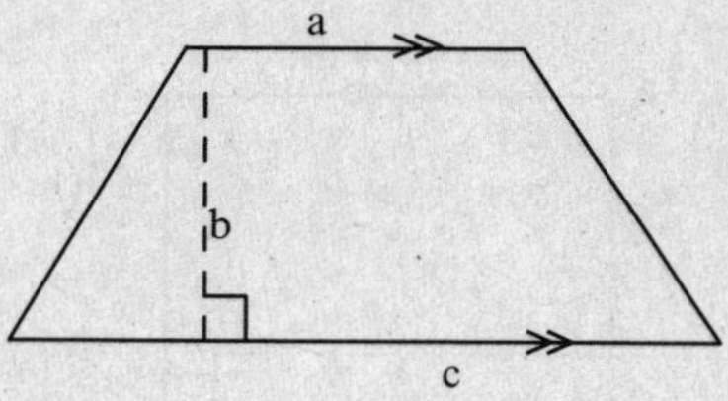

3. Calculate the area of each of the following diagrams:

3.1

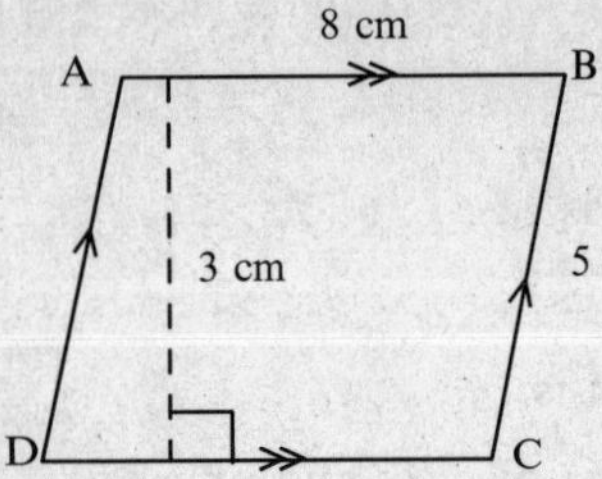

3.2

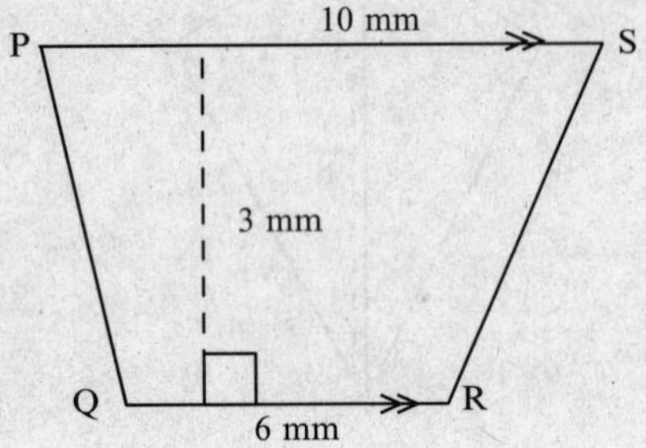

3.3

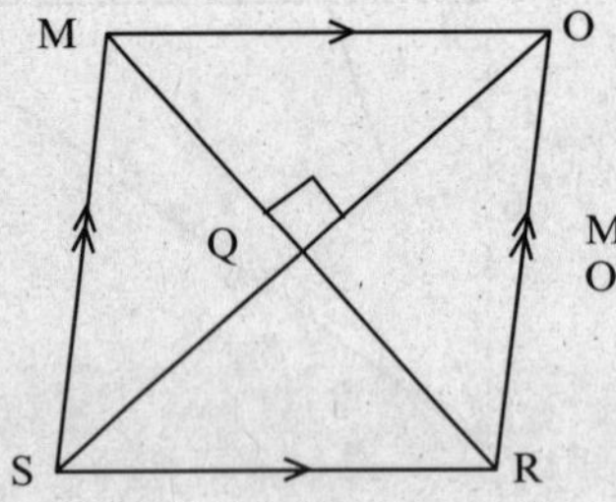

MR = 15 cm
OS = 25 cm

3.4

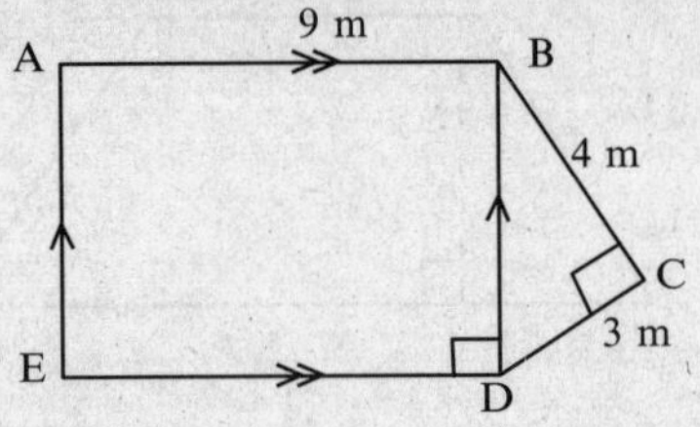

3.5

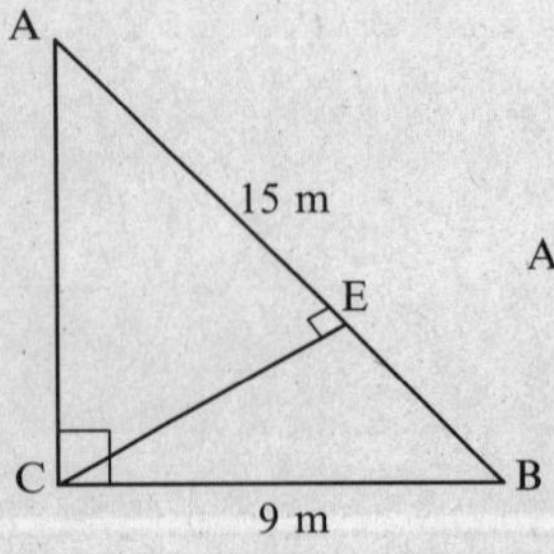

AB = 15 m

3.6

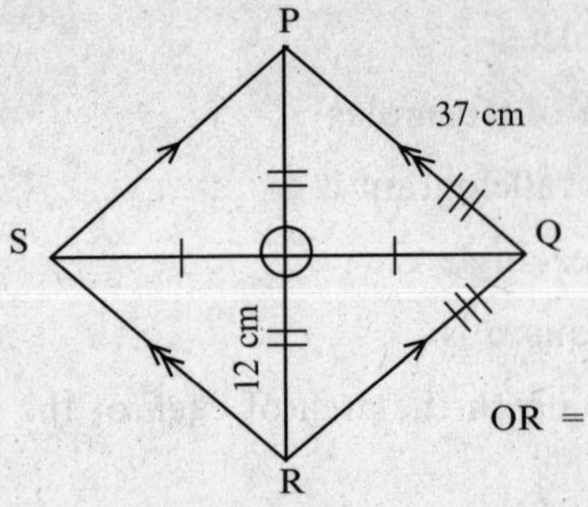

OR = 12 cm

3.7

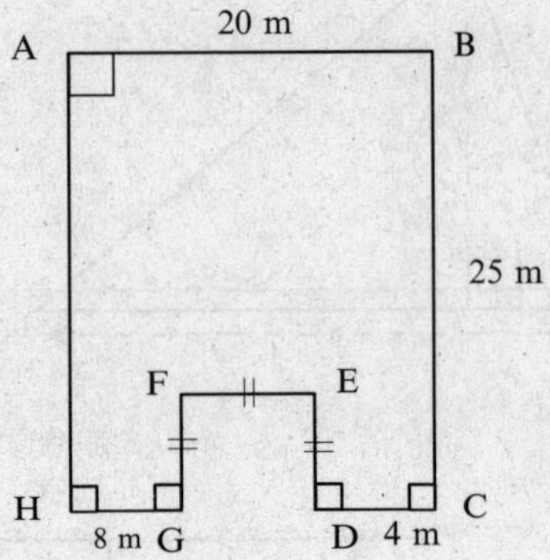

3.8

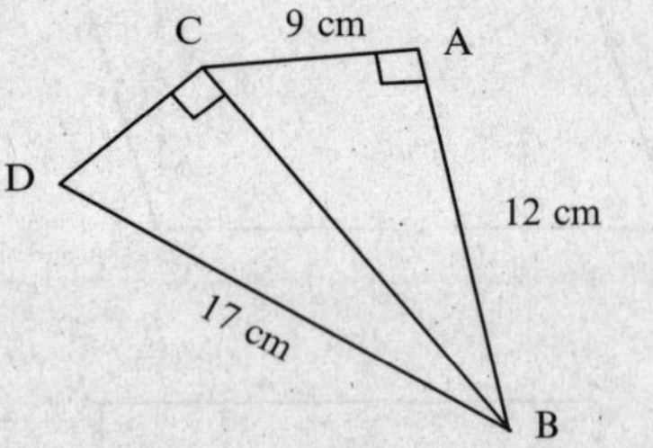

3.9

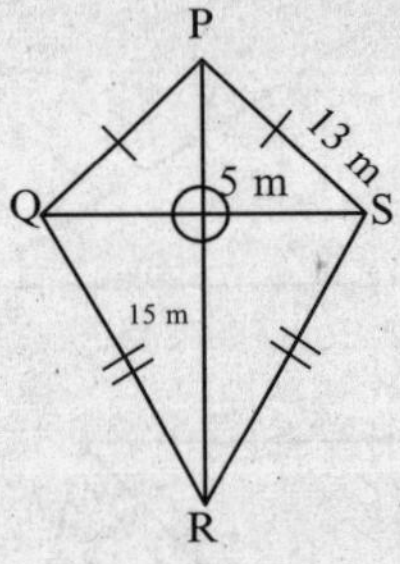

OP = 5 m
OR = 15 m

3.10

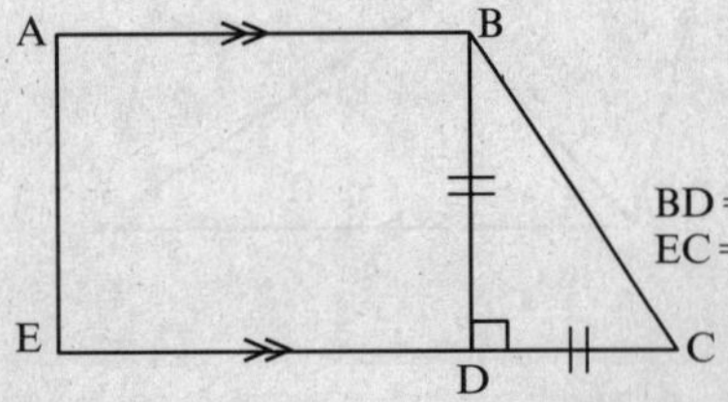

BD = 6 cm
EC = 14 cm

3.11

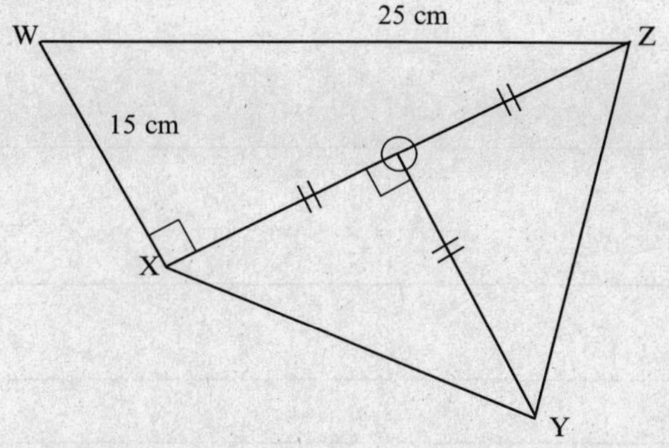

3.12

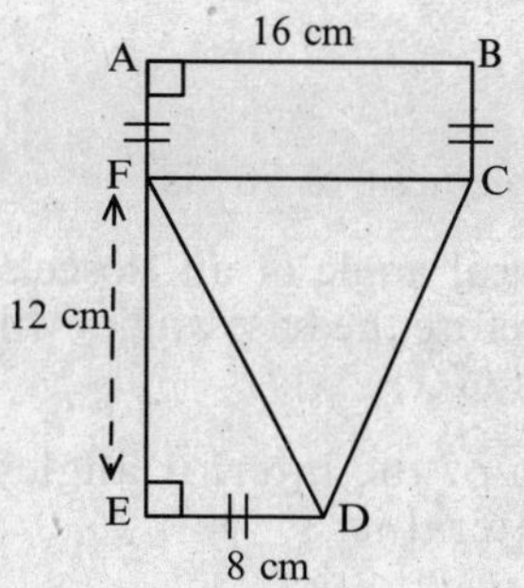

4. ABCD is a rectangle.
CD = 7 cm and the area of the rectangle is 63 cm^2
Calculate the area of trapezium AECD.

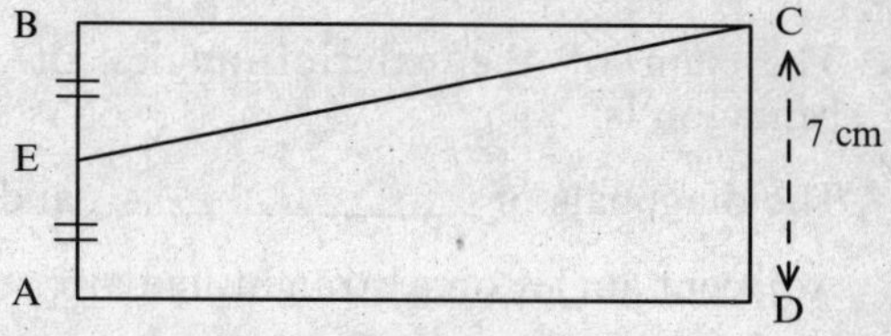

5. Use the diagram and calculate

5.1 AE
5.2 Area of $\triangle AED$
5.3 Area of AEBCD.

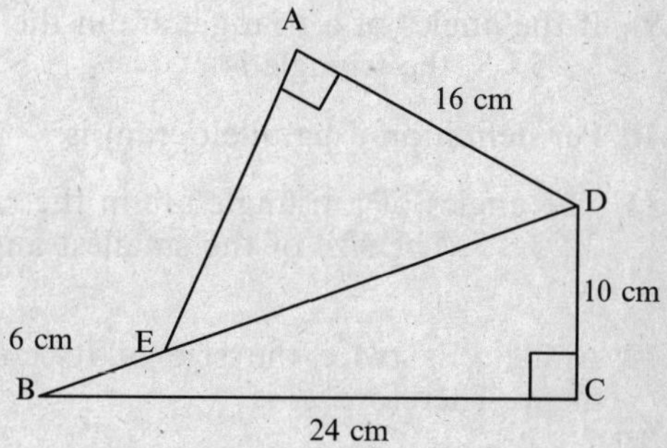

6. ABCD is a rhombus with AO = 4 cm.
The area of the rhombus is 48 cm^2.

Calculate BD.

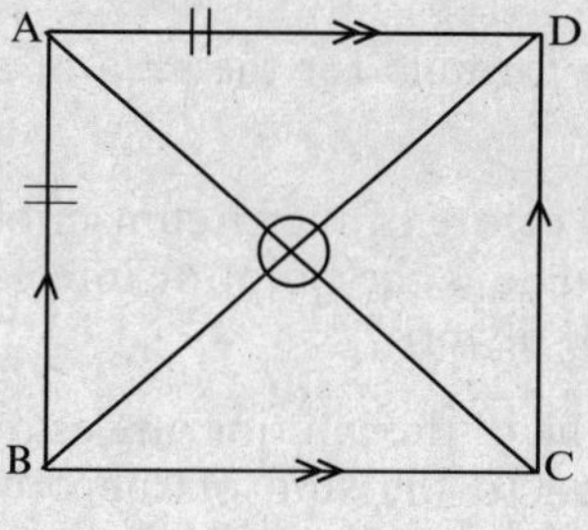

7. The area of the kite is 64 cm^2.

Determine the value of x.

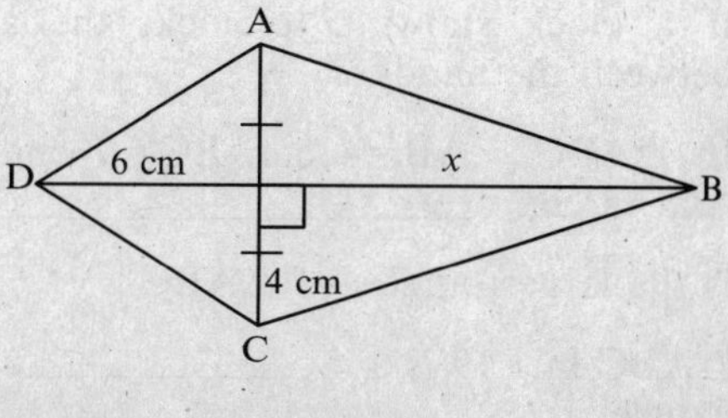

Exercise 4: Mixed geometry questions

1. Complete:

1.1 The supplement of 70° is ______________________°

1.2 The vertical angle of an isosceles triangle is x, therefore the base angles will each be: (in terms of x) ______________________°

1.3 The sum of the interior angles of a quadrilateral is: ______________________

1.4 The complement of $(90 - x)$ is: ______________________

1.5 The sum of the exterior angles of a pentagon is: ______________________

1.6 The diagonals of ______________ and ______________________ are equal.

1.7 Adjacent angles on a straight line are: ______________________

1.8 If the sum of the interior angles of a polygon is 720°, then the polygon has ______________________ sides.

1.9 If the angles of a triangle are in the ratio 5 : 5 : 8, the triangle is ______________________ triangle.

1.10 Per definition a parallelogram is: ______________________

1.11 The angles of a triangle are in the ratio 2 : 3 : 5. The size of the smallest angle is ______________________°

1.12 Angle x is twice the size of its complement. Therefore x is ______________________°

1.13 The sum of the interior angles of a polygon with n sides is ______________________.

1.14 The formula for the area of a circle is ______________________.

1.15 The centre of the circumscribed circle of a triangle is the point of intersection of the (Std. 6 work) ______________________.

1.16 If one of the interior angles of a triangle is equal to the sum of the other 2 interior angles, then the triangle ______________________.

1.17 The supplement of $(90° - x)$ is ______________________.

1.18 If a clock shows 5 o'clock, the angle between the hands is ______________________°

1.19 In △ABC AB = 4 cm; BC = 7 cm and AC = 5 cm. ______________________

is the largest angles of △ABC.

1.20 △ABC in 1.19 is a ______________ triangle.

2. Formulate the theorem of Pythagoras in words.

3. Why are similar triangles not always congruent too?

4. In the diagram

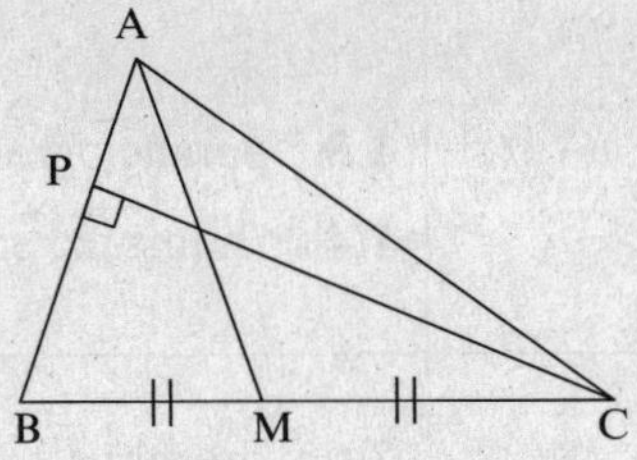

AM is a ____________

CP is a ____________

5. **Are the following statements TRUE or FALSE?**

5.1 The exterior angle of an equalateral triangle is 120°.

5.2 The exterior angles of a triangle can possibly be equal to one of the opposite interior angles.

5.3 The diagonals of a kite bisect one another at 90°.

5.4 A triangle with sides in the ratio 3 : 4 : 5, is a right-angled triangle.

5.5 The diagonals of a parallelogram bisect the area of the parallelogram.

5.6 In a right-angled triangle, the other 2 angles are supplementary.

5.7 Congruent triangles are always similar.

6. The diagonals of a rhombus are 8 cm and 6 cm. Calculate the area of a square which has the same perimeter as this rhombus.

7. **What conclusion can be drawn from each of the following statements?**

(eg. In $\triangle ABC$ $AB = AC$. **Conclusion** $\hat{B} = \hat{C}$.)

7.1 In parallelogram ABCD, $AC = BD$.

7.2 The sum of the interior angles of a polygon is 720°.

7.3 In $\triangle ABC$ $a = 6$ *cm*; $b = 5$ *cm*; $c = 8$ *cm*.

7.4 In $\triangle ABC$ and $\triangle PQR$ $a = p$; $b = r$; and $c = q$.

7.5 In $\triangle ABC$ $\hat{A} : \hat{B} : \hat{C} = 1 : 3 : 2$.

SOLUTIONS

Exercise 1: Quadrilaterals

1.1 rectangle; square

1.2 rhombus; square; kite

1.3 parallelogram; rectangle; rhombus; square

1.4 rhombus; square

2.1 kite

2.2 square

2.3 trapezium

2.4 parallelogram

2.5 rhombus

2.6 rhombus

2.7 rectangle

2.8 rhombus

2.9 square

2.10 kite

3. ABCD is a rhombus

$\therefore$ **a = 90°** (diagonals of a rhombus bisect at 90°)
b = 30° (diagonals of a rhombus bisect the angles)

DEFG is a parallelogram ($\|^{m}$)

c = 70° (co-interior angles; DG$\|$EF)
d = 110° (Opposite angles of $\|^{m}$ equal)

PQRS is a rectangle

e = 5 cm (opposite sides of rectangle equal)
$f^2 = 12^2 + 5^2$ (Pythagoras)
$= 144 + 25 = 169$
$f = \sqrt{169} = 13\ cm$

ABCD is a kite
$\therefore$ $A\hat{O}B = 90°$ (diagonals of a kite intersect at 90°)
$\therefore$ $g = 90° - 25°$ (3 $\angle^{s}$ of $\triangle AOB = 180$)
g = 65°

4.1 PQRS is a square

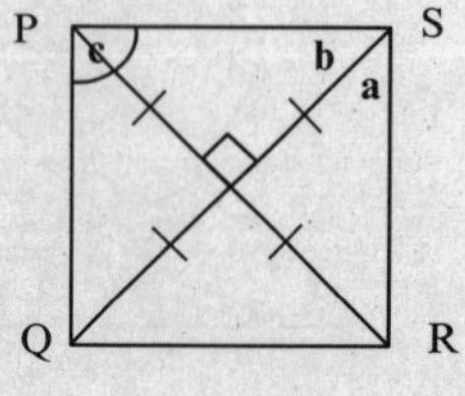

$a = 45°$ (diagonals of square bisect angles)
$b = 45°$

$c = 90°$ (PQRS is a square)

4.2 ABCD is a kite

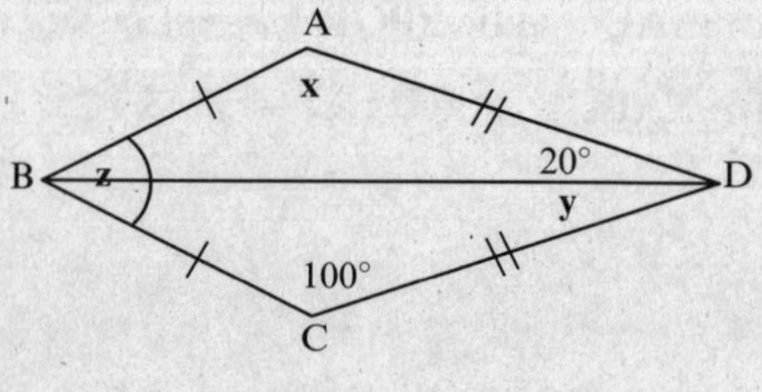

$x = 100°$ ($\triangle ABD \equiv \triangle CBD$ (S; S; S)
$y = 20°$ ($\triangle ABD \equiv \triangle CBD$ (S; S; S))
$z = 120°$ (4 angles of ABCD $= 360°$)

4.3 DEFG is a $||^m$

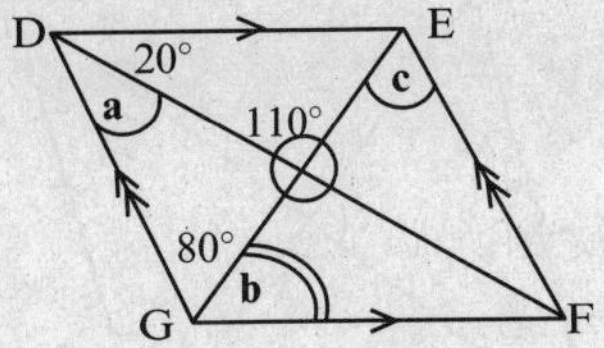

$a = 110° - 80°$ (Ext. $\angle$ of DOG)
$= 30°$

$b = 180° - (20° + 30° + 80°)$
$= 50°$ (co-interior $\angle^s$; DE$\|$GF)

$c = 80°$ (Alternate $\angle^s$; DG$\|$EF)

4.4

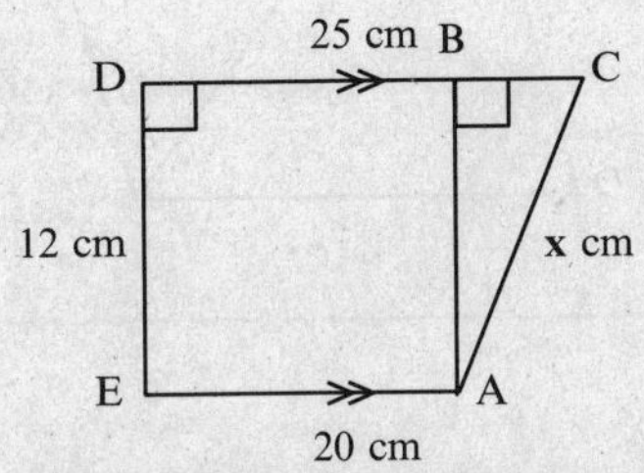

AB = 12 cm (ABDE is a rectangle)
BC = 5 cm

$x^2 = 12^2 + 5^2$
$= 144 + 25$
$= 169$
$x = 13$ cm

4.5 ABCD is a rhombus

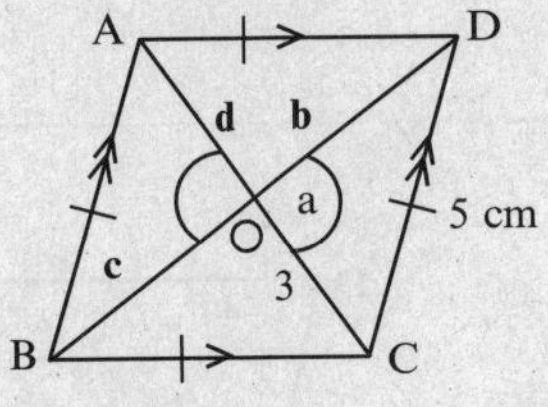

a = 90° (diagonals bisect $\perp$)
$b^2 = 5^2 - 3^2$ (Pyth.)
$= 25 - 9$
$= 16$
$\therefore$ b = 4 cm
c = 4 cm (diagonals of rhombus bisect one another)
d = 3 cm (diagonals of rhombus bisect one another)

4.6

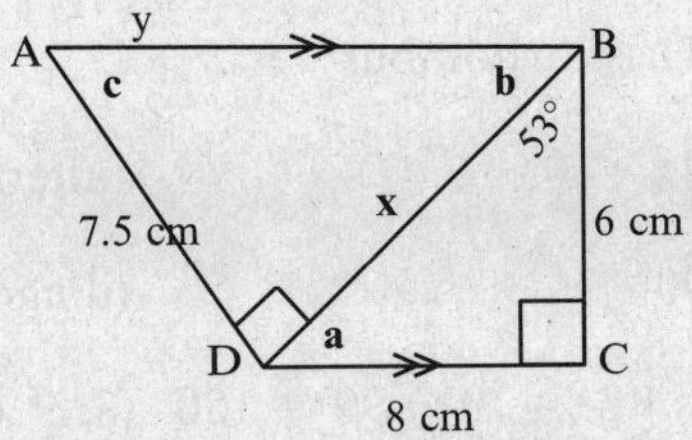

$a = 90° - 53°$ (3 $\angle^s$ of $\triangle$BDC)
$= 37°$
$b = 37°$ (alternate $\angle^s$; AB$\|$DC)
$c = 90° - 37°$ (3 $\angle^s$; of $\triangle$ABD)
$= 53°$
$x^2 = 8^2 + 6^2$ (Pyth.; $\hat{C} = 90°$)
$= 64 + 36$
$= 100$
$x = 10$ cm
$y^2 = 7{,}5^2 + 10^2$ (Pyth.; $\hat{D}_1 = 90°$)
$= 156{,}25$
$y = 12{,}5$ cm

4.7 DEFG is a rhombus.

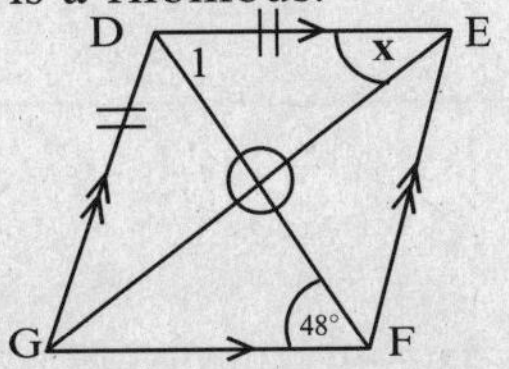

$\hat{D}_1 = 40°$ (alternate $\angle^s$; DE$\|$GF)
$x = 180° - (90° + 40°)$
(diagonals of rhombus intersect at 90°
$= 50°$

4.8 ABCD is a rectangle

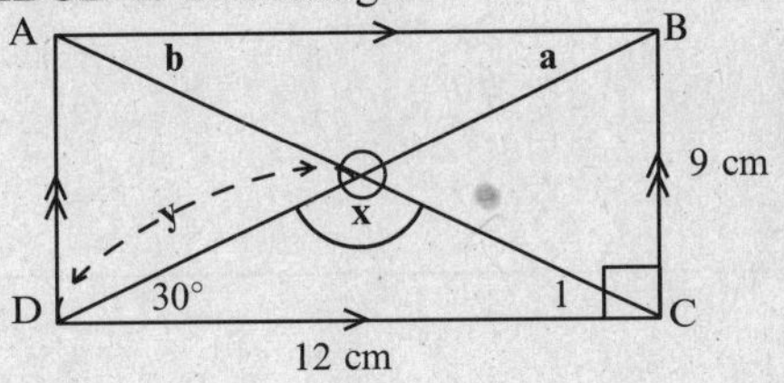

$a = 30°$ (alternate $\angle^s$ AB$\|$DC)
$\hat{c}_1 = 30°$ (isosceles $\triangle$DOC; diagonals of rectangle are equal and bisect)
$b = 30°$ (alternate $\angle^s$; AB$\|$DC)
$x = 180° -$ $(30° + 30°)$ (3 $\angle^s$ of $\triangle$DOC)
$= 120°$
$BD^2 = 9^2 + 12^2$ (Pyth.; $\hat{C}$ = 90°)
$= 225$
BD = 15 cm
y = 7,5 cm (diagonals of rectangle bisect one another)

5. ABCD is a rhombus

$\hat{D}_1 = 3x - 10°$ (alternate $\angle^s$; AB∥BC)

$A\hat{O}D = 90°$ (diagonals intersect ⊥)

$\therefore x + 10° + 3x - 10° + 90° = 180°$ (3 $\angle^s$ of △AOD)

$4x = 90°$

$x = 22,5°$

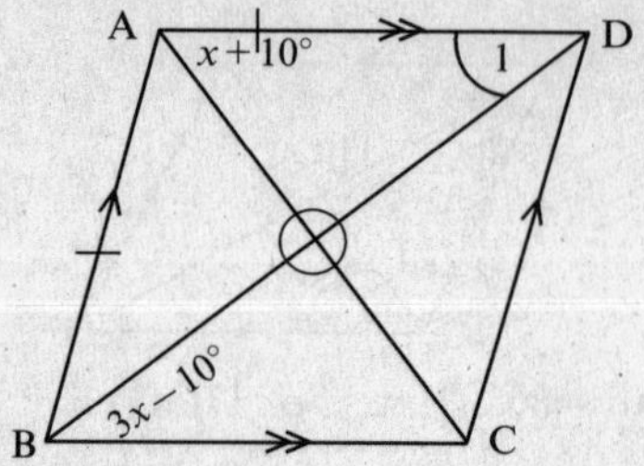

6. $\hat{F}_1 = \dfrac{180° - 38°}{2}$ (3 $\angle^s$ of △BEF; BF = BE)

$= 71°$

$\hat{A}_1 = 71°$ (corresponding $\angle^s$; EF∥AC)

$x = 71°$ (alternate $\angle^s$; AB∥DC)

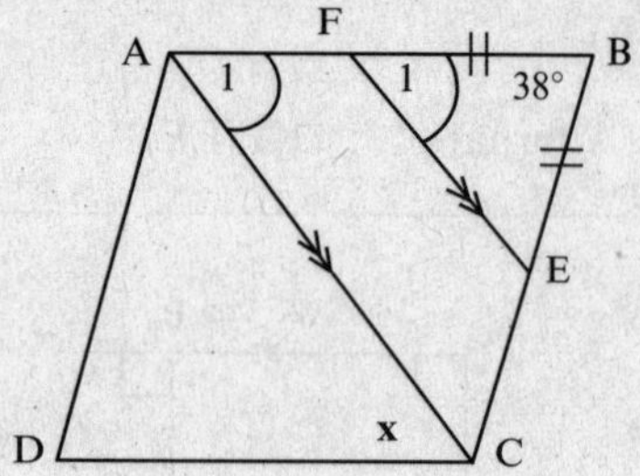

7. $2x + 20° = 5x - 40°$ (opposite $\angle^s$ of $||^m$ =)

$\therefore 20° + 40° = 5x - 2x$

$60° = 3x$

$20° = x$

$\hat{A} = 2x + 20°$

$= 60°$

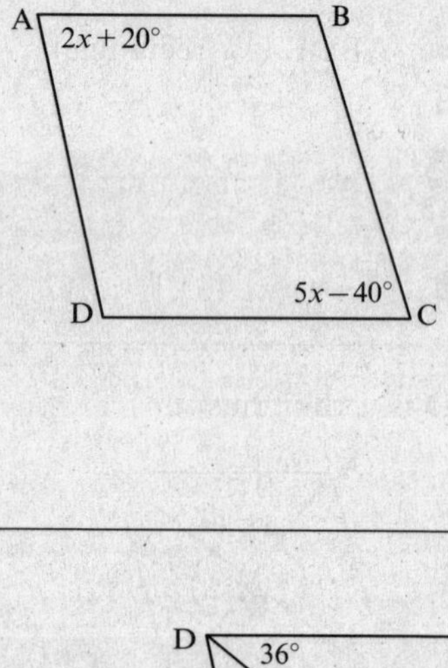

8. $\hat{D}_1 = 36°$ (Diagonals of rhombus bisect one another)

$x = 180° - (36° + 36°)$ (Co-interior $\angle^s$; DE∥GF)

$x = 108°$

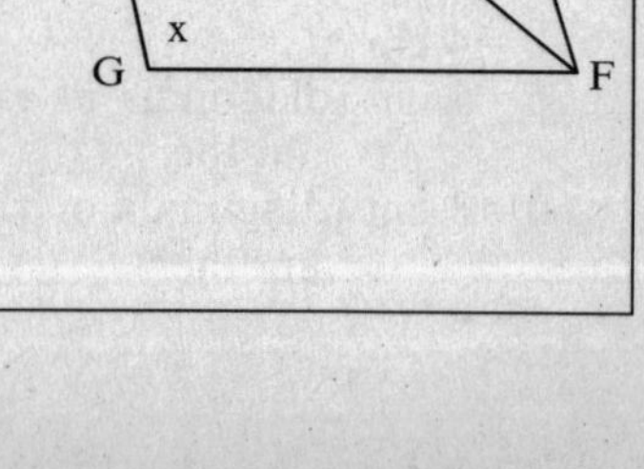

Exercise 2: Polygons Solutions

1.1 Sum interior $\angle^s$ of 5 angles

$= (5-2)180°$

$= 3 \times 180°$

$= 540°$

$\therefore (x + 2x - 20°) + (3x - 25° + x + 2x) = 540°$

$9x = 540° + 20° + 25°$

$x = 65°$

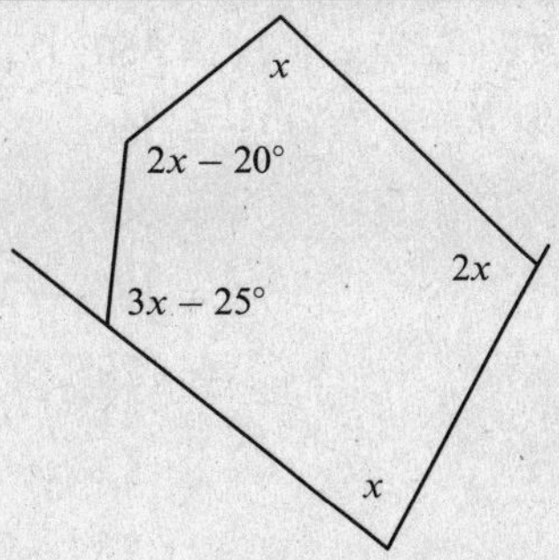

1.2 $\hat{A}_1 = 42°$ (alternate $\angle^s$; BC∥AE)

$A_2 = 138°$

Sum interior $\angle^s = 360°$

$\therefore 42° + 2x + 30° + 90° + 3x - 10° + 138° = 360°$

$5x = 360° - 138° + 10° - 90° - 30° - 42°$

$= 70°$

$x = 14°$

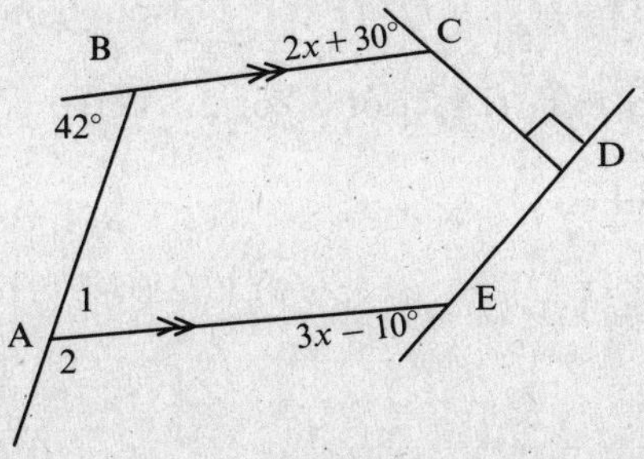

1.3 Regular polygon:

Every corner: $\dfrac{(n-2)(180°)}{n}$

$x = \dfrac{(5-2)(180°)}{5}$

$= 108°$

$y = 180° - 108°$ (straight line, ext. $\angle$)

$= 72°$

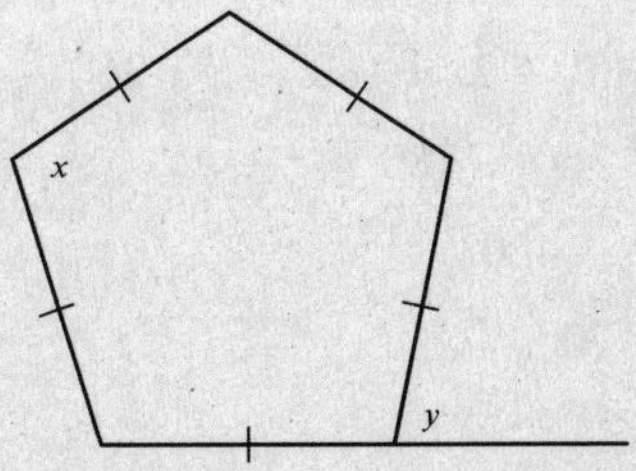

1.4 Regular hexagon:

Every interior $\angle$: $\dfrac{(n-2)(180°)}{n}$

$\therefore \hat{A} = \dfrac{(6-2)(180°)}{6}$

$= 120°$

$x = \dfrac{180° - 120°}{2}$ Isosceles △ABF

$= 30°$

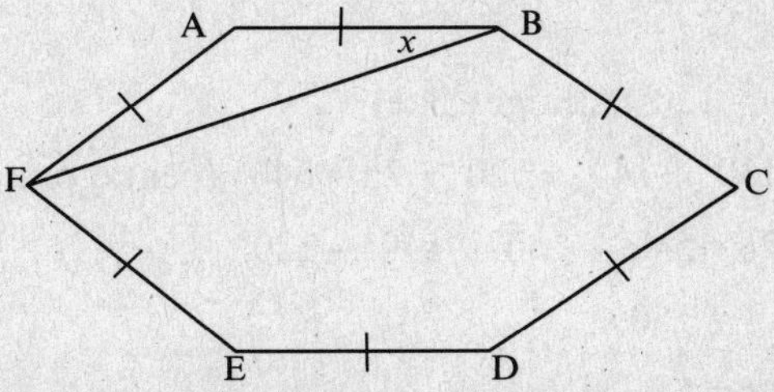

1.5 $5x + 3x + x + x + 2x = 360°$ (Sum external $\angle^s = 360°$)

$12x = 360°$

$x = 30°$

$y = 180° - 2x$ (straight line)

$= 120°$

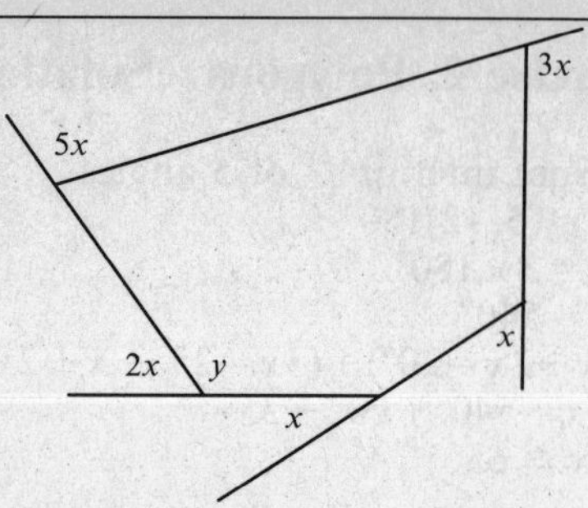

1.6 $\hat{S}_1 = \hat{T}_1 = \frac{360°}{5} = 72°$ (Ext. $\angle^s$ of polygons = 360°)

$x = \hat{S}_1 + \hat{T}_1$ (Exterior $\angle^s$ of $\triangle$STU)

$= 144°$

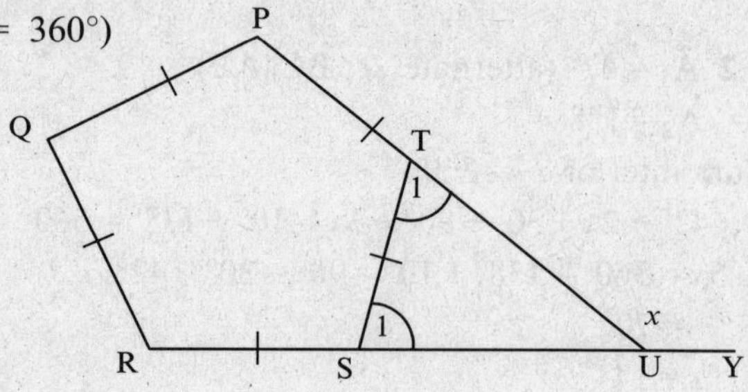

1.7 $60° + 180° - 2x + 180° - x + 360° - 3x = 360°$ (Quadrilateral)

$-6x = 360° - 780°$

$-6x = -420°$

$x = 70°$

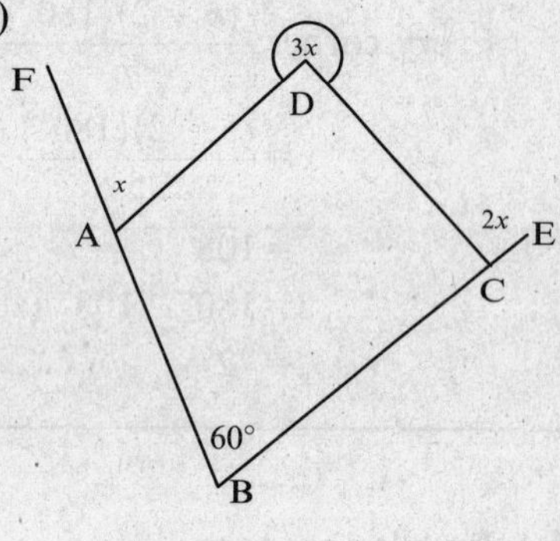

1.8 $a = 120°$ (straight line)

$130° + 140° + 120° + 2b = 540°$ (pentagon: $(5-2)180°$)

$2b = 540° - 130° - 140° - 120°$

$= 150°$

$b = 75°$

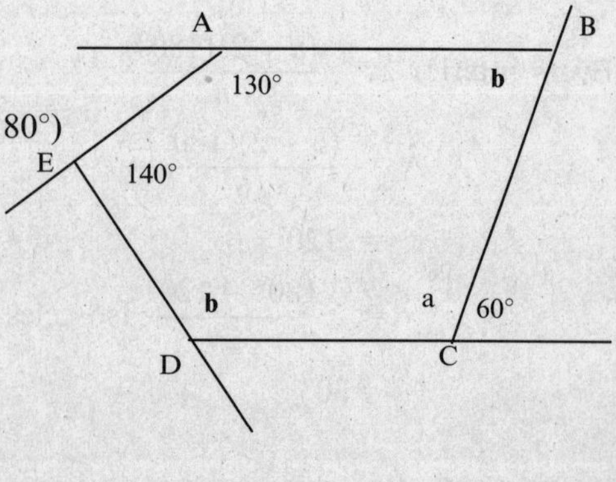

2.1 Sum of interior $\angle^s$ of hexagon:

$$(n-2)180° = (6-2)180° = 720°$$

2.2 Interior $\angle^s$ of regular pentagon

$$\frac{(n-2)180°}{n} = \frac{(5-2)180°}{5} = 108°$$

2.3 Interior $\angle^s$ of regular nonagon:

$$\frac{(n-2)180°}{n} = \frac{(9-2)180°}{9} = 140°$$

2.4 Every exterior $\angle$ of octagon:

$$\frac{360°}{8} = 45°$$

2.5 Sum of exterior $\angle^s$: 360°

$$\therefore 36° \times n = 360°$$
$$n = 10 \text{ sye}$$

2.6 Every interior $\angle$: $180° - 36°$ (straight line) $= 144°$

2.7 Every interior $\angle$: 135°

$\therefore$ Every exterior $\angle$: 45°

$$\therefore 45° \times n = 360°$$
$$n = 8 \text{ sides}$$

2.8 $x + x + 10° + 2x - 40° + 2x - 60° + x = 540°$ (5 sided polygon)

$$7x = 540° - 10° + 40° + 60° = 630°$$
$$x = 90°$$

Exercise 3: Areas. Solutions

1.1 $b \times h$

1.2 $l \times b$

1.3 side × side

1.4 $b\ x\ h$ or $\frac{1}{2}$ product of diagonals

1.5 $\frac{1}{2}$ product of diagonals

1.6 $\frac{h}{2}$ (sum of parallel sides)

2.1 $\text{Area}_{\triangle} = \frac{1}{2}\text{a.c}$

2.2 $\text{Area}_{\|m} = c \times b = bc$

2.3 $\text{Area}_{\square} = a \times b = ab$

2.4 $\text{Area}_{\text{rhombus}} = \frac{1}{2}.2a \times 2b = 2ab$

2.5 $\text{Area}_{\text{kite}} = \frac{1}{2}2a.2b = 2ab$

2.6 $\text{Area}_{\triangle} = \frac{1}{2}.ab$

2.7 $\text{Area}_{\text{trapezium}} = \frac{b}{2}(a + c)$

3.1 Parallelogram:

$$\text{Area}_{\|_m} = b \times h$$
$$= 8 \times 3$$
$$= 24\ cm^2$$

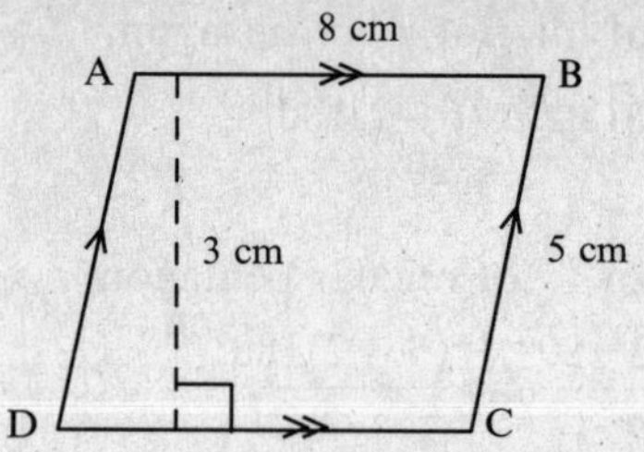

3.2 Trapezium:

$$\text{Area}_{\text{trapezium}} = \frac{h}{2}\,(\text{sum of } \|\text{–sides})$$
$$= \frac{3}{2}(10 + 6)$$
$$= 24\ \text{mm}^2$$

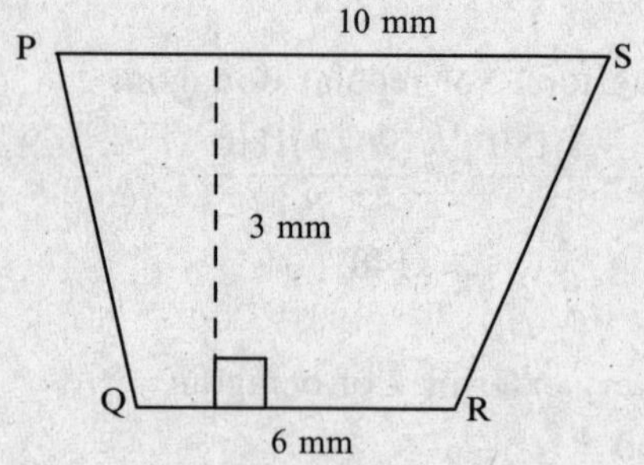

3.3 MORS is a rhombus

$$\text{Area}_{\text{rhombus}} = \tfrac{1}{2}\ \text{product of diagonals}$$
$$= \tfrac{1}{2}.15 \times 25$$
$$= 187{,}5\ \text{cm}^2$$

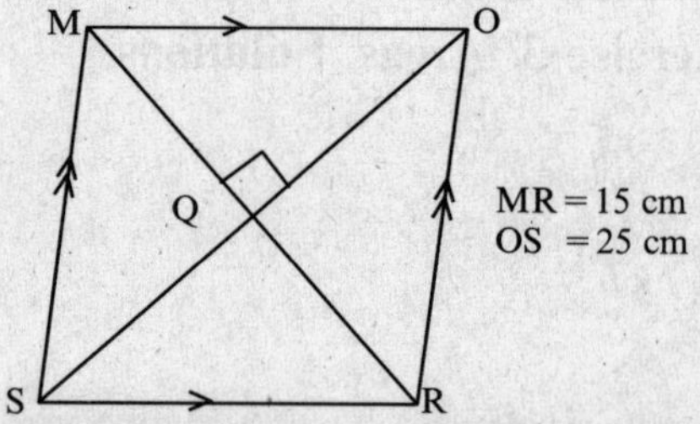

3.4 $BD^2 = 4^2 + 3^2 = 25$ (Pyth.)

$\therefore\ BD = 5$

Rectangle + triangle area

$\text{Area} = l \times b + \tfrac{1}{2}b.h$

$= 9 \times 5 + \tfrac{1}{2}.4 \times 3$

$= 45 + 6$

$= 51\ \text{m}^2$

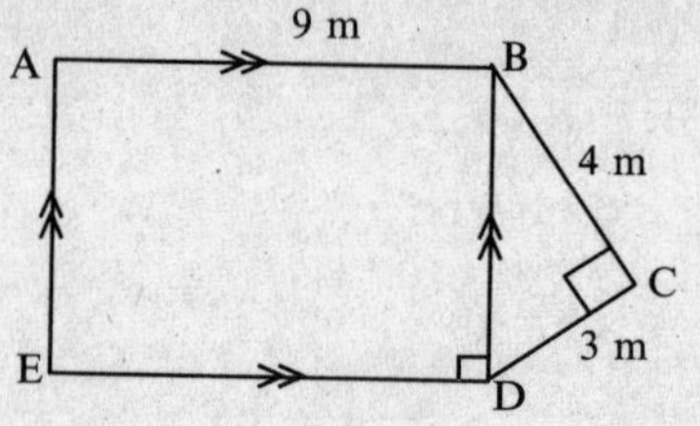

3.5 $AC^2 = 15^2 - 9^2$ (Pyth.)

$= 144$

$AC = 12$

$\text{Area}_{\triangle} = \tfrac{1}{2}.b.h$

$= \tfrac{1}{2}.9 \times 12$

$= 54\ \text{m}^2$

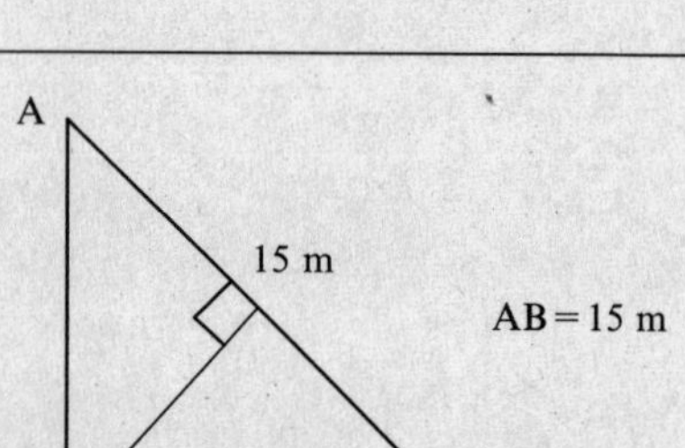

3.6 PQRS is a rhombus.

$Q0^2 = 37^2 - 12^2$ (Pyth. $\hat{0} = 90°$; Rhombus)

$= 1\,225$

$Q0 = 35$ cm

$\text{Area}_{\text{rhombus}} = \frac{1}{2}$ product of diagonals

$= \frac{1}{2}.24 \times 70$

$= 840 \text{ cm}^2$

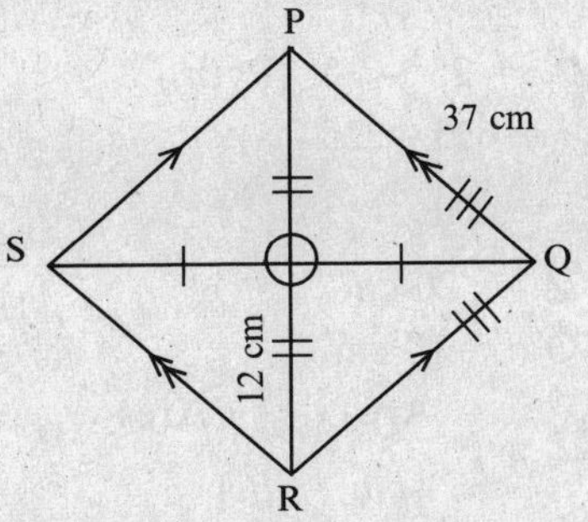

3.7 $\text{Area}_{\text{diagram}} = \text{Area}_{\text{rectangle}}$ ABCH – $\text{Area}_{\text{square}}$ DEFG

$= (l \times b) - (\text{side} \times \text{side})$

$= 25 \times 20 - 8 \times 8$

$= 436 \text{ m}^2$

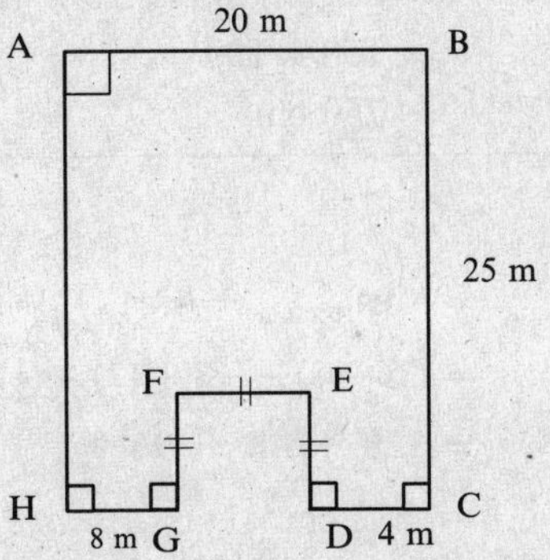

3.8 $BC^2 = 9^2 + 12^2$ (Pyth.; $\hat{A} = 90°$)

$= 225$

$BC = 15$ cm

$\therefore\ \text{Area}_{\text{diagram}} = \frac{1}{2} b_1 h_1 + \frac{1}{2} b_2 h_2$

$= \frac{1}{2}.9 \times 12 + \frac{1}{2}.8 \times 15$

$= 54 + 60$

$= 114 \text{ cm}^2$

$DC^2 = 17^2 - 15^2$

$= 289 - 225$

$= 64$

$DC = 8$ cm

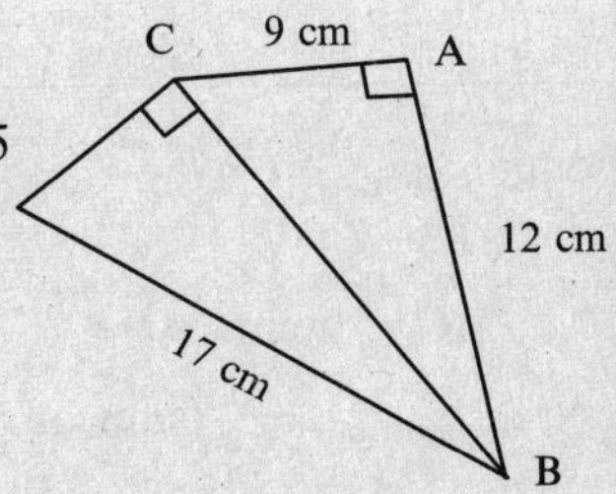

3.9 $OS^2 = 13^2 - 5^2$ (Pyth.)

$= 169 - 25^2$

$= 144$

$OS = 12$

$\text{Area}_{\text{kite}} = \frac{1}{2}$ product of diagonals

$= \frac{1}{2}.24 \times 20$

$= 240 \text{ m}^2$

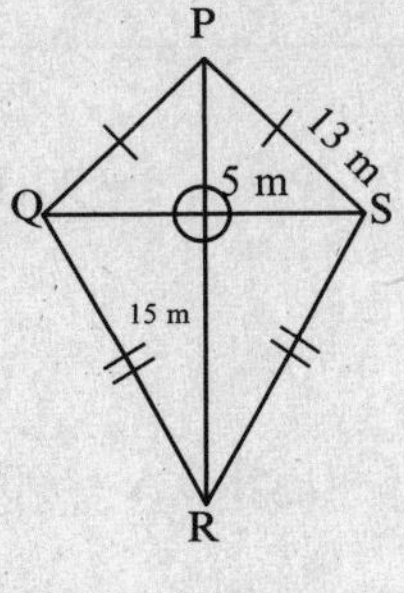

3.10 $\text{Area}_{\text{trapezium}} = \frac{h}{2}$ (sum of $\|$-sides)

$= \frac{6}{2}(8 + 14)$

$= 66 \text{ cm}^2$

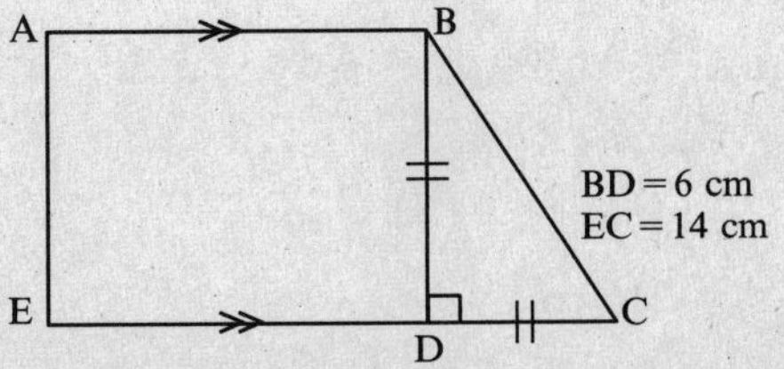

3.11 $XZ^2 = 25^2 - 15^2$ (Pyth.)

$= 625 - 225$

$= 400$

$\therefore XZ = 20$ cm

$XO = OY = 10$ cm

$\text{Area}_{\text{diagram}} = \text{Area}_{\triangle_{WXZ}} + \text{Area}_{\triangle_{XYZ}}$

$= \frac{1}{2}b.h + \frac{1}{2}b.h$

$= \frac{1}{2}.15 \times 20 + \frac{1}{2}.20 \times 10$

$= 150 + 100$

$= 250 \text{ cm}^2$

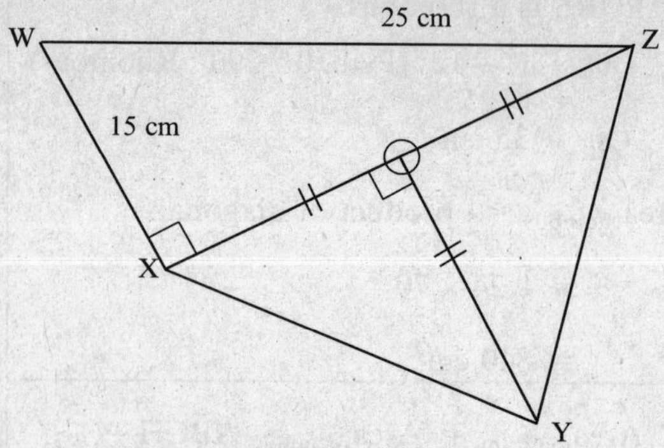

3.12

$\text{Area}_{\text{diagram}} = \text{Area}_{\text{rectangle}} + \text{Area}_{\triangle_{DEF}} + \text{Area}_{\triangle_{RCD}}$

$= (16 \times 8) + (\frac{1}{2}8 \times 12) + (\frac{1}{2}16 \times 12)$

$= 128 + 48 + 96$

$= 272 \text{ cm}^2$

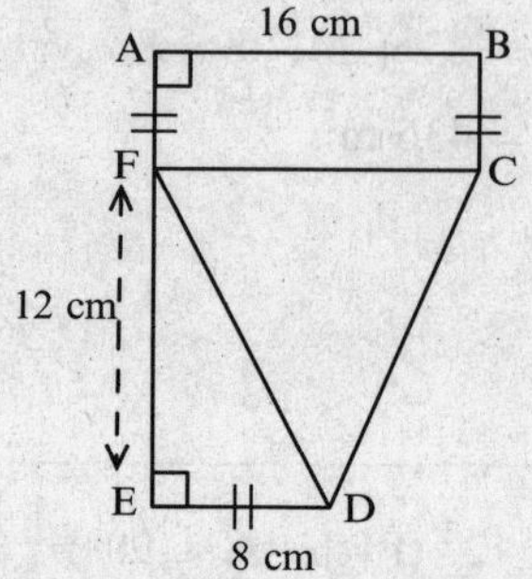

4. $\text{Area}_{\text{rectangle}} = l \times b$

$\therefore 63 = l \times 7$

$\therefore BC = 9$ cm

$\text{Area}_{\text{trapezium}} = \frac{h}{2}$ (sum of $\|$–sides)

$= \frac{9}{2}(7 + 3,5)$

$= 47,25 \text{ cm}^2$

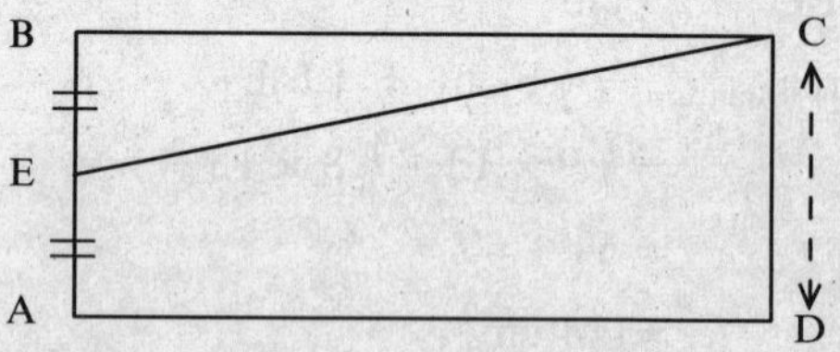

5. $BD^2 = 24^2 + 10^2$ (Pyth.; $\hat{c} = 90°$)

$= 676$

$BD = 26$ cm

$\therefore DE = 20$ cm

$\therefore AE^2 = 20^7 - 16^2$ (Pyth.; $\hat{A} = 90°$)

$= 144$

$AE = 12$

$\text{Area}_{\text{diagram}} = \text{Area}_{\triangle_{ADE}} + \text{Area}_{\triangle_{BCD}}$

$= \frac{1}{2}b.h + \frac{1}{2}b.h$

$= \frac{1}{2}16 \times 12 + \frac{1}{2}24 \times 10$

$= 216 \text{ cm}^2$

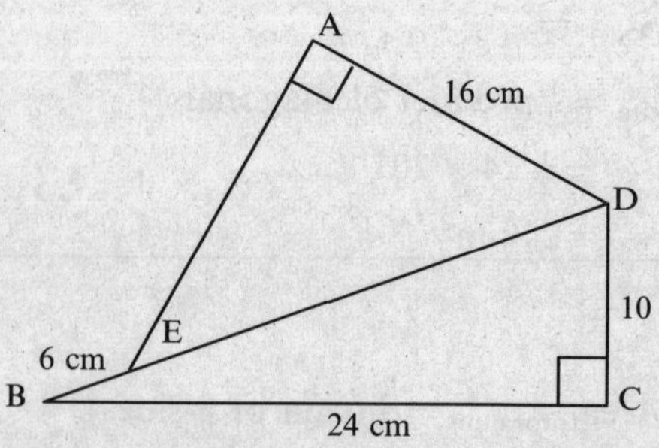

6. $\text{Area}_{\text{rhombus}} = \frac{1}{2}\text{AC.BD}$

$\therefore\ 48\ \text{cm}^2 = \frac{1}{2}.8.\text{BD}$

$\text{BD} = 12\ \text{cm}$

7. $\text{Area}_{\text{kite}} = \frac{1}{2}$ product of diagonals

$= \frac{1}{2}\ \text{AC.BD}$

$64 = \frac{1}{2}8.\text{BD}$

$\therefore\ \text{BD} = 16\ \text{cm}$

$x = 10\ \text{cm}$

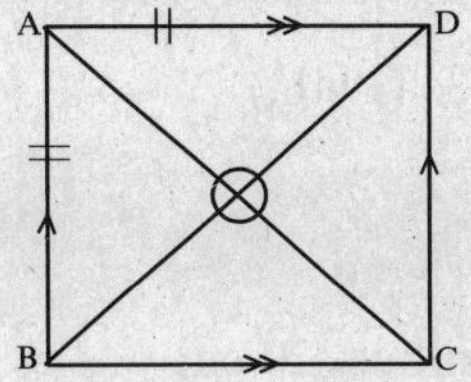

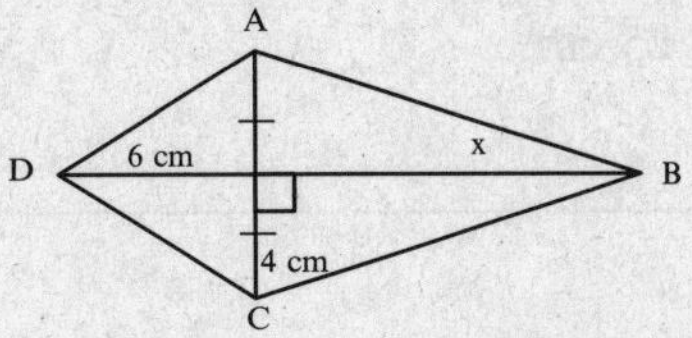

Exercise 4: Mixed Geometry. Solutions

1.1 110°

1.2 $\dfrac{180° - x}{2}$

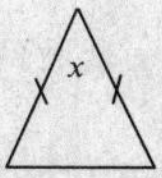

1.3 360°

1.4 x

1.5 360°

1.6 rectangle and square

1.7 supplementary

1.8 6 sides

1.9 isosceles triangle

1.10 a quadrilateral with 2 pr opposite sides parallel

1.11 Smallest ∠ : $\frac{2}{10} \times 180° = 36°$

1.12 60°

1.13 $(n-2)180°$

1.14 πr^2

1.15 perpendicular bisectors of sides

1.16 right-angled

1.17 $90° + x$

1.18 150°

1.19 Â (opposite longest side)

1.20 obtuse-angled $(49 > 25 + 16)$

2. In a right-angled triangle the square on the hypotenuse is equal to the sum of the squares on the other 2 sides.

3. Their sides are not equal, but their ratios are equal.

4. AM: Median; CP: Altitude (Height)

5.1 true

5.2 false

5.3 false (they intersect at right-angles and the one bisects the other, but **not one another**)

5.4 true

5.5 true

5.6 false (complementary)

5.7 true

6. $AB^2 = 4^2 + 3^2$ (Pyth.)

$= 25$

$\therefore$ AB = 5 cm

Perimeter rhombus: 20 cm

Perimeter square: 20 cm (each side 5 cm)

$Area_{square}$: 25 cm^2

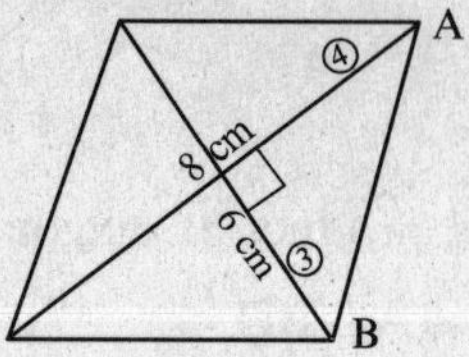

7.1

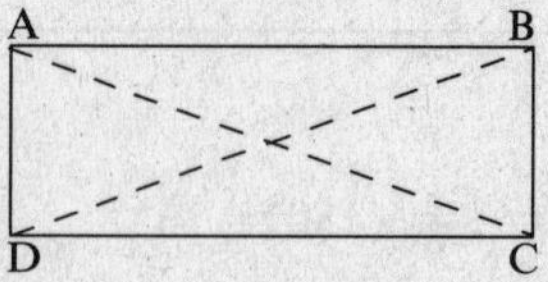

AC = BD

$\therefore$ ABCD is a rectangle

7.2 Polygon is : hexagon

7.3 $c^2 = 64$

$a^2 + b^2 = 36 + 25 = 61$ } $\therefore \triangle$ ABC is obtused

$c^2 > a^2 + b^2$

7.4 $\triangle$ ABC $\equiv$ $\triangle$PRQ (S;S;S)

7.5 $\hat{B} = \frac{3}{6} \times 180°$

$= 90°$

$\therefore \triangle$ ABC is right-angled

16 *Multiple Choice Questions*

Choose the correct answer from the 4 possibilities:

1. $(-2)^4 =$

A 8
B -8
C 16
D -16

2. $a^4 \times a^4 =$

A a^{16}
B a
C a°
D a^8

3. $(a-b)^2 =$

A $a^2-2ab-b^2$
B $a^2-2ab+b^2$
C a^2+b^2
D a^2-b^2

4. $(4x^2y^3)^3 =$

A $12x^6y^9$
B $4x^5y^9$
C $64x^6y^9$
D $4^3x^5y^6$

5. $\dfrac{x^2-2y}{z} = \ldots\ldots$ if $x = -1$; $y = -1$ and $z = 1\frac{1}{2}$

A 2
B $\frac{2}{3}$
C $\frac{9}{12}$
D -2

6. The gradient (m) of the line $2y - x = -6$ is:

A -1
B 6
C -3
D $\frac{1}{2}$

7. Which of the following points lie on the line: $2x + y = 4$

A $(-1;\ 5)$
B $(1;\ -2)$
C $(1;\ -6)$
D $(\frac{1}{2};\ -3)$

8. $\sqrt{1\frac{9}{16}x^{16}} =$

A $1\frac{3}{4}x^4$
B $\frac{5}{4}x^4$
C $1\frac{1}{4}x^8$
D $1\frac{3}{4}x^8$

9. $\sqrt{\dfrac{20x^3y^5}{5x^5y}} =$

A $\dfrac{4y^4}{x^2}$

B $\dfrac{2y^2}{x}$

C $\dfrac{10y^4y^3}{5}$

D $2xy^2$

10. The value of x is:

A 18°
B 36°
C 60°
D 30°

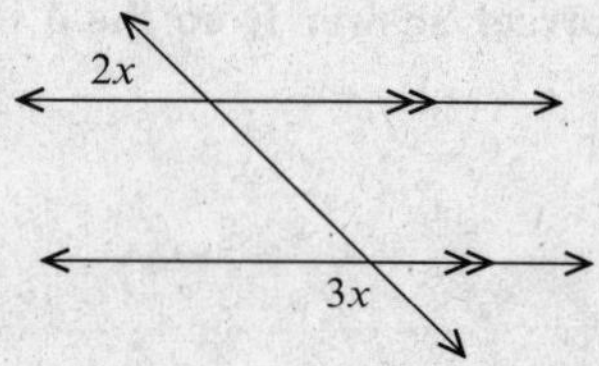

11. In △ABC Â = B̂ and Ĉ = 50°; therefore Â =

A 40°
B 50°
C 60°
D 65°

12. The value of x is:

A 60°
B 90°
C 120°
D 130°

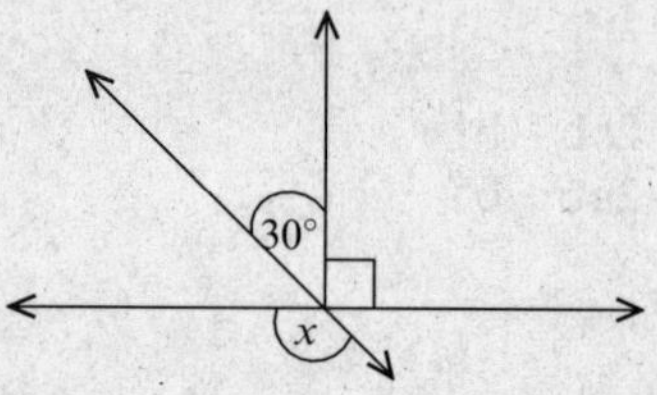

13. ABCD is a parallelogram with AD = BC = AC
BÂD =

A 50°
B 80°
C 115°
D 100°

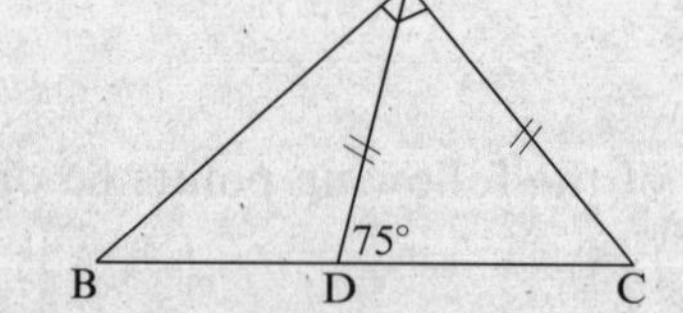

14. BÂC = 90° and AD = AC.
BÂD =

A 30°
B 75°
C 45°
D 60°

15. The ratio of the angles of a triangle is 2:3:5. The smallest angle is:

A 5°
B 18°
C 36°
D 10°

16. The size of x is:

A 30°
B 40°
C 50°
D 60°

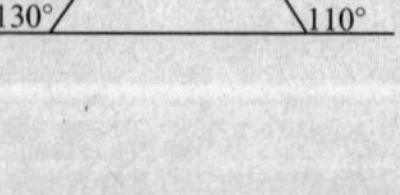

17. The area of the diagram can be calculated by:
A AF.DC
B $\frac{1}{2}$ DC(AD + BC)
C AC.DB
D $\frac{1}{2}$ AF(AB + DC)

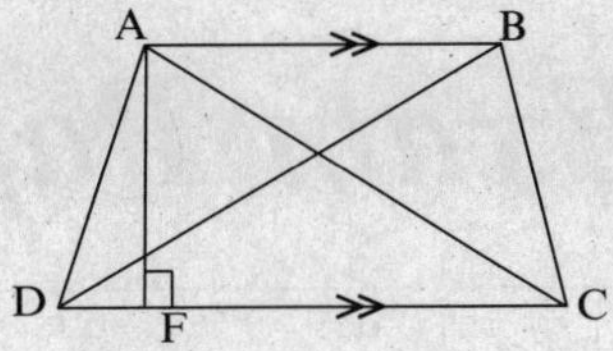

18. The diagonals of a . . . . are equal.
A rhombus
B rectangle
C parallelogram
D trapesium

19. One of the interior angles of a triangle is equal to the sum of the other 2 interior angles; therefore the triangle is:
A Equilateral
B Obtuse
C Right-angled
D Isosceles

20. In the diagram AM is the
A altitude.
B median.
C bisector.
D perpendicular bisector.

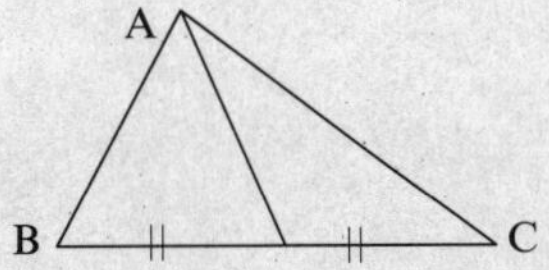

Solutions: Multiple Choice

1. C	2. D	3. B	4. C	5. A
6. D	7. A	8. C	9. B	10. B
11. D	12. C	13. C	14. D	15. C
16. D	17. D	18. B	19. C	20. B

17 *Exam Papers*

Std 7 Mathematics: First Paper

June

Time: 2 hours
Total: 120
Chapters: 1; 2; 3; 4.

Question 1

(Calculators may not be used in this question.)

1.1 Simplify and give answers with positive indices:

1.1.1 $2(a+b)^0$

1.1.2 b^{-7}

1.1.3 -8^0

1.1.4 $\sqrt[3]{3^{-6}}$

1.1.5 $(\frac{1}{2})^{-3}.2^{-2}$

1.1.6 $-(-3x)^2$

1.1.7 $2^{-3}+(x+y)^0+\dfrac{2^3.2^{-4}}{2^{-2}}$

1.1.8 $a^b \times a^c \times 3a^d$

(15)

1.2 If $a=-2$, determine the value of:
$-2a^3-(-2a)^3$ (6)

21

Question 2

2.1 Given the expression: $2x^4 - 3x(x + y) - 4xy - 2$

2.1.1 Simplify the expression and rewrite in descending powers of x.
Now answer the following questions:

2.1.2 How many terms are there?

2.1.3 What is the index of x in the third term?

2.1.4 What is the numerical coefficient of the 2nd term?

2.1.5 Determine the numerical value of the expression if $x = 2$ and $y = -3$.
(Show your calculation) (8)

2.2 If $a = -3$, $b = -1$ and $c = 2$, calculate the value of: $\frac{-3a}{b - c}$ (3)

2.3 Determine the perimeter of a rectangle if the length is $(2x^2 + 3x - 5)cm$ and the breadth $(x^2 - x - 8)cm$. (5)

[16]

Question 3

3.1 Simplify:

3.1.1 $(3x - 2y)(3x + 2y)$

3.1.2 $-6ab^2(3ab - 2bc)$

3.1.3 $\frac{1}{4}a(12a^3 - 4a^2 + a)$

3.1.4 $(2a - 3b)^2$

3.1.5 $(3a - 4b)(a + b) - 2(a - b)^2 - 2(3a^2 - b^2)$

3.1.6 $(2x - 3)(4x^2 - 5x + 1)$ (20)

3.2 Divide $2x^3 + 7x^2 - 19x + 6$ by $(2x - 3)$ (6)

3.3 Subract $6xy - 2x^2 + 3y^2$ from the product of $(x + 2y)$ and $(5x - y)$ (6)

[32]

Question 4

4.1 Write in scientific notation:

4.1.1 65432000

4.1.2 0,005353 (4)

4.2 Represent the following values of x on the number line:

4.2.1 $-4 \leq x < 2$; $x \in R$

4.2.2 $x < -\frac{1}{2}$ of $x > 2\frac{1}{2}$; $x \in R$ (4)

4.3 If $x \in Z$ (Integers), solve for x and represent the solution on a number line:

$x + 1 \leq -3$ (4)

12

Question 5

5. Factorise the following:

5.1 $3a^2b + 3ab^2$

5.2 $(b + a)^2 - (b + a)$

5.3 $4x^2(x - 2) + 2x(x - 2)$

5.4 $4a^2 - 9b^2$

5.5 $x^2 + 7x + 10$

5.6 $x^2 + x - 6$

5.7 $2a^2 - 22a - 24$

5.8 $18x^2 - 2$

5.9 $4(a + b)^2 - 9$

5.10 $3x^3 - 21x^2 + 36x$

5.11 $(x^2 - x) - (2 - 2x)$

39

Total: 120

Memorandum – First Paper
June

Question 1 (Memo)

1.1.1 2

1.1.2 $\dfrac{1}{b^7}$

1.1.3 -1

1.1.4 $3^{-2} = \frac{1}{9}$

1.1.5 $8.\frac{1}{4} = 2$

1.1.6 $-9x^2$

1.1.7 $\frac{1}{8} + 1 + 2^1 = 3\frac{1}{8}$

1.1.8 $3a^{b+c+d}$ (15)

1.2 $-2(-2)^3 - (-2(-2))^3$
$= -2(-8) - (4)^3$
$= 16 - 64$
$= -48$ (6)

21

Question 2 (Memo)

2.1.1 $2x^4 - 3x^2 - 3xy - 4xy - 2$

$= 2x^4 - 3x^2 - 7xy - 2$

2.1.2 4 terms

2.1.3 1

2.1.4 -3

2.1.5 $2(2)^4 - 3(2)^2 - 7(2)(-3) - 2$

$= 32 - 12 + 42 - 2$

$= 60$ (8)

2.2 $\dfrac{-3(-3)}{-1-2} = \dfrac{9}{-3} = -3$ (3)

2.3 Perimeter:

$2l + 2b$

$= 2(2x^2 + 3x - 5) + 2(x^2 - x - 8)$

$= 4x^2 + 6x - 10 + 2x^2 - 2x - 16$

$= (6x^2 + 4x - 26)cm$ (5)

16

Question 3 (Memo)

3.1.1 $9x^2 - 4y^2$

3.1.2 $-18a^2b^3 + 12ab^3c$

3.1.3 $3a^4 - a^3 + \frac{1}{4}a^2$

3.1.4 $4a^2 - 12ab + 9b^2$

3.1.5 $3a^2 - ab - 4b^2 - 2(a^2 - 2ab + b^2) - 6a^2 + 2b^2$
$= 3a^2 - ab - 4b^2 - 2a^2 + 4ab - 2b^2 - 6a^2 + 2b^2$
$= -5a^2 + 3ab - 4b^2$

3.1.6 $8x^3 - 22x^2 + 17x - 3$

(20)

3.2 $2x^3 + 7x^2 - 19x + 6$

$= (2x - 3)(x^2 + 5x - 2)$

(6)

3.3 $5x^2 + 9xy - 2y^2$

(+) (−) (−)

$-2x^2 + 6xy + 3y^2$

$7x^2 + 3xy - 5y^2$

(6)

32

Question 4 (Memo)

4.1.1 $6{,}5432 \times 10^7$

4.1.2 $5{,}353 \times 10^{-3}$

(4)

4.2.1

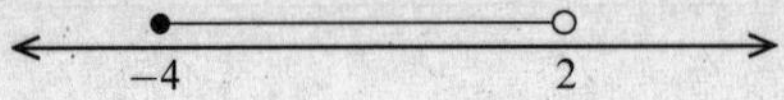

4.2.2

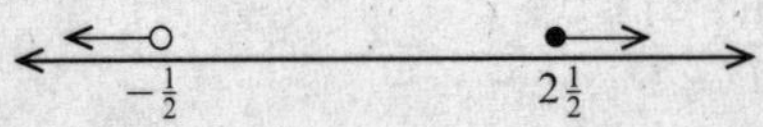

(4)

4.3 $x + 1 \leq -3$

$x \leq -4$

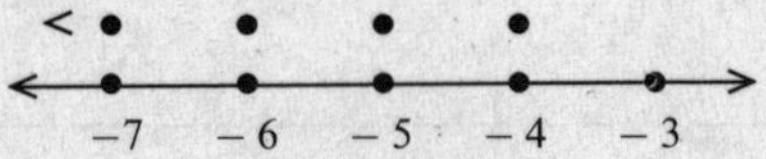

(4)

12

Question 5 (Memo)

5.1 $3ab(a + b)$ (2)

5.2 $(b + a)(b + a - 1)$ (3)

5.3 $2x(x - 2)(2x + 1)$ (3)

5.4 $(2a - 3b)(2a + 3b)$ (2)

5.5 $(x + 5)(x + 2)$ (4)

5.6 $(x + 3)(x - 2)$ (4)

5.7 $2(a^2 - 11a - 12)$ (4)

$= 2(a - 12)(a + 1)$ (4)

5.8 $2(9x^2 - 1)$

$2(3x - 1)(3x + 1)$ (4)

5.9 $\left[2(a + b) - 3\right]\left[2(a + b) + 3\right]$

$= \left[2a + 2b - 3\right]\left[2a + 2b + 3\right]$ (4)

5.10 $3x(x^2 - 7x + 12)$ (4)

$3x(x - 4)(x - 3)$ (4)

5.11 $x(x - 1) - 2(1 - x)$

$= x(x - 1) + 2(x - 1)$

$= (x - 1)(x + 2)$ (5)

Total: 120

Std 7 Mathematics: Second Paper
June

Time: 90 min
Total: 75

Question 1

Complete by only writing down the missing parts:

(*a*) The complement of x° is .... degrees.

(*b*) Every angle of an equilateral triangle is equal to .... degrees.

(*c*) If one of the base angles of an isosceles triangle is 40°, the size of the vertical angle is .... degrees.

(*d*) Vertically opposite angles are equal if .... intersect one another.

(*e*) The sum of two interior angles of a triangle is equal to ....

(*f*) Determine the size of $\angle x$ if x is an angle which is twice the size of its complement.

6

Question 2

(*a*) Calculate x and $\angle C\hat{O}D$, with reasons.

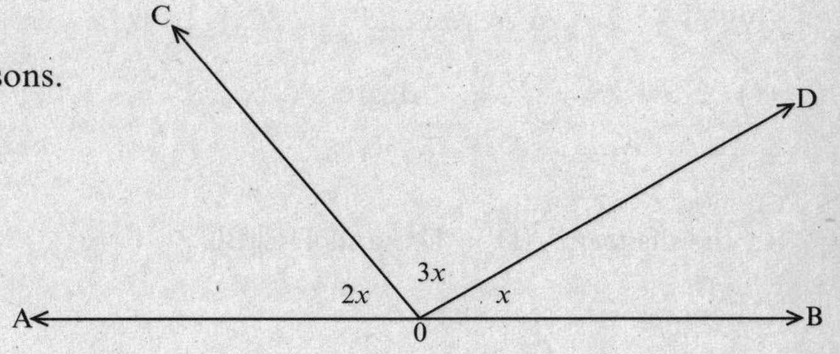

(5)

(*b*) From the diagram, write down:

(i) All the pairs of vertically opposite angles.

(ii) All the pairs of complementary angles.

(iii) An angle adjacent to $\angle c$.

(iv) A pair of adjacent supplementary angles.

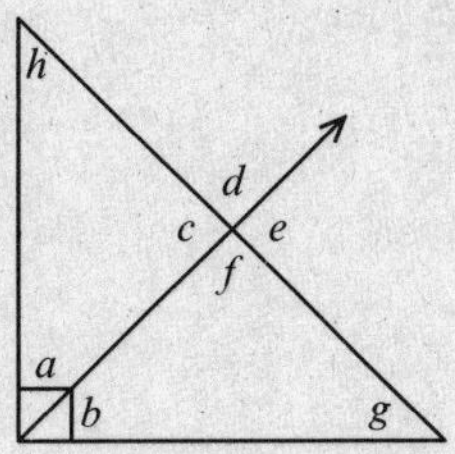

(6)

11

Question 3

(*a*) In the diagram ∠BAC = 120° and ∠ACE = 115°:

Calculate ∠1 with reasons.

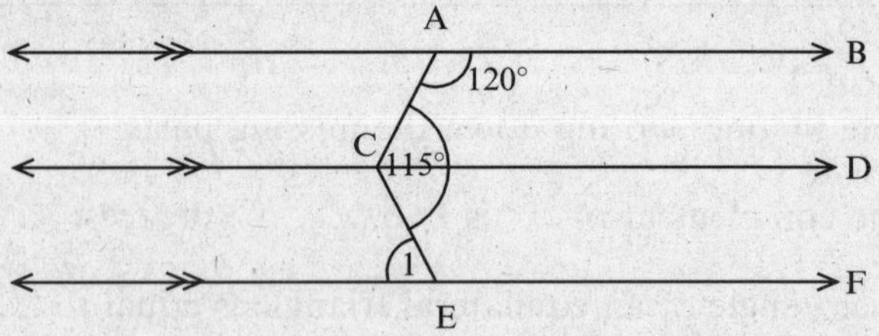

(5)

(*b*) Given △HEF with FH extended to A. FD is drawn parallel to EH and D and H are joined and extended to B. It is also given that HF is the bisector of ∠EHD.

Complete the following and supply reasons where required:

(i) ∠2 = ... (..........)

(ii) ∠4 + ∠5 + ∠6 = ... degrees (..........)

(iii) ∠1 + ∠2 + ∠3 + ∠4 + ∠9 = ... degrees

(iv) ∠1 + ∠2 = ∠6 + ∠ .... (..........)

(v) 2∠3 + ∠8 = ... degrees (..........)

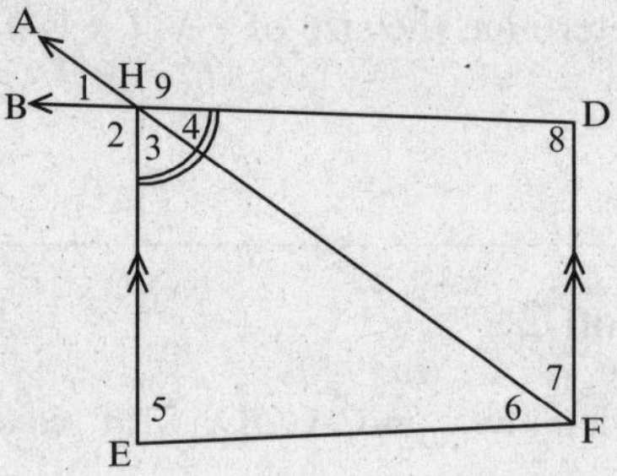

(9)

(*c*) In the diagram AD = DC and DE||BC.

Calculate with reasons:

(*a*) ∠BDC

(*b*) ∠EDC

(*c*) ∠ADC

(*d*) ∠DCA

(*e*) ∠DEA

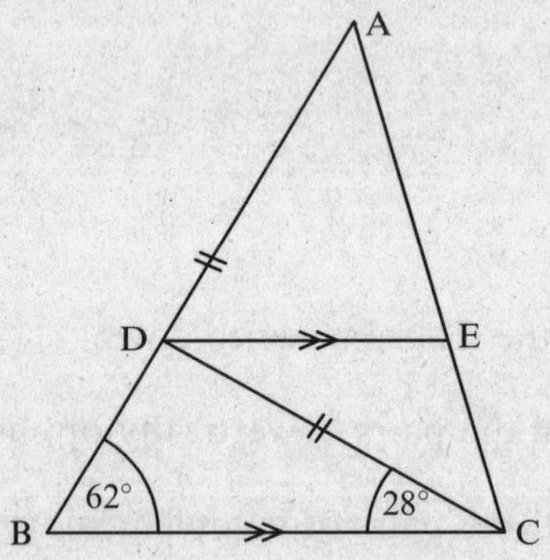

(10)

24

Question 4

(*a*) Prove the theorem: The exterior angle of a triangle is equal to the sum of the opposite interior angles.

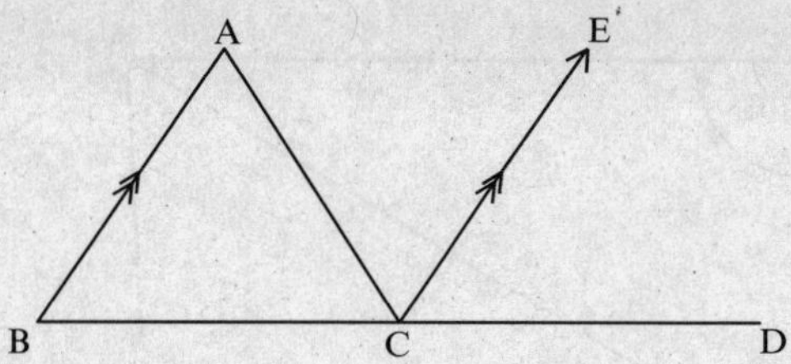

(5)

(*b*) Calculate with reasons ∠1, ∠2 and ∠3 from the diagram.

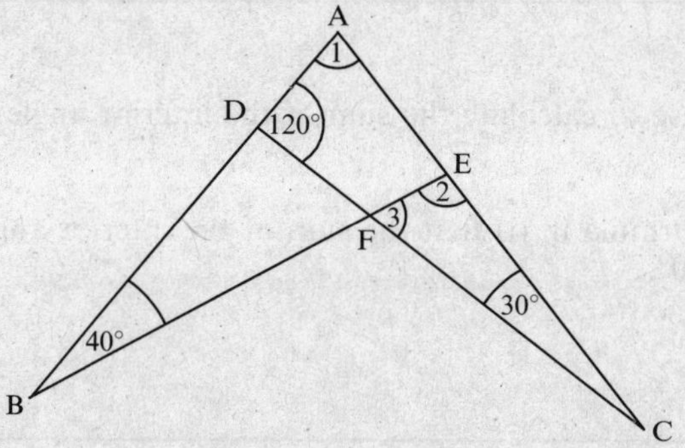

(6)

(*c*) In △EFG EF = EG. H is a point on EG so that FH = FG, ∠E = 50° and ∠EFH = x. Determine, with reasons, the value of x and ∠EHF.

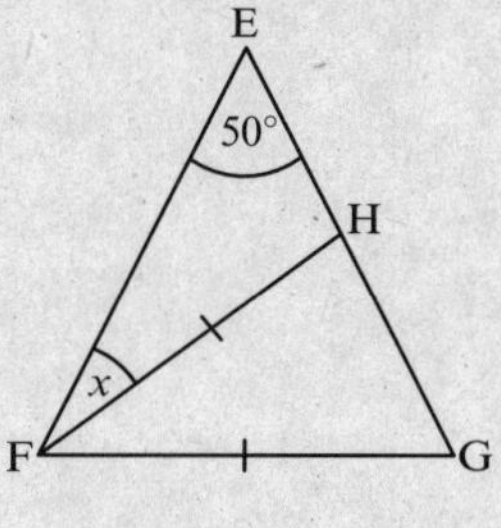

(8)

19

Question 5

(*a*) In the diagram AC∥DF, ∠CBP = x = ∠PBE and ∠BEP = y = ∠PEF.

Prove that

(i) $x + y = 90°$

(ii) ∠BPE = 90°

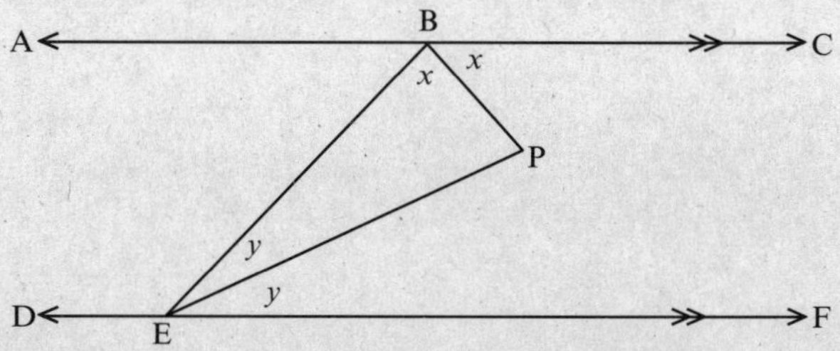

(7)

(*b*) In the accompanying diagram AB = AC.

Prove that AP||TB and AT||PB:

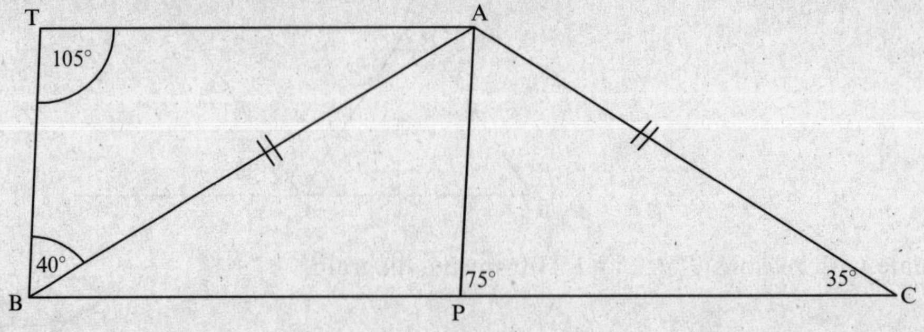

(6)

(*c*) (i) Which formula do we use to calculate the sum of the interior angles of a polygon with *n* sides? (1)

(ii) Now prove, using the formula in (i) that the sum of the interior angles of a pentagon is equal to 540°. (1)

15

Total: 75

Memorandum – Second Paper
June

Question 1 (Memo)

(*a*) $90° - x°$

(*b*) 60°

(*c*) 100°

(*d*) 2 lines

(*e*) opposite exterior angle

(*f*) 60°

6

Question 2 (Memo)

(*a*) $2x + 3x + x = 180°$ (*AOB*: straight line)

$6x = 180°$

$\therefore x = 30°$

$\angle COD = 3x = 90°$

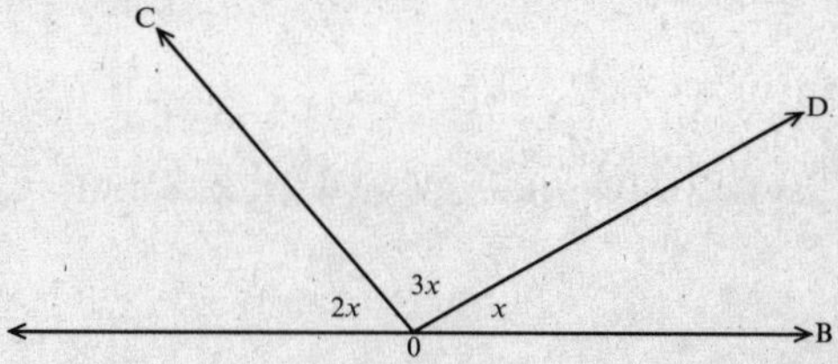

(5)

(*b*) (i) $\angle c$ and $\angle e$ $\angle d$ and $\angle f$

(ii) $\angle a$ and $\angle b$ $\angle h$ and $\angle g$

(iii) $\angle f$ or $\angle d$

(iv) ($\angle c$ and $\angle f$) or ($\angle c$ and $\angle d$) or ($\angle f$ and $\angle e$) or ($\angle d$ and $\angle e$).

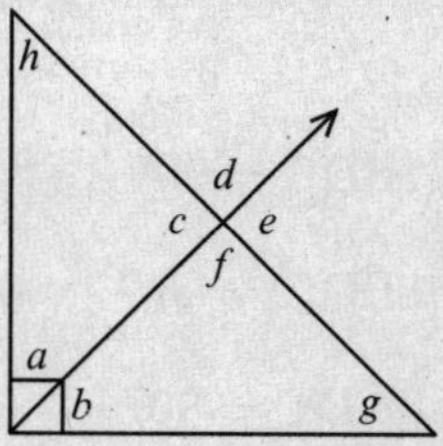

(6)

11

Question 3 (Memo)

(*a*)

$\angle ACD = 60°$ (co-interior $\angle^s$ AB∥CD)

$\angle ACD + \angle DCE = \angle ACE$

$60° + \angle DCE = 115°$

$\therefore \angle DCE = 55°$

$\angle 1 = \angle DCE$ (alternate $\angle^s$ CD∥EF)

$= 55°$

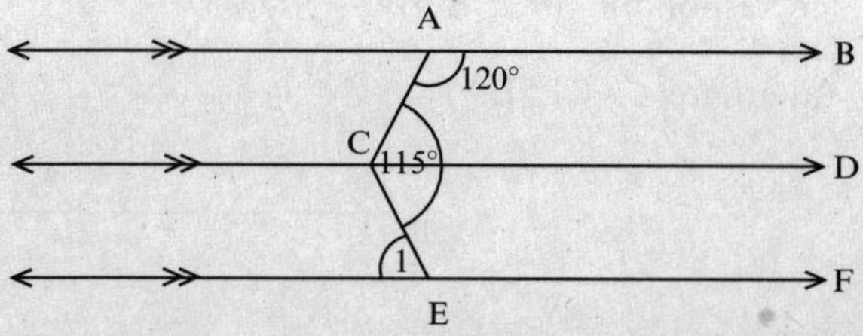

(5)

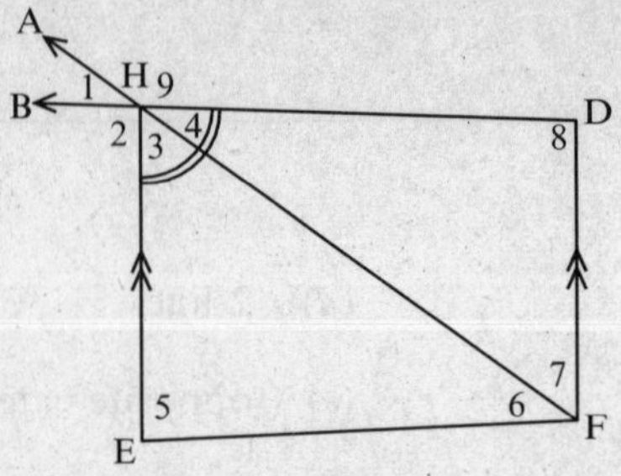

(*b*) (i) $\angle 2 = \angle 8$ (corresponding $\angle^s$ FD$\parallel$EH)

(ii) $\angle 4 + \angle 5 + \angle 6 = \angle 3 + \angle 5 + \angle 6 = 180°$ (HF bisects E$\hat{H}$D; 3 $\angle^s$ of $\triangle$ HEF)

(iii) $\angle 1 + \angle 2 + \angle 3 + \angle 4 + \angle 9 = 360°$ (revolution)

(iv) $\angle 1 + \angle 2 = \angle 6 + \angle 5$ (ext. $\angle$ = opp. interior $\angle^s \triangle$HEF)

(v) $2\angle 3 + \angle 8 = \angle 3 + \angle 4 + \angle 8 = 180°$ ($\angle 3 = \angle 4$)

(co-interior $\angle^s$ HE$\parallel$FD)

(9)

(*c*) $\angle$BDC = 90° (3 $\angle^s$ of $\triangle$ BDC = 180°)

$\angle$EDC = 28° (alternate $\angle^s$, DE$\parallel$BC)

$\angle$ADC = 90° (straight line)

$\angle$DCA = 45° (DA = DC and 3$\angle^s \triangle$ADC)

$\angle$DEA = $\angle$ BCA (corresponding $\angle^s$, DE$\parallel$BC)

= 45° + 28°

= 73°

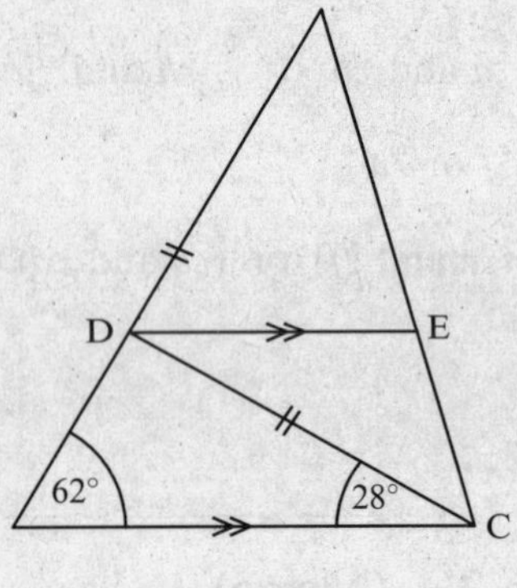

(10)

24

Question 4 (Memo)

(*a*) Proof: $\angle x_1 = \angle x_2$ (corresponding $\angle^s$, CE$\parallel$BA)

$\angle y_1 = \angle y_2$ (alternate $\angle^s$, CE$\parallel$BA)

$\therefore \angle x_1 + \angle y_1 = \angle x_2 + \angle y_2$

$\therefore \angle ACD = \angle A + \angle B$

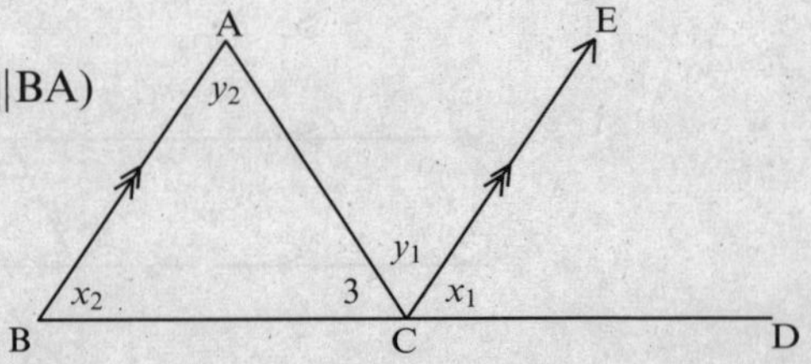

(5)

(*b*) $\angle 1 = 180° - (120° + 30°)$ ($3\angle^s$ of $\triangle$ ADC = 180°)

$= 180° - 150°$

$= 30°$

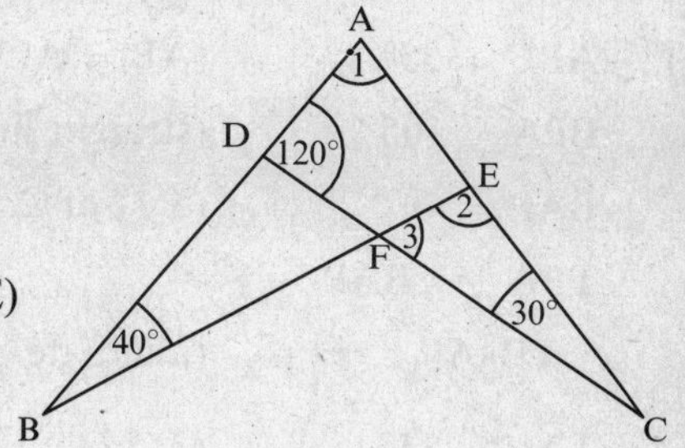

$\angle 2 = \angle 1 + 40°$ (ext. $\angle$ = opp. interior $\angle^s$ $\triangle$ ABE)

$= 30° + 40°$

$= 70°$

$\angle 3 = 180° - (70 + 30)$ ($3\angle^s$ of $\triangle$ FEC)

$= 80°$

(6)

(*c*) $\angle EFG = \angle G$ (EF = EG)

$\angle EFG = \angle G = \dfrac{180° - 50°}{2}$ ($3\angle^s$ of $\triangle$ = 180°)

$= 65°$

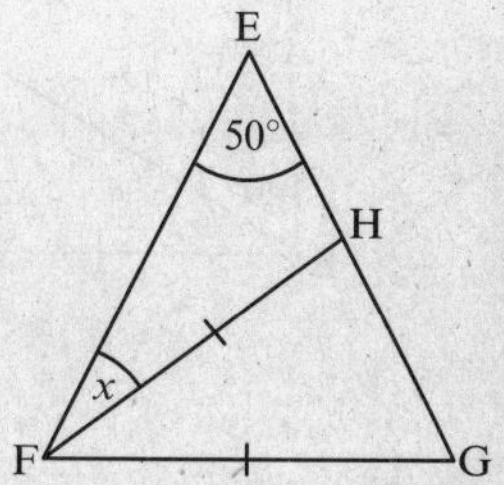

$\therefore \angle FHG = 65°$ (FH = FG)

$\therefore \angle x = 65° - 50°$ (ext. $\angle$ = opp. interior $\angle^s$ $\triangle$ EFH)

$= 15°$

$\angle EHF = 180° - \angle FHG$ (straight line)

$= 180° - 65°$

$= 115°$

(8)

19

Question 5 (Memo)

(*a*) (i) $\angle CBE + \angle BEF = 180°$ (Co-interior $\angle^s$, AC||DF)

$\therefore 2x + 2y = 180°$

$\therefore x + y = 90°$

(ii) $\angle BPE = 180° - (x + y)$ ($3\angle^s$ of $\triangle$BPE = 180°)

$= 180° - 90°$

$= 90°$

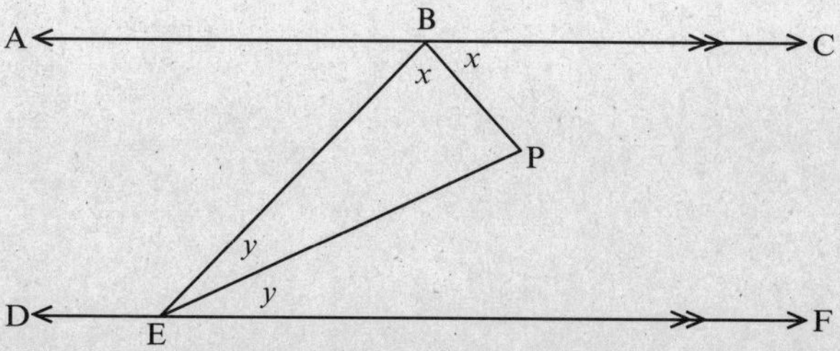

(7)

(*b*) ∠ABC = 35° (AB = AC)

∠BPA = 105° (straight line BPC)

∠BAP = 40° (3 ∠^{s} of △ BAP = 180°)

∠TBA = ∠BAP

∴ TB∥AP (alternate ∠^{s} proved equal)

∠TAB = 35° (3∠^{s} of △ TAB = 180°)

∴ AT∥BP (alternate ∠^{s} proved equal)

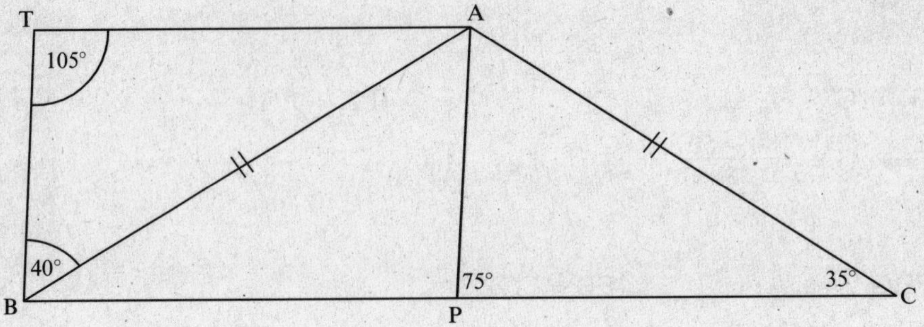

(6)

(*c*) (i) (n − 2) 180° (1)

(ii) (5 − 2) 180° = 3 × 180° = 540° (1)

15

Total: 75

Std 7 Mathematics: First Paper
November

Time: 2 hours
Total: 150

Question 1

Simplify without using a calculator.

1.1 $(-3x^2y^3)^3$ (3)

1.2 $\sqrt{9(x-2)^2y^4}$ (3)

1.3 $(2\sqrt{7})^2 - \sqrt{100-36}$ (2)

1.4 $-2(-3x^2)^2$ (2)

1.5 $(x^2y^{-1})^3$ and write with positive indices. (2)

1.6 $\dfrac{x^{-4}y^3}{2x^{-1}y}$ and write with positive indices. (2)

[14]

Question 2

2.1 Simplify:

2.1.1 $6a^2 - 2a^2(a-5)$

2.1.2 $2x^2(x-5) + x(3x^2-2)$

2.1.3 $(2x+1)(3x-2) - (x-3)(x+5)$

2.1.4 $-2(2x-3)^2 + (5x-2)(5x+2)$ (16)

2.2 Divide $6x^3 + 5x^2 - 5x - 3$ by $2x + 1$ (6)

[22]

Question 3

3.1 Write down all the numbers which satisfy the conditions given below:

3.1.1 $x < 3$ and $x \in N$

3.1.2 $-4 < x \leq 2$ and $x \in Z$ (4)

3.2 Represent the following on the number line:

3.2.1 $x > 5$ and $x \in Z$

3.2.2 $x \leq -2$ or $x > 1{,}5$ and $x \in R$ (5)

3.3 Write in scientific notation: 0,003 456 (2)

11

Question 4

4.1 Factorise

4.1.1 $9x^2 - 4y^2$

4.1.2 $m^2 - m - 12$

4.1.3 $x^2 - 7x + 10$

4.1.4 $3b^3 - 15b^2 + 18b$

4.1.5 $x^2(a + b) - y^2(b + a)$

4.1.6 $(x - 2y)(m - n) - y(n - m)$

4.1.7 $3 - \frac{12}{x^2}$

4.1.8 $3b(2a + b) - 2a - b$ (25)

4.2 Simplify the following fractions:

4.2.1 $\frac{3x^2 - 6x}{3x}$

4.2.2 $\frac{3m - 3p}{a^2 + 3a + 2} \div \frac{m - p}{a + 2}$

4.2.3 $\frac{2x - 1}{3x} - \frac{x - 1}{2x} - \frac{1}{2}$ (17)

42

Question 5

5.1 Solve for x:

5.1.1 $2(x-2)+(x-2)=5$

5.1.2 $\dfrac{x+3}{4}-\dfrac{x+2}{8}=\dfrac{x}{2}-1$

5.1.3 $\dfrac{3(x+6)}{4}-\dfrac{x}{3}=2-\dfrac{2x-3}{6}$ (14)

5.2

5.2.1 There are 13 single desks in a classroom and a certain number of double desks. The classroom has seating for 31 pupils. How many double desks are there in the room? (3)

5.2.2 Each base angle of an isosceles triangle is 30° less than the vertical angle. Calculate the size of the 3 angles of the triangle. (Show your calculations.) (3)

5.2.3 The perimeter of the diagram is 48 cm. Calculate the area of the diagram.

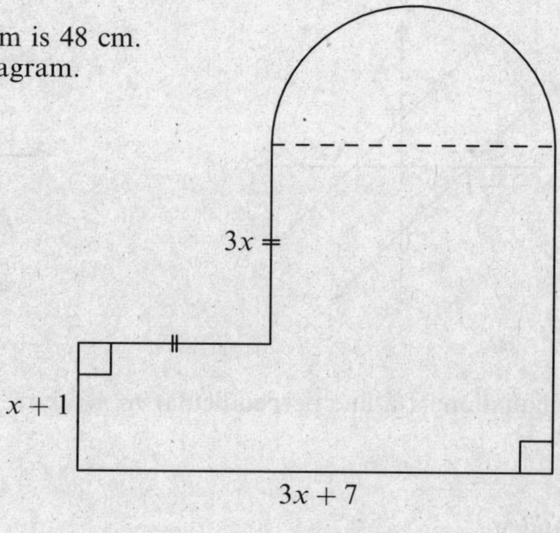

(7)

27

Question 6

6.1 On the same set of axes, draw the graphs of: (Clearly show the intercepts with the axes, and name the graphs with (A) or (B).)

6.1.1 $y = 2x - 3$ ....(A)

6.1.2 $-3x = 6y + 6$ ....(B) (6)

6.2 Determine **the value** of a if $y = 4x + 2$ is parallel to $y = ax + 7$. (2)

6.3 If the graph of $y = 2x + 6$ passes through the points $(3; a)$ and $(b; 2)$, calculate the values of a and b. (4)

6.4.1 Determine the equations of f, g, h and k in the diagrams below:

6.4.2 Hence determine the coordinates of A, B and C.

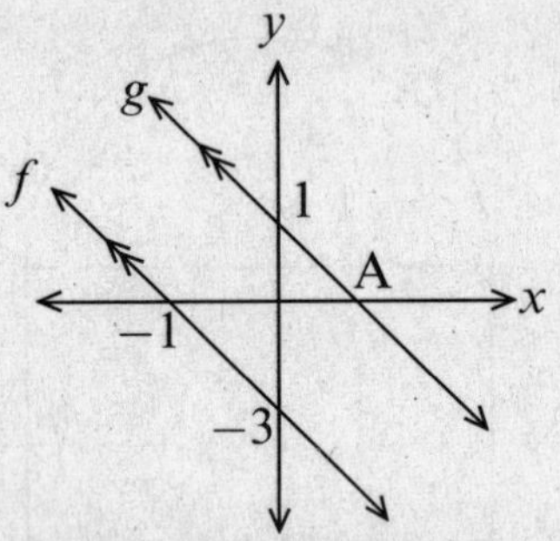

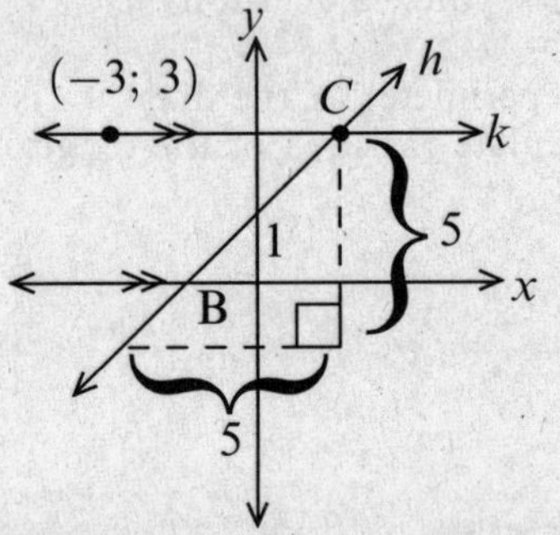

6.4.3 What is the equation of a line perpendicular to k (above) and passing through the point (3, 5) (19)

6.5 Solve for x and y:
$5 = -x + 2$
$y = 2x - 1$ (3)

34

Total: 150

Memorandum – First Paper November

Question 1 (Memo)

1.1 $-27x^6y^9$

1.2 $3(x-2)y^2$

1.3 $28-\sqrt{64}=20$

1.4 $-2\times 9x^4=-18x^4$

1.5 $x^6y^{-3}=\dfrac{x^6}{y^3}$

1.6 $\dfrac{y^2}{2x^3}$

14

Question 2 (Memo)

2.1.1 $6a^2-2a^3+10a^2$
$=16a^2-2a^3$

2.1.2 $2x^3-10x^2+3x^3-2x$
$=5x^3-10x^2-2x$

2.1.3 $6x^2-x-2-(x^2+2x-15)$
$=5x^2-3x+13$

2.1.4 $-2(4x^2-12x+9)+25x^2-4$
$=17x^2+24x-22$ (16)

2.2 $6x^3+5x^2-5x-3$
$=(2x+1)(3x^2+x-3)$ (6)

22

Question 3 (Memo)

3.1.1 $x=1;2$

3.1.2 $-3;-2;-1;0;1;2$

3.2.1

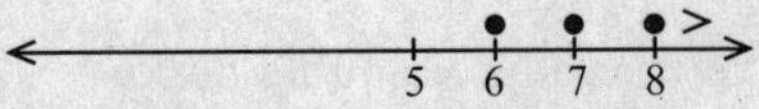

3.2.2

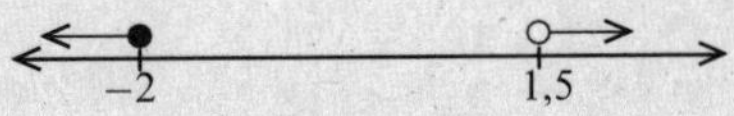

3.3 $3{,}456\times 10^{-3}$

11

Question 4 (Memo)

4.1.1 $(3x-2y)(3x+2y)$

4.1.2 $(m-4)(m+3)$

4.1.3 $(x-5)(x-2)$

4.1.4 $3b(b^2-5b+6)$
$=3b(b-3)(b-2)$

4.1.5 $(a+b)(x^2-y^2)$
$(a+b)(x-y)(x+y)$
$(m-n)(x-y)$

4.1.6 $(m-n)(x-2y\ \mathbf{+}y)$

4.1.7 $3\left(1-\dfrac{4}{x^2}\right)$
$=3\left(1-\dfrac{2}{x}\right)\left(1+\dfrac{2}{x}\right)$

4.1.8 $3b(2a+b)-(2a+b)$
$(2a+b)(3b-1)$ (25)

4.2.1 $\dfrac{3x(x-2)}{3x}$

$= x - 2$

4.2.2 $\dfrac{3\cancel{(m-p)}}{\cancel{(a+2)}(a+1)} \times \dfrac{\cancel{a+2}}{\cancel{m-p}}$

$= \dfrac{3}{a+1}$

4.2.3 $\dfrac{2(2x-1) - 3(x-1) - 3x}{6x}$

$= \dfrac{4x - 2 - 3x + 3 - 3x}{6}$

$= \dfrac{-2x+1}{6}$ (17)

42

Question 5 (Memo)

5.1.1 $2x - 4 + x - 2 = 5$

$3x = 11$

$x = 3\frac{2}{3}$

5.1.2 **×8:** $2(x+3) - (x+2) = 4x - 8$

$2x + 6 - x - 2 = 4x - 8$

$-3x = -12$

$x = 4$

5.1.3 **×12:** $9(x+6) - 4x = 24 - 2(2x-3)$

$9x + 54 - 4x = 24 - 4x + 6$

$9x = -24$

$x = -2\frac{2}{3}$ (14)

5.2.1 Let there be x double desks

$\therefore \quad 2x + 13 = 31$

$2x = 18$

$x = 9$

$\therefore$ 9 double desks. (3)

5.2.2 Let vertical angle be: x

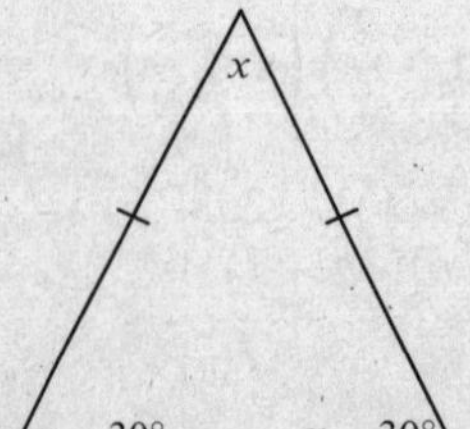

$\therefore \quad x + x - 30 + x - 30 = 180°$

$3x = 240$

$x = 80$

Angles: 80°; 50°; 50°. (3)

5.2.3

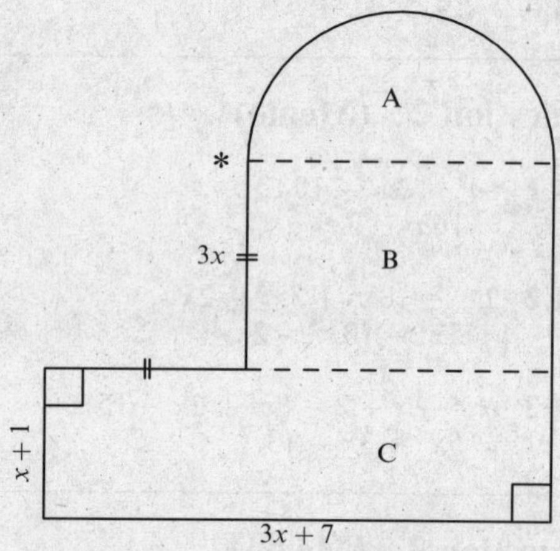

Circumference of $\frac{1}{2}\odot$: (radius $\frac{7}{2}$)

$\frac{1}{2}.2\,\pi\,r$

$= \pi\,r$

$= \pi\,(\frac{7}{2})$

$\simeq 11cm$

$3x+3x+x+1+3x+7+x+1+3x+11 = 48$

(Begin at *; anti-clockwise)

$14x = 28$

$x = 2$

Area: $\frac{1}{2}\,\pi\,r^2 + \text{Area}_B + \text{Area}_C$

$= \left(\frac{1}{2}.\frac{22}{7} \times 3.5^2\right) + (3x \times 7) +$

$\left((x+1) \times (3x+7)\right)$

$= 19{,}25 + 6 \times 7 + 3 \times 13$

$= 19{,}25 + 42 + 39$

$= 100{,}25\ \text{cm}^2$

(7)

27

Question 6 (Memo)

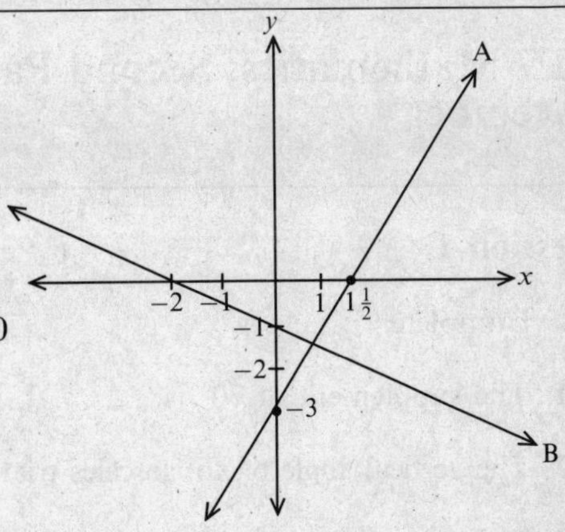

6.1.1 $y = 2x - 3$: y-intercepts: -3

x-intercept: $y = 0$: $2x - 3 = 0$
$x = 1\frac{1}{2}$

6.1.2 $-3x = 6y + 6$: y-intercept: $x = 0$: $6y + 6 = 0$
$6y = -6$
$y = -1$

x-intercept: $y = 0$: $-3x = 6$
$x = -2$

(6)

6.2 $a = 4$ (2)

6.3 $y = 2x + 6 \Rightarrow a = 2(3) + 6 \Rightarrow a = 12$

$y = 2x + 6 \Rightarrow 2 = 2b + 6 \Rightarrow 26 = -4 \Rightarrow b = -2$ (4)

6.4.1 f: $y = -3x - 3$

g: $y = -3x + 1$

h: $y = x + 1$

k: $y = 3$

6.4.2 A $(\frac{1}{3}; 0)$

B $(-1; 0)$

C $(2; 3)$

To find C:

$y = x + 1$

$3 = x + 1$

$2 = x$

(19)

6.4.3 $x = 3$ (1)

6.5 $-x + 2 = 5$ (1)
$y = 2x - 1$ (2)

From (1):
$-x + 2 = 5$
$-x = 3$
$x = -3$

Substitute $x = -3$ into (2):
$y = 2x - 1$
$= 2(-3) - 1$
$= -6 - 1$
$= -7$
$(-3; -7)$.

(3)

34

Total: 150

Std 7 Mathematics: Second Paper
November

Time: 2 hours
Total: 120

Question 1

1.1 Complete:

1.1.1 The supplement of 70° is ....°.

1.1.2 The vertical angle of an isosceles triangle is x°; each base angle is thus equal to ....°.

1.1.3 The sum of the interior angles of a quadrilateral is ....°.

1.1.4 The complement of $(90 - x)$ is ....°.

1.1.5 The sum of the exterior angles of a pentagon is ....°.

(5)

1.2 Choose the correct answer from each of the five possible answers:

1.2.1 The diagonals of a .... are equal.
A: Rhombus
B: Rectangle
C: Trapezium
D: Parallelogram

1.2.2 ABCD is a trapezium with AB||DC and AF⊥DC. The area of the trapezium can be calculated by:
A: AF.DC
B: $\frac{1}{2}$ DC(AD + BC)
C: AC.DB
D: $\frac{1}{2}$ (AB + DC) AF

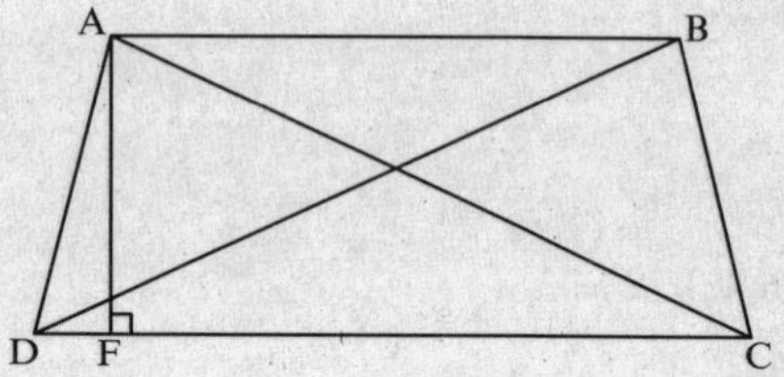

1.2.3 In the diagram, AM is:
A: angle of elevation
B: perpendicular line
C: median
D: perpendicular bisector

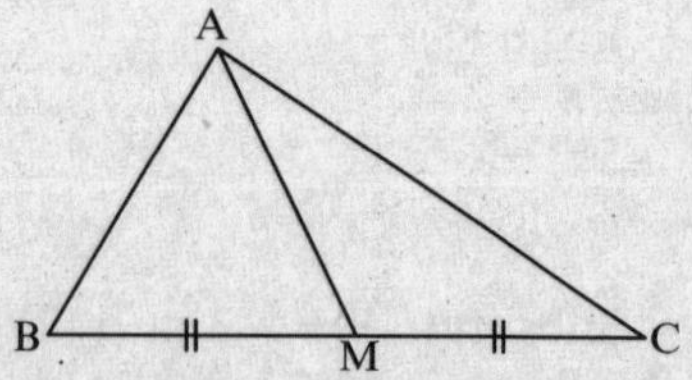

1.2.4 If one interior angle of a triangle is equal to the sum of the other two interior angles, then the triangle is:
A: Equalateral
B: Obtuse-angled
C: Right-angled
D: Acute-angled

1.2.5 Which of the following is a Pythagorean triple?
A: 5 : 1 : 2
B: 49 : 16 : 9
C: 2 : 1 : 1
D: 13 : 12 : 5

(10)

15

Question 2

2.1 In the diagram AB||DC and AD||BC.

Calculate a, b, c and d

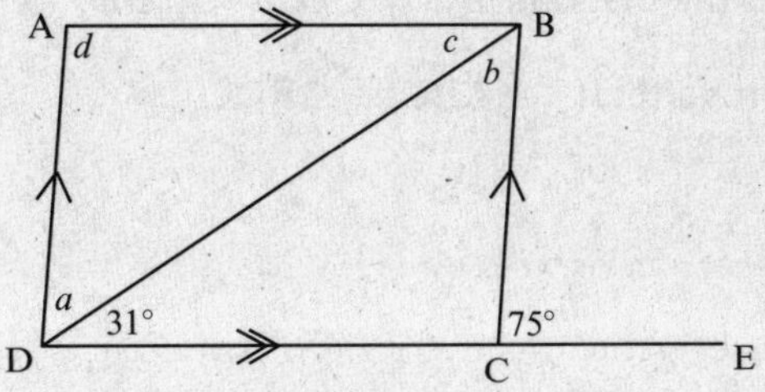

(8)

2.2 In the diagram AB = BE = BC

2.2.1 Express each of the following in terms of x:
(*a*) ∠CEB
(*b*) ∠ABE
(*c*) ∠AEB

2.2.2 Prove that: CE⊥AD (give reasons)

2.2.3 If CE biscets $A\hat{C}D$, determine the relationship between BE and CD. Explain.

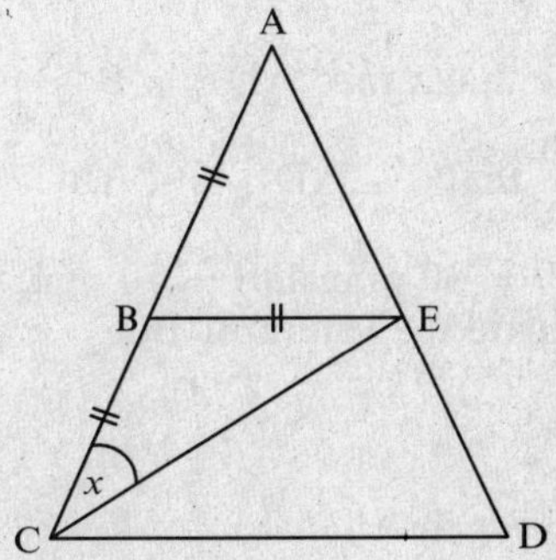

(8)

2.3 Use the accompanying diagram and determine x in terms of a and b:

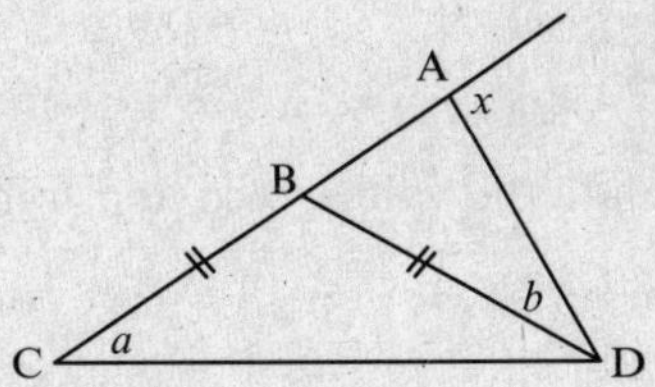

(4)

20

Question 3

3.1 In the diagram $A\hat{B}C = A\hat{C}B$ and $BD = DC$.

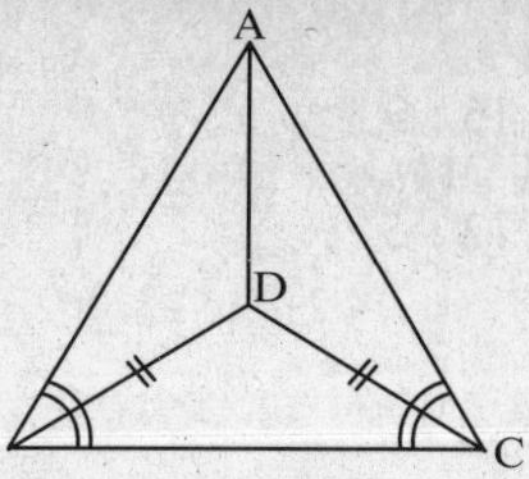

3.1.1 Prove that: $\triangle ABD \equiv \triangle ACD$.

3.1.2 Complete: $A\hat{B}D = \ldots\ldots\ldots\ldots\ldots\ldots$

(5)

3.2 In the diagram $BC = CD$; $AE \parallel BD$; $CA = CE$ and $E\hat{C}D = E\hat{A}C$.

Prove that: $\triangle ABC \equiv \triangle EDC$

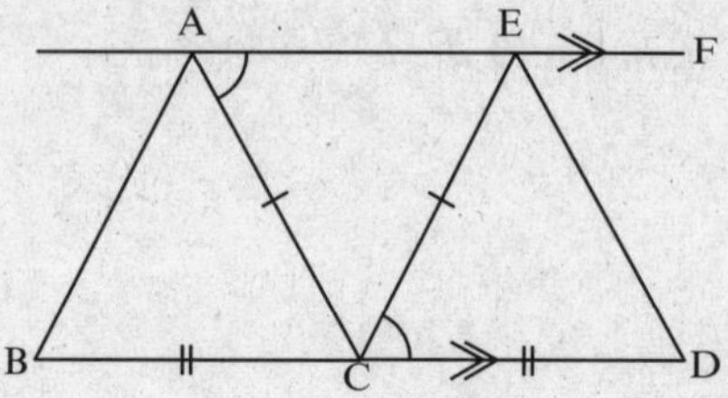

(5)

3.3 In the diagram: $A\hat{D}E = \hat{B}$

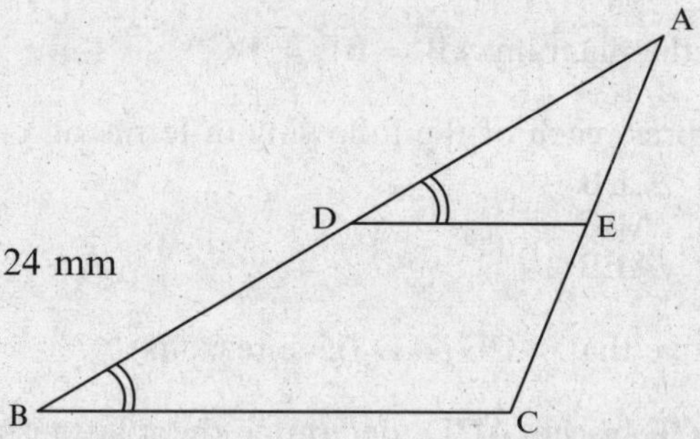

3.3.1 Prove that: $\triangle ADE \parallel\!| \triangle ABC$

3.3.2 If $AD = 30$ mm; $DB = 50$ mm; $DE = 24$ mm
Calculate the length of BC.

(6)

16

Question 4

4.1 Calculate the values of a and b in the accompanying diagram.

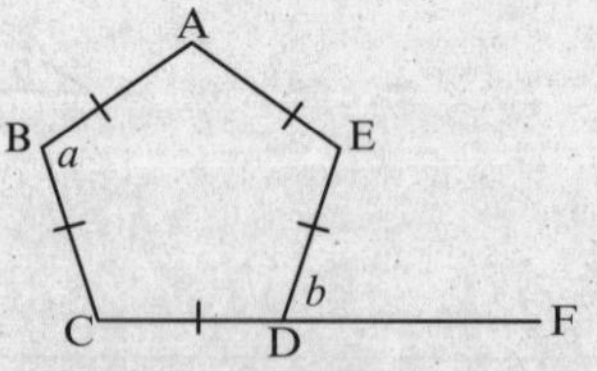

(6)

4.2 In a regular polygon, every exterior angle is 36°.
(*a*) How many sides does the polygon have?
(*b*) What is the size of each of the interior angles of the polygon?

(2)

8

Question 5

5.1 ABCD is a rectangle.
E is the midpoint of AB.
CD = 7cm.
The area of the rectangle is $63cm^2$.

Calculate:
(*a*) BC
(*b*) The area of trapezium EADC.

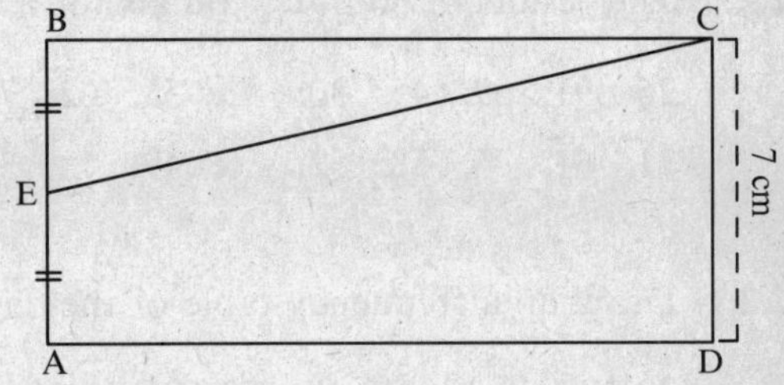

(5)

5.2 MNQP is a kite.
MO = 5; MP = 13; OQ = 25; OP = x

5.2.1 Calculate the length of x.

5.2.2 Calculate the area of the kite.

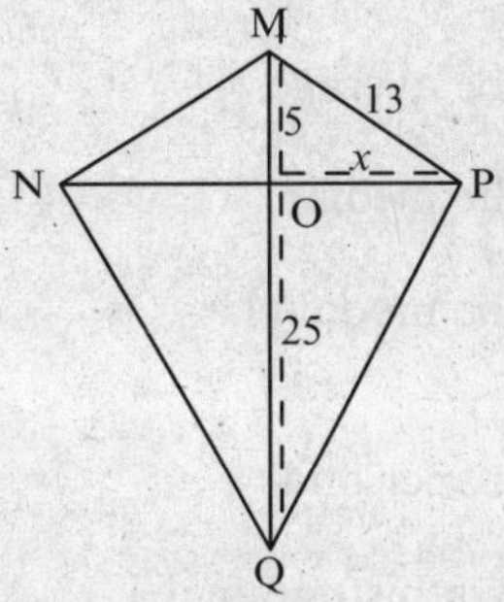

(6)

11

Question 6

6.1 Write down the theorem of Pythagoras (in words). (4)

6.2 In the diagram $\hat{A} = \hat{C} = 90°$.
BC = 24 cm; AD = 16 cm; DC = 10 cm; BE = 6 cm.

Calculate:

6.2.1 AE

6.2.2 Area △AED.

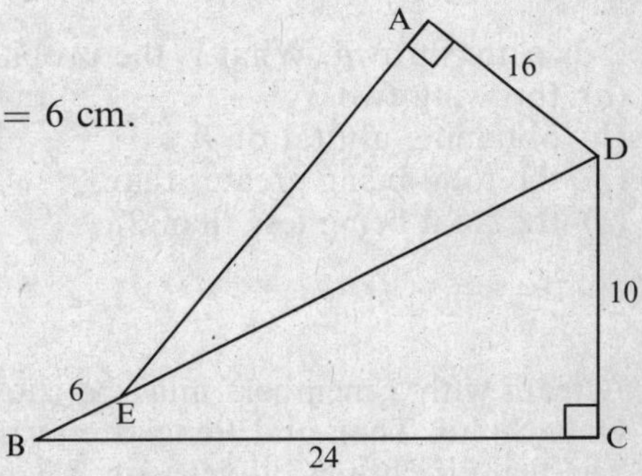

(8)

12

Question 7

7.1 50 teachers from a school were interviewed to determine which TV-channel they preferred. The results were:
SABC 1: 30%
SABC 2: 10%
SABC 3: 20%
M-NET: 40%

Draw a pie chart to represent the given information. (First determine the degrees of each sector) How many teachers (of the 50) prefer M-Net?

(6)

7.2 The results of 30 Std. 7 pupils for a Maths test, are as follows:

26; 91; 59; 45; 30; 81; 53; 62; 70; 53; 49; 52; 66; 56; 88;
31; 61; 75; 58; 52; 36; 49; 48; 34; 64; 72; 42; 39; 52; 69.

7.2.1 Draw up a frequency table of the data, using the following intervals:

20-29; 30-39; 40-49; 50-59; 60-69; 70-79; 80-89; 90-99.

7.2.2 Calculate:

7.2.2.1 the average.

7.2.2.2 the median.

7.2.2.3 the modal class.

Use graph paper and:

7.2.3 Draw a histogram.

(8)

[14]

Question 8

8.1 2 dice are thrown. What is the probability of:
(*a*) throwing two 3^s;
(*b*) obtaining a total of 7;
(*c*) the total being greater than 7;
(*d*) the total being less than 8?

(7)

8.2 A team with 2 members must be chosen from a Std. 7-class.
Only Sipho; Thali and Bengi are available.
The team is chosen by drawing 2 names from a hat.

8.2.1 Draw a tree-diagram to show all the possible outcomes.

8.2.2 What is the probability that both Sipho and Thali are chosen?

8.2.3 What is the probability that Sipho will be part of the team?

(8)

[15]

Question 9

9.1 Simon invests R800 at 10% p.a. simple interest. After 2 years the interest rate changes to 12% per annum. What amount will he have available after 5 years?

(4)

9.2 Cardo borrows R12 000 from Joan at the beginning of 1992 at 10% compound interest. At the end of each year hy pays back R500 on the loan. What amount does he still owe at the end of 1995?

(5)

9

Total: 120

Memorandum – Second Paper
November

Question 1 (Memo)

1.1.1 110°

1.1.2 $\frac{180°-x}{2}$

1.1.3 360°

1.1.4 x

1.1.5 360° (5)

1.2.1 B

1.2.2 D

1.2.3 C

1.2.4 C

1.2.5 D (10)

15

Question 2 (Memo)

2.1 $a = 75° - 31°$ (corresponding ∠ˢ AD∥BC)
$= 44°$

$b = a$ (alternate ∠ˢ AD∥BC)
$= 44°$

$c = 31°$ (alternate ∠ˢ AB∥DC)

$d = 180° - (44° + 31°)$ (3 ∠ˢ of △ABD)

$= 105°$

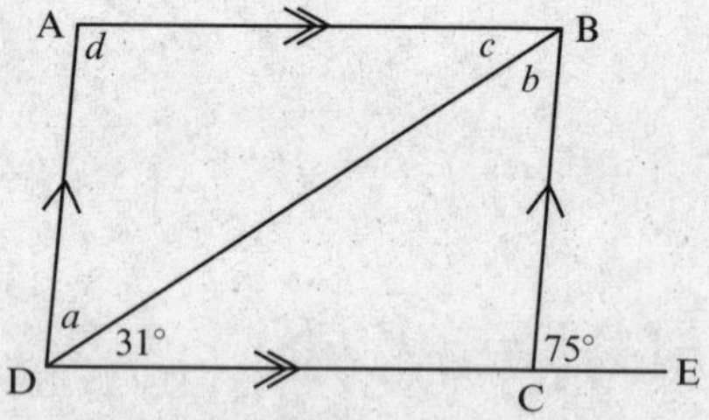

(8)

2.2

2.2.1 (*a*) $C\hat{E}B = x$ (isosceles △ ; BE = BC)

(*b*) $A\hat{B}E = 2x$ (ext. ∠ of △CBE)

(*c*) $A\hat{E}B = \frac{180 - 2x}{2}$ (3 ∠ˢ of △ABE)
$= 90 - x$

2.2.2 $C\hat{E}D = 180° - (x + 90 - x)$ (straight line $A\hat{E}D$)
$= 90°$

$\therefore$ CE⊥AD

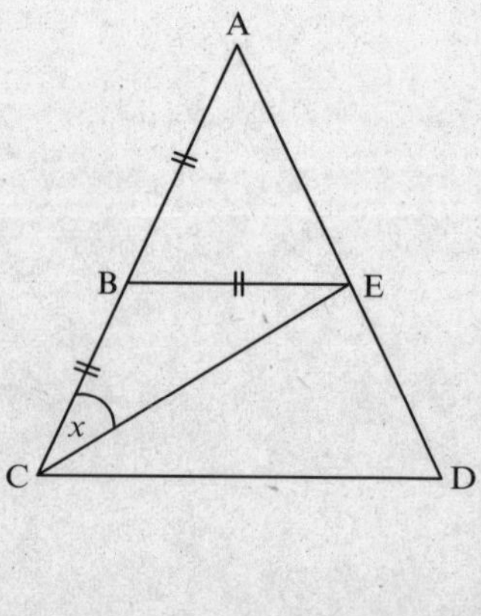

△ ABC ≡ △ YZX (S, ∠, S)

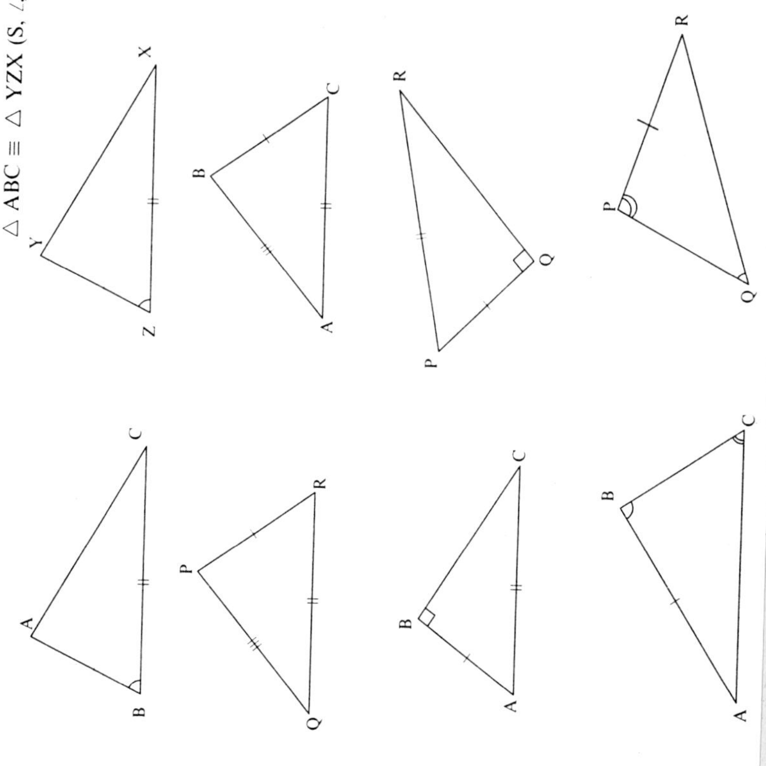

(8)

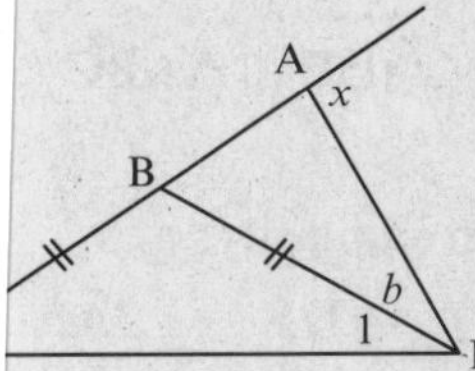

(4)

20

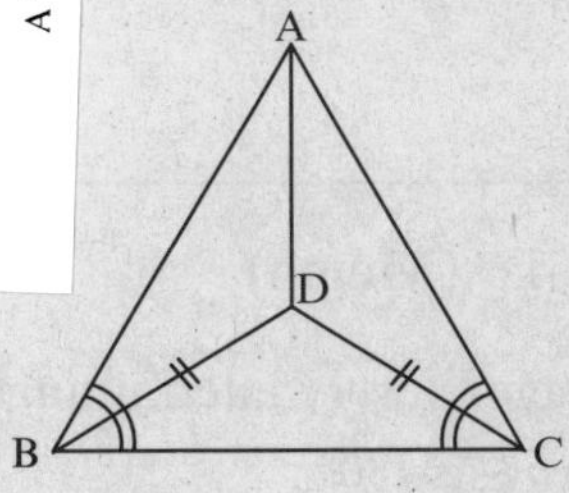

($A\hat{B}C = A\hat{C}B$)

congruency)

(5)

3.

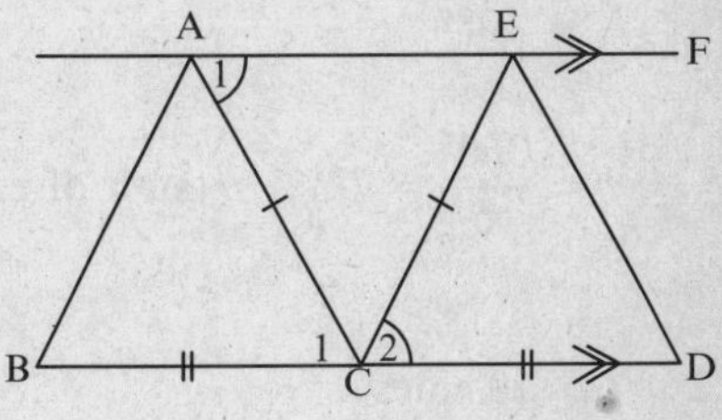

(given)

(given)

(alternate $\angle^s$ AF||BD)

2 (given)

∴ △EDC (S, ∠, S)

(5)

3.3

3.3.1 In $\triangle^s$ ADE and ABC:

1. $\hat{A}$ is common

2. $\hat{D}_1 = \hat{B}$ (given)

$\therefore$ $\triangle ADE \,|||\, \triangle ABC$ ($\angle$, $\angle$,)

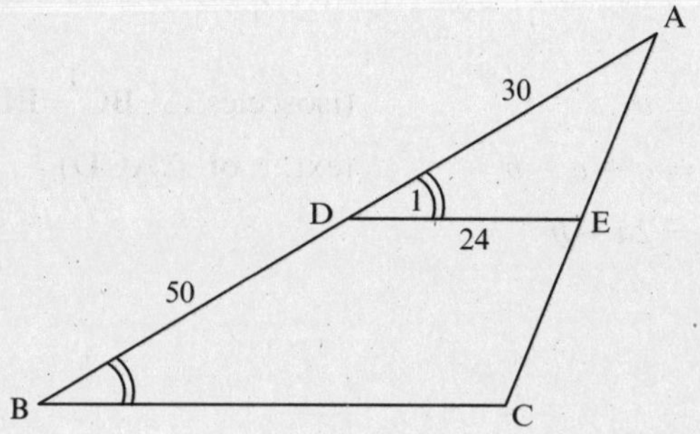

3.3.2 From similarity:

$$\frac{AD}{AB} = \frac{DE}{BC} = \frac{AE}{AC}$$

$$\therefore \frac{30}{80} = \frac{24}{BC}$$

$30BC = 80 \times 24$

$= 1920$

$\therefore BC = 64$ mm

(6)

16

Question 4 (Memo)

4.1 Pentagon (every interior angle)

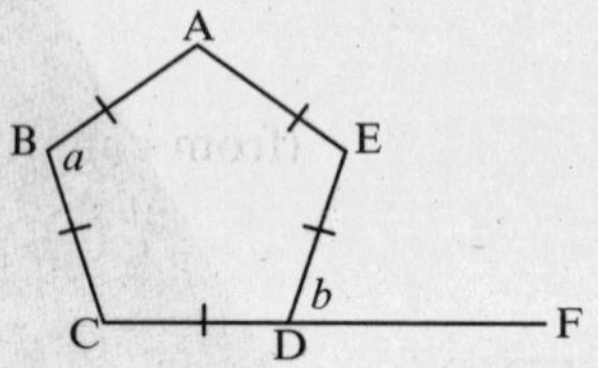

$$a = \frac{(n - z)180°}{n}$$

$$= \frac{(5 - 2)180°}{5}$$

$$= \frac{540}{5}$$

$= 108°$

$b = \frac{360°}{5} = 72°$ (sum of exterior $\angle^s = 360°$)

(6)

4.2 (*a*) 10 sides

(*b*) Every interior angle:
$180° - 36° = 144°$ (straight line)

(2)

8

Question 5 (Memo)

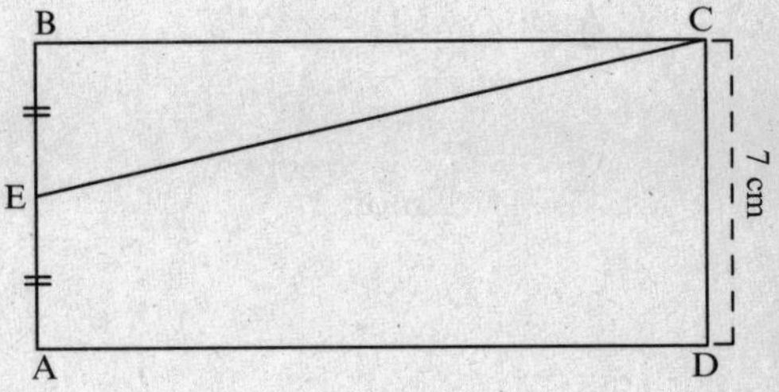

5.1 1. BC (length) $= \frac{63}{7} = 9$ cm

2. Area trap.: $\frac{h}{2}$ (sum of $\|$-sides)

$= \frac{9}{2}(7 + 3,5)$

$= 47,25 \text{ cm}^2$

(5)

5.2

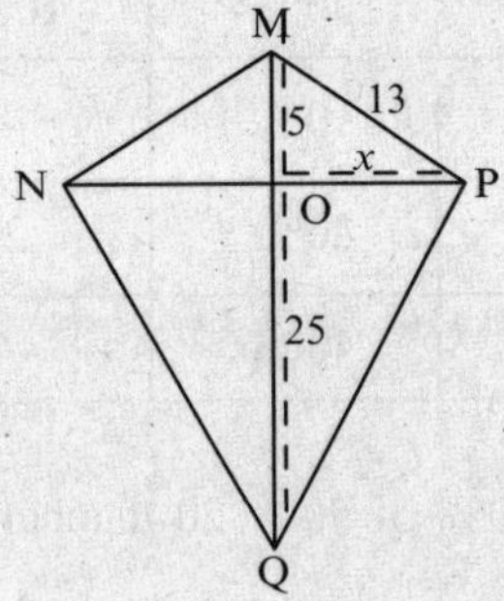

5.2.1 $x^2 = 13^2 - 5^2$ (Pyth.)

$= 169 - 25$

$= 144$

$x = 12$

5.2.2 Area kite: $\frac{1}{2}$ product diagonals

$= \frac{1}{2} . 30 . 24$

$= 360 \text{ units}^2$

(6)

11

Question 6 (Memo)

6.1 In a **right-angled triangle** the **square** on the **hypotenuse** is equal to the **sum** of the **square** on the other 2 sides. (4)

6.2

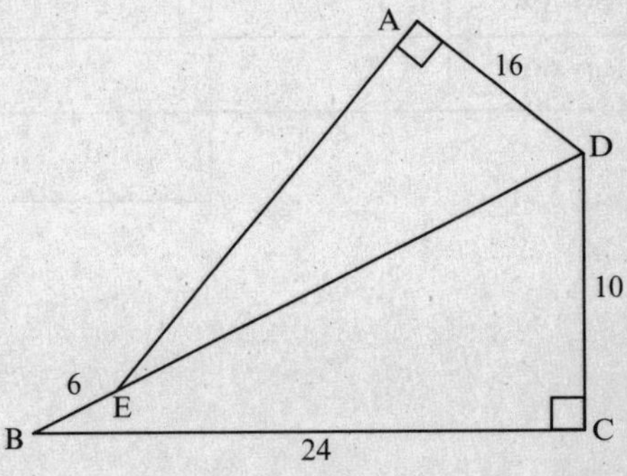

6.2.1 $BD^2 = 24^2 + 10^2$ (Pyth; $\hat{C} = 90°$)

$= 576 + 100$

$= 676$

$\therefore$ BD $= 26$ cm

$\therefore$ ED $= 20$ cm

$\therefore$ $AE^2 = 20^2 - 16^2$ (Pyth; $\hat{A} = 90°$)

$= 400 - 256$

$= 144$

$\therefore$ AE $= 12$ cm

6.2.2 Area $\triangle AED = \frac{1}{2}\, b.h$

$= \frac{1}{2}\, 16.12$

$= 96\ \text{cm}^2$

(8)

12

Question 7 (Memo)

7.1

SABC 1	30%	108°
SABC 2	10%	36°
SABC 3	20%	72°
M-NET	40%	144°

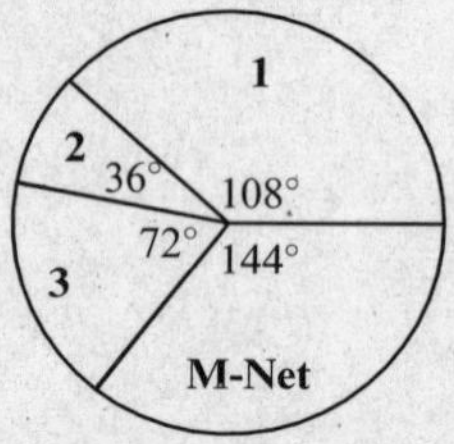

M-NET: 40% of 50 = 20 teachers.

(6)

7.2

Interval	Tally	Frequency
20-29	I	1
30-39	~~IIII~~	5
40-49	~~IIII~~	5
50-59	~~IIII~~ I II	8
60-69	~~IIII~~	5
70-79	I II	3
80-89	II	2
90-99	I	1
		30

7.2.2.1 $\overline{x} = \dfrac{1663}{30}$

$= 55,4$

7.2.2.2 Median: 53

7.2.2.3 Modal class: 50-59

7.2.3

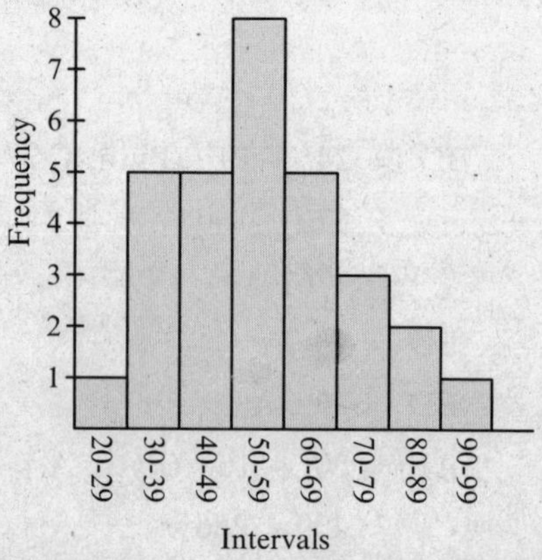

(8)

14